Political Ecology of Tourism

Why has political ecology been assigned so little attention in tourism studies, despite its broad and critical interrogation of environment and politics? As the first full-length treatment of a political ecology of tourism, the collection addresses this lacuna and calls for the further establishment of this emerging interdisciplinary subfield.

Drawing on recent trends in geography, anthropology, and environmental and tourism studies, *Political Ecology of Tourism: Communities, power and the environment* employs a political ecology approach to the analysis of tourism through three interrelated themes: communities and power, conservation and control, and development and conflict. While geographically broad in scope—with chapters that span Central and South America to Africa, and South, Southeast, and East Asia to Europe and Greenland—the collection illustrates how tourism-related environmental challenges are shared across prodigious geographical distances, while also attending to the nuanced ways they materialize in local contexts and therefore demand the historically situated, place-based and multi-scalar approach of political ecology.

This collection advances our understanding of the role of political, economic and environmental concerns in tourism practice. It offers readers a political ecology framework from which to address tourism-related issues and themes such as development, identity politics, environmental subjectivities, environmental degradation, land and resources conflict, and indigenous ecologies. Finally, the collection is bookended by a pair of essays from two of the most distinguished scholars working in the subfield: Rosaleen Duffy (foreword) and James Igoe (afterword).

This collection will be valuable reading for scholars and practitioners alike who share a critical interest in the intersection of tourism, politics and the environment

Mary Mostafanezhad is an Assistant Professor in the Department of Geography at the University of Hawai'i at Mānoa.

Roger Norum is a Postdoctoral Research Fellow at the University of Leeds School of English and a researcher on the HERA-funded project Arctic Encounters: Contemporary Travel/Writing in the European High North and the Marie Sklodowska-Curie Innovative Training Network in Environmental Humanities. Trained in social anthropology, his research focuses on sociality, temporality, travel and the environment.

Eric J. Shelton works with environmental NGOs in New Zealand and strives to situate nature-based tourism within environmental philosophy.

Anna Thompson-Carr is a Senior Lecturer in the Department of Tourism at the University of Otago, NZ. She has conducted research and published in high-quality tourism journals on visitors' experiences of cultural values for landscapes in New Zealand with a focus on integrating cultural values within interpretation.

Contemporary geographies of leisure, tourism and mobility
Series Editor: C. Michael Hall
Professor at the Department of Management, College of Business and Economics, University of Canterbury, Christchurch, New Zealand

For a complete list of titles in this series, please visit www.routledge.com

The aim of this series is to explore and communicate the intersections and relationships between leisure, tourism and human mobility within the social sciences.

It will incorporate both traditional and new perspectives on leisure and tourism from contemporary geography, e.g. notions of identity, representation and culture, while also providing for perspectives from cognate areas such as anthropology, cultural studies, gastronomy and food studies, marketing, policy studies and political economy, regional and urban planning, and sociology, within the development of an integrated field of leisure and tourism studies.

Also, increasingly, tourism and leisure are regarded as steps in a continuum of human mobility. Inclusion of mobility in the series offers the prospect to examine the relationship between tourism and migration, the sojourner, educational travel, and second home and retirement travel phenomena.

The series comprises two strands:

Contemporary geographies of leisure, tourism and mobility aims to address the needs of students and academics, and the titles will be published in hardback and paperback. Titles include:

1 **The Moralisation of Tourism**
   Sun, sand . . . and saving the world?
   Jim Butcher

2 **The Ethics of Tourism Development**
   Mick Smith and Rosaleen Duffy

3 **Tourism in the Caribbean**
   Trends, development, prospects
   Edited by David Timothy Duval

4 **Qualitative Research in Tourism**
   Ontologies, epistemologies and methodologies
   Edited by Jenny Phillimore and Lisa Goodson

5 **The Media and the Tourist Imagination**
   Converging cultures
   Edited by David Crouch, Rhona Jackson and Felix Thompson

6 **Tourism and Global Environmental Change**
   Ecological, social, economic and political interrelationships
   Edited by Stefan Gössling and C. Michael Hall

7 **Cultural Heritage of Tourism in the Developing World**
   Edited by Dallen J. Timothy and Gyan Nyaupane

8 **Understanding and Managing Tourism Impacts**
An integrated approach
C. Michael Hall and Alan Lew

9 **An Introduction to Visual Research Methods in Tourism**
Edited by Tijana Rakic and Donna Chambers

10 **Tourism and Climate Change**
Impacts, adaptation and mitigation
C. Michael Hall, Stefan Gössling and Daniel Scott

11 **Tourism and Citizenship**
Raoul V. Bianchi and Marcus L. Stephenson

Routledge studies in contemporary geographies of leisure, tourism and mobility is a forum for innovative new research intended for research students and academics, and the titles will be available in hardback only. Titles include:

40 **Scuba Diving Tourism**
Edited by Kay Dimmcock and Ghazali Musa

41 **Contested Spatialities Lifestyle Migration and Residential Tourism**
Michael Janoschka and Heiko Haas

42 **Contemporary Issues in Cultural Heritage Tourism**
Edited by Jamie Kaminski, Angela M. Benson and David Arnold

43 **Understanding and Governing Sustainable Tourism Mobility**
Edited by Scott Cohen, James Higham, Paul Peeters and Stefan Gossling

44 **Green Growth and Travelism**
Concept, policy and practice for sustainable tourism
Edited by Terry DeLacy, Min Jiang, Geoffrey Lipman and Shaun Vorster

45 **Tourism, Religion and Pilgrimage in Jerusalem**
Kobi Cohen-Hattab and Noam Shoval

46 **Trust, Tourism Development and Planning**
Edited by Robin Nunkoo and Stephen L.J. Smith

47 **A Hospitable World?**
Organising work and workers in hotels and tourist resorts
Edited by David Jordhus-Lier and Anders Underthun

48 **Tourism in Pacific Islands**
Current issues and future challenges
Edited by Stephen Pratt and David Harrison

49 **Social Memory and Heritage Tourism Methodologies**
Edited by Stephen P. Hanna, Amy E. Potter, E. Arnold Modlin, Perry Carter, and David L. Butler

50 **Affective Tourism**
Dark routes in conflict
Dorina Maria Buda

51 **Scientific Tourism**
Edited by Susan L. Slocum, Carol Kline and Andrew Holden

52 **Volunteer Tourism and Development**
The lifestyle politics of international development
Jim Butcher and Peter Smith

53 **Imagining the West through Film and Tourism**
Warwick Frost and Jennifer Laing

54 **The Practice of Sustainable Tourism**
Resolving the paradox
Edited by Michael Hughes, David Weaver and Christof Pforr

55 **Mountaineering Tourism**
Edited by Ghazali Musa, James Higham and Anna Thompson

56 **Tourism and Development in Sub-Saharan Africa**
Current issues and local realities
Marina Novelli

57 **Tourism and the Anthropocene**
Edited by Martin Gren and Edward H. Huijbens

58 **The Politics and Power of Tourism in Palestine**
Edited by Rami K. Isaac, C. Michael Hall and Freya Higgins-Desbiolles

59 **Political Ecology of Tourism**
Community, power and the environment
Edited by Mary Mostafanezhad, Eric Jacob Shelton, Roger Norum and Anna Thompson-Carr

Forthcoming:

International Tourism and Cooperation and the Gulf Cooperation Council States
Developments, challenges and opportunities
Edited by Marcus Stephenson and Ala Al-Hamarneh

Protest and Resistance in the Tourist City
Edited by Johannes Novy and Claire Colomb

Women and Sex Tourism Landscapes
Erin Sanders-McDonagh

Research Volunteer Tourism
Angela M Benson

Managing and Interpreting D-day's Sites of Memory
War graves, museums and tour guides
Edited by Geoffrey Bird, Sean Claxton and Keir Reeves

Co-Creation in Tourist Experiences
Nina Prebensen, Joseph Chen and Muzaffer Uysal

Authentic and Inauthentic Places
Jane Lovell and Chris Bull

Political Ecology of Tourism
Community, power and the environment

Edited by
Mary Mostafanezhad,
Roger Norum, Eric J. Shelton and
Anna Thompson-Carr

LONDON AND NEW YORK

First published 2016
by Routledge
2 Park Square, Milton Park, Abingdon, Oxon OX14 4RN

and by Routledge
711 Third Avenue, New York, NY 10017

Routledge is an imprint of the Taylor & Francis Group, an informa business

© 2016 selection and editorial matter, Mary Mostafanezhad, Roger Norum, Eric J. Shelton and Anna Thompson-Carr; individual chapters: the contributors.

The right of Mary Mostafanezhad, Roger Norum, Eric J. Shelton and Anna Thompson-Carr to be identified as the authors of the editorial material, and of the authors for their individual chapters, has been asserted in accordance with sections 77 and 78 of the Copyright, Designs and Patents Act 1988.

All rights reserved. No part of this book may be reprinted or reproduced or utilised in any form or by any electronic, mechanical, or other means, now known or hereafter invented, including photocopying and recording, or in any information storage or retrieval system, without permission in writing from the publishers.

Trademark notice: Product or corporate names may be trademarks or registered trademarks, and are used only for identification and explanation without intent to infringe.

British Library Cataloguing in Publication Data
A catalogue record for this book is available from the British Library

Library of Congress Cataloging in Publication Data
A catalog record for this book has been requested

ISBN: 978-1-138-85944-9 (hbk)
ISBN: 978-1-315-71722-7 (ebk)

Typeset in Times New Roman
by Swales & Willis Ltd, Exeter, Devon, UK

Contents

List of figures	x
List of contributors	xi
Foreword	xvi
ROSALEEN DUFFY	
Introduction	1
MARY MOSTAFANEZHAD, ROGER NORUM, ERIC J. SHELTON AND ANNA THOMPSON-CARR	

PART I
Communities and power 23

	Introduction to Communities and power	25
	ANNA THOMPSON-CARR	
1	A gendered political ecology of tourism and water	31
	STROMA COLE	
2	Ngarrindjeri authority: a sovereignty approach to tourism	50
	RON NICHOLLS, FREYA HIGGINS-DESBIOLLES AND GRANT RIGNEY	
3	Co-management of natural resources in protected areas in 'postcolonial' Africa	70
	CHENGETO CHADEROPA	
4	'Few people know that Krishna was the first environmentalist': religiously motivated conservation as a response to pilgrimage pressures in Vrindavan, India	92
	TAMARA LUTHY	
5	Festive environmentalism: a carnivalesque reading of eco-voluntourism at the Roskilde Festival	108
	METTE FOG OLWIG AND LENE BULL CHRISTIANSEN	

PART II
Conservation and control 129

Introduction to Conservation and control 131
ERIC J. SHELTON

6 Unsettling the moral economy of tourism on Chile's Easter Island 134
FORREST WADE YOUNG

7 Rethinking ecotourism in environmental discourse in Shangri-La: an antiessentialist political ecology perspective 151
JUNDAN (JASMINE) ZHANG

8 (Re)creating forest natures: assemblage and political ecologies of ecotourism in Japan's central highlands 169
ERIC J. CUNNINGHAM

9 Ecotourism or eco-utilitarianism: exploring the new debates in ecotourism 188
STEPHEN WEARING AND MICHAEL WEARING

PART III
Development and conflict 207

Introduction to Development and conflict 209
MARY MOSTAFANEZHAD

10 Political ecologies and economies of tourism development in Kaokoland, north-west Namibia 213
JARKKO SAARINEN

11 Cleaning up the streets, Sandinista-style: the aesthetics of garbage and the urban political ecology of tourism development in Nicaragua 231
JOSH FISHER

12 The political ecology of tourism development on Mount Kilimanjaro 251
MEGAN HOLROYD

13 'Absolutely not smelly': the political ecology of disengaged slum tours in Mumbai, India 270
KEVIN HANNAM AND ANYA DIEKMANN

14 **Composing Greenlandic tourism futures: an integrated political ecology and actor-network theory approach**   284
CARINA REN, LILL RASTAD BJØRST AND
DIANNE DREDGE

Conclusion: Towards future intersections of tourism studies and political ecology   302
ROGER NORUM, MARY MOSTAFANEZHAD, ERIC J. SHELTON
AND ANNA THOMPSON-CARR

Afterword   309
JAMES IGOE

Index   317

Figures

3.1	Old Makuleke/Pafuri Triangle, now the location of the 26,500-hectare Makuleke Contractual Park (MCP)	72
3.2	Income earned by Makuleke community	79
3.3	Lodges operation in the MCP	82
3.4	Existing institutional arrangements	83
3.5	Proposed new governance arrangement	84
3.6	Makuleke residents as members of the Community Property Association	86
4.1	The pamphlet cover: 'Dham Seva, Save a Dham'	102
5.1	Bicycle-powered Ferris wheel at the 2009 Roskilde Festival	115
5.2	2015 Roskilde Festival poster listing the organizations that received donations in 2014	117
5.3	The *fisk* booth at Roskilde Festival 2009	120
5.4	Karen Mukupa being interviewed by MTV at the *fisk* booth. Karen Mukupa is wearing clothing items from the *SKIFt* collection	122
7.1	Images of Shangri-La	152
8.1	Depiction of 'Mountain Girl' (yama gāru)	181
10.1	Sharing the benefits from tourism: distributing maize flour to households after the tourist group has left the community	222
10.2	The Opuwo Country Lodge's swimming pool and a view to dry 'Himbaland'	223
12.1	Map of Tanzania's protected areas	255
12.2	Population in 1988	261
14.1	Narsaq seen from the hillside	290
14.2	Children in the Narsaq museum	291
14.3	Screenshot of *Greenland Minerals and Energy – Kvanefjeld*	294

Contributors

Lill Rastad Bjørst is Assistant Professor at Aalborg University, Denmark and holds a PhD in Arctic studies from the University of Copenhagen. Her doctoral research focused on the climate change debate in Greenland, and since 2012 she has explored the political debate about Greenland's uranium. She is assistant research coordinator at CIRCLA and academic coordinator for Arctic studies – a specialization of the CCG programme at Aalborg University. Her research interests include Inuit culture and society; climate change and sustainability; mining and industrialization; postcolonialism and tourism.

Chengeto Chaderopa is a Senior Lecturer in the Department of Tourism and Hospitality Management at Waiariki Institute of Technology, New Zealand. He lectures in tourism, hotel management, and services marketing. Chengeto has also taught tourism at universities in South Africa and Zimbabwe. His research interests include sustainability in the tourism and hospitality industry, transnational parks and local communities, power, political ecology and local communities' environmental narratives. Finally, Chengeto has an interest in service quality and tourism destination marketing.

Lene Bull Christiansen holds a PhD in international development studies from Roskilde University, Denmark, where she is an Associate Professor at the Institute of Culture and Identity. Her PhD research dealt with gender in Zimbabwean cultural politics. Her current work deals with development communication, celebrity and nationalism in Denmark. She is a core member of the Research Network on Celebrity and North-South Relations and heads the Research Cluster on celebrities as new global actors at Roskilde University.

Stroma Cole is a Senior Lecturer in tourism geography at the University of the West of England. Stroma combines her academic career with action research and consultancy, most recently looking at tourism and the abuse of water rights in Labuan Bajo, Indonesia with a grant from the British Academy. She was the Chair of Tourism Concern (2007–2011), and is now a Director of Equality in Tourism. With research interests in responsible tourism development in less developed countries, and the link between tourism and human rights, Stroma is an activist researcher critiquing the consequences of tourism development.

Eric J. Cunningham is Assistant Professor in the Japanese Studies and Environmental Studies programmes at Earlham College in Richmond, Indiana. His research focuses on intersections between capitalism and other life projects within highland ecologies in Japan.

Anya Diekmann is Professor of Tourism at the Université Libre de Bruxelles (Belgium). Her research and publications include work on social tourism and cultural tourism with a particular focus on heritage, urban, (slum and ethnic) tourism in India and Europe. Among others, she co-authored with Kevin Hannam *Tourism and India: A Critical Introduction* and co-edited with Melanie Smith *Ethnic and Minority Cultures as Tourist Attractions*.

Dianne Dredge is Professor in the Department of Culture and Global Studies, Aalborg University, Copenhagen, Denmark and Chair of the Tourism Education Futures Initiative. Originally trained as an environmental and urban planner, Dianne has 20 years of practical experience working with communities, government and non-government organizations, and industry stakeholders in a variety of locations including Australia, Mexico and China. Her research interests include tourism development processes, collaborative governance, tourism policy networks, policy knowledge dynamics and tourism education.

Rosaleen Duffy is Professor of Political Ecology, in the Department of Development Studies, SOAS University of London. Her research takes an interdisciplinary approach, drawing on international politics, geography and development studies. She focuses on global environmental governance, neoliberalism and nature, tourism, ecotourism and biodiversity conservation. Her most recent books are *Nature Crime: How We're Getting Conservation Wrong* and *Nature Unbound: Conservation, Capitalism and the Future of Protected Areas* with Dan Brockington and James Igoe.

Josh Fisher is Assistant Professor of Anthropology and serves as Director of both Environmental Studies and Anthropology at High Point University, North Carolina. He researches and writes on questions of how different forms of 'value' are produced in development, alternative development, urban environmental policy and employment generation in Central America. He has served as the technical director of the employment-generation project called the Organization Workshop in Ciudad Sandino, Nicaragua, and he is also a member of the board of directors for a Nicaragua-based NGO called the Center for Development in Central America.

Mette Fog Olwig is a human geographer and an Assistant Professor in International Development Studies at the Department of Society and Globalisation, Roskilde University, Denmark. In her research she has applied a multi-sited, multi-level ethnographic perspective on the social and power dimensions of the political ecology of development policy in relation to climate change discourse in Ghana. Her current research focuses on land grabbing in Tanzania, natural disasters, community dynamics, power relations and climate change in Vietnam as well as development communication and consumption through fair trade, benefit events and eco/social tourism globally.

Kevin Hannam is Professor of Tourism Mobilities at Leeds Beckett University, UK. Previously he was Associate Dean for Research in the Faculty of Business and Law at the University of Sunderland, UK. He is founding co-editor of the journal *Mobilities*, co-author of the books *Understanding Tourism* and *Tourism and India* (Routledge) and co-editor of the *Routledge Handbook of Mobilities Research* and *Moral Encounters in Tourism*. He has a PhD in geography from the University of Portsmouth, UK and is a Fellow of the Royal Geographical Society.

Freya Higgins-Desbiolles is a Senior Lecturer in Tourism in the School of Management, University of South Australia. She has researched and taught on Indigenous tourism for more than a decade; she is a critical scholar employing an Indigenous-rights approach based on Indigenous community engagement and collaborations. Her work through the course Tourism and Indigenous Peoples has received numerous awards for community engagement, teaching and research, including an Australian national award – the Australian Teaching and Learning Council's Citation Award for Outstanding Contributions to Student Learning in 2009.

Megan Holroyd is a PhD Candidate in Geography at the University of Kansas. She has spent time working and researching in Tanzania. She holds a Master of Arts in international studies from the University of Kansas and a Bachelor of Science in sociology from Pittsburg State University. She has taught world regional geography, human geography and African studies courses at the University of Kansas.

James Igoe is an Associate Professor in the Department of Anthropology at the University of Virginia. He is the author of *Conservation and Globalization: a Study of Indigenous Communities and National Parks from East Africa to South Dakota*. He has conducted extensive ethnographic fieldwork related to the environment, identity, and community well-being in Tanzania, the Pine Ridge Reservation (South Dakota) and New Orleans (Louisiana). He is one of the key organizers of a network of scholars, practitioners and community activists concerned with the commodification of nature and culture in the context of global biodiversity conservation.

Tamara Luthy is a PhD student in the Department of Anthropology at the University of Hawai`i at Mānoa, working toward a concurrent MS in botany.

Mary Mostafanezhad is an Assistant Professor in the Department of Geography and a faculty affiliate in Thai Studies at the University of Hawai`i at Mānoa. Mary's current research is situated at the intersection of critical geopolitics and cultural and development studies and explores popular humanitarianism in several contexts, including tourism, fair trade, celebrity humanitarianism and corporate social responsibility. Additionally, Mary is a board member for the Association of American Geographers Cultural and Political Ecology and Recreation, Tourism and Sport Specialty Groups and a co-founder of the American Anthropological Association Anthropology of Tourism Interest Group and the Critical Tourism Studies Asia-Pacific Consortium.

xiv *Contributors*

Ron Nicholls is a Lecturer and Open Universities Coordinator in the David Unaipon College of Indigenous Education and Research at the University of South Australia. His research focuses on global and national Indigenous issues, alternative worldviews, experiential learning, and peace studies. Ron has also worked as a professional musician and from 1980-1995 held the position of Lecturer in Music at the Centre of Aboriginal Studies in Music, University of Adelaide. His recent publications and presentations have focused on the necessity of forging innovative ways of being and the movement toward a post-enlightenment world.

Roger Norum is Postdoctoral Research Fellow at the University of Leeds and Teaching Fellow in Norwegian at University College London. He is a researcher on the HERA-funded project *Arctic Encounters: Contemporary Travel/Writing in the European High North* and is also Research Director of the Marie Skłodowska-Curie Innovative Training Network in Environmental Humanities (2015–2018). Trained in social anthropology, his research currently focuses on mobility, tourism and the global travel-writing industry. He is the co-author, with Alejandro Reig, of *Migraciones* and is co-convenor of ANTHROMOB, the Anthropology and Mobility Network of the European Association of Social Anthropologists.

Carina Ren is Associate Professor at the Tourism Research Unit at Aalborg University, Denmark. In her research, she ethnographically explores encounters, controversies and multiplicity in tourism taking inspiration from material and relational approaches such as ANT. She is a co-editor of *Actor-Network Theory and Tourism: Ordering, Materiality and Multiplicity* and *Tourism Encounters and Controversies Ontological Politics of Tourism Development*.

Grant Rigney is a Ngarrindjeri Regional Authority Board Member at Ngarrindjeri Regional Authority in South Australia.

Jarkko Saarinen is Professor of Human Geography at the University of Oulu, Finland and Distinguished Visitor Professor at the School of Tourism and Hospitality, University of Johannesburg, South Africa. His research interests include tourism and development, sustainability management, tourism-community relations, tourism and climate change, community-based natural resource management and wilderness studies. He is currently the Vice President of the International Geographical Union and Associate Editor of the *Journal of Ecotourism*. Previously he has been Professor of Tourism at the University of Botswana. His recent publications include the book: *Tourism and Millennium Development Goals* (co-edited with Rogerson and Manwa)

Eric J. Shelton is a trustee of the Yellow-Eyed Penguin Trust, an environmental NGO operating from Dunedin, New Zealand, and working to produce protected whole-of-ecosystem habitat primarily for sea birds. Also, Eric writes about the craft of conservation within neoliberalism and the place of nature-based tourism.

Contributors xv

Anna Thompson-Carr is a Senior Lecturer at the Department of Tourism, University of Otago, New Zealand. She is on the editorial boards for *Tourism in Marine Environments* and the *Journal of Heritage Tourism*. Her research interests focus on the interdisciplinary aspects of sustainable ecotourism, adventure tourism, wilderness management and cultural landscapes. Prior to academia Anna was co-owner of two adventure tourism businesses and she is a founding Co-Director of the Centre for Recreation Research (University of Otago). She recently co-edited a book on *Mountaineering Tourism* with James Higham and Ghazali Musa.

Michael Wearing is a Senior Lecturer in The School of Social Sciences, Faculty of Arts and Social Sciences, University of New South Wales (UNSW) Sydney, Australia since 1996. He completed a PhD in sociology at The Social Policy Research Centre, UNSW. In the last 25 years he has taught and published in the areas of youth studies, social policy, sociology and political sociology while an academic at Sydney University and then UNSW. He is the author of several books and other publications in areas of socio-cultural aspects of tourism, youth and community services, youth studies and social policy. His current research interests are in the environment and ecotourism, marginal youth studies and comparative social policy.

Stephen Wearing is an Associate Professor at the University of Technology Sydney. His research and projects are in the area of leisure and tourism studies, with a PhD focused on sustainable forms of tourism. Stephen has made seminal contributions in many areas including ecotourism, volunteer tourism and community development; the importance of community-based approaches in the leisure, recreation and tourism sector has formed the focus of his research. He is a prolific scholar and his work ranks among the top professors in the world for publications in sustainable tourism and leading research journals in that field.

Forrest Wade Young is a Lecturer in the Departments of Anthropology and Political Science at the University of Hawai`i at Mānoa. He holds a PhD in anthropology and MA degrees in anthropology, philosophy and political science. He regionally specializes in Indigenous politics of Pacific Island worlds, and engages research in political and legal anthropology, as well as linguistic anthropology.

Jundan (Jasmine) Zhang recently completed her PhD dissertation in the Department of Tourism, University of Otago, New Zealand. Her research involves working with ideas of 'nature' and nature–society relationship in the context of global tourism. In the past she has conducted long-term research in Shangri-La County, Southwest China. Currently her interests of research include poststructuralist political ecology, environmental discourse, environmental subjectivity, tourism as 'worldmaking', continental philosophy and Chinese philosophy.

Foreword

Rosaleen Duffy

This is a very important book, one which develops a much-needed fresh perspective for tourism studies. The editors are quite correct to point out that the debates on tourism and on political ecology have developed separately, and fusing them together provides an excellent way to develop critical interrogations of tourism. A political ecology frame produces a nuanced understanding of the uneven effects of tourism across the world. Since political ecology is not a strictly unified theoretical approach, it allows for a rich and detailed exploration of a range of issues that are often overlooked, and hence under-researched by scholars of tourism as well as tourism-focused scholars in cognate disciplines such as anthropology, geography and environmental studies. Discussions in this book are not constrained by the narrow frameworks and research questions that often guide tourism studies. For example, instead of asking, How can tourism be made more sustainable?, a political ecology approach prompts us to ask: How do claims to sustainability produce or deepen inequalities? What labor relations does ecotourism produce? What gender relations are produced by sustainable tourism? Or even, for that matter: What do we mean by 'sustainability' in the first place?

It is surprising that no one has thought to produce a collection like this before. Tourism is one of the biggest industries in the world, and demands a much greater degree of theoretical and empirical interrogation than it is given at present. Tourism has continued to grow, despite the ongoing effects of austerity, financial crises and security concerns; the expansion of the industry has exceeded expectations, even from its positive promoters, such as the United Nations World Tourism Organization (UNWTO). One simple indicator of its sheer size and continued growth is the number of trips taken by tourists. In 2012 international tourist arrivals (overnight visitors) reached the 1 billion mark, and in 2014 this rose to a record 1.138 billion. By 2030 UNWTO expects 1.8 billion international tourist arrivals.[1] Tourism is now a central part of the global economy, and one which is only likely to increase in size and significance. Tourism has powerful backers: international organizations, corporations, wealthy individuals, national governments and NGOs. For example, it has been promoted by major global organizations, including the World Bank and the United Nations; indeed the United Nations Environment Programme has even identified tourism as a central means of developing a 'green economy', as laid out at the Rio+20 conference and related report, *The Future We Want*.

With such levels of growth, and powerful underwriters and supporters, tourism has the capacity to transform and reshape labor relations, environmental systems, economies, political systems and societies. For example, international tourism is heavily reliant on aviation to move tourists around the globe – with obvious effects on climate change. Yet climate change itself poses threats to many of the very attractions the tourism industry relies on – the development of Arctic tourism to view wildlife or tours of low-lying coral atolls are obvious examples. The industry itself has responded and, as early as 2003 the UNWTO Djerba Declaration on Tourism and Climate Change noted the potential risk posed by planetary environmental shifts. Despite this, key industry stakeholders (including the UNWTO and large tourism corporations such as TUI) have managed to evade binding regulations and have successfully argued in favor of forms of regulation that do not inhibit the industry's growth.

Furthermore, tourism is often promoted as a growing industry that makes an important economic contribution, especially to marginalized communities in rural areas. Taking a political ecology approach, we can dig a bit deeper: What sort of contribution does tourism really make? Why are its benefits spread unevenly? And have communities necessarily needed to give up access and use rights to certain natural resources to pave the way for tourism development? It is important to remember that once we start to develop tourism in a particular place, it has the capacity to displace other options for realizing economic development and environmental conservation goals. We need only to examine the rapid development of tourism in 'emerging destinations' such as Myanmar and Vietnam to understand the power of the industry to fundamentally reshape environments, and restructure economic, political and social relations.

Tourism is one expression of global capitalism, and it can produce rapid transformations; but we should not limit our interrogations by asking whether this is for better or worse. Rather, it is more compelling and constructive to focus on how benefits are distributed and why distribution has taken that form. We should not assume tourism is a completely negative force either, and this book provides a welcome space to discuss what its benefits might be and who is able to access them – but it does so in a way that is markedly different from the usual arguments anchored in assumptions about the claimed economic benefits of the industry. Of course, it is vital to place tourism within the context of wider capitalist relations; tourism has been a decisive force in the extension and intensification of neoliberalism. However, many of these chapters serve as important reminders that seeing tourism only through the lens of capitalism can cut off potentially productive lines of enquiry centered on subjects such as discourses of tourism, the agency of nature and post-human geographies, the centrality of the body or the ethics of tourism. Political ecology helps us consider multiple answers – and raise further questions – left out by other modes of enquiry into the complex worlds involved in tourism production and practice. This book is a very welcome and timely step in that direction.

Note

1 UNWTO (2014) UNWTO Tourism Highlights: 2014 Edition.

Reference

The Future We Want (2012) Available from: http://www.uncsd2012.org/content/documents/727The%20Future%20We%20Want%2019%20June%201230pm.pdf [Accessed 30 August 2015].

Introduction

*Mary Mostafanezhad, Roger Norum,
Eric J. Shelton and Anna Thompson-Carr*

As we write this introduction, there are more planes flying tourists across the globe than ever before. Sea levels are rising faster than ever anticipated. Simultaneously, climate change threatens the lives and livelihoods of millions of people internationally as the global tourism industry continues to grow at unprecedented rates. Many small island nations – often economically dependent on international tourism – have been compelled to purchase land on higher ground to relocate their citizens. Kiribati in Micronesia, for example, has acquired 20 square kilometers of land in Fiji, some 1,200 miles away. Meanwhile, the government of Fiji continues to explore new ways of increasing international tourist arrivals, while simultaneously striving to account for and address the integral links between tourism and climate change (in 2007 the country merged its Ministries of the Environment and Tourism). Tripadvisor.com's Best Hotel in the World award put the Maldives, an Indian Ocean island nation, back into the international tourism spotlight in 2015 (tripadvisor.com). And yet, it is expected that, by the year 2100, many of the islands comprising the Maldives will be submerged. The Maldivian government now puts the revenue it collects from tourism towards the purchase of higher ground abroad on which to eventually relocate its residents, who are among the world's most at-risk climate refugees-in-waiting (Kothari, 2014). Emergent markets such as 'last chance' (Eijgelaar *et al.*, 2010; Hall *et al.*, 2010; Lemelin *et al.*, 2013) and 'apocalyptic' (Tucker and Shelton, 2014) tourism are indicative of a growing societal consciousness about global environmental crises. More tellingly perhaps, is how tourism has emerged as an increasingly popular practice through which to address to this emerging sentiment.

Tourism-as-environmental-end-game has in many ways become a seemingly logical outcome of the global nature of late capitalism. Newly emerging economies such as Myanmar (Burma) rapidly tap into the global tourism market as the government simultaneously displaces millions of residents for expeditious tourism development along coastlines, engendering a perhaps new kind of refugee: the 'tourism refugee'. In Kenya, meanwhile, tens of thousands of Maasai people have been dispossessed of their homes through the allocation of game reserves for tourists; they continue to battle with the government over access to their ancestral lands, which they need to graze cattle, their primary livelihood, just as they are compelled to commoditize their local culture for tourist consumption (Bruner, 2001). These events represent the spread of tourism habitat and the myriad economic, ecological, social and political processes that accompany what is often understood as neoliberal engagement with conservation (Brockington *et al.*, 2008, p. 131).

In this introduction we lay out a framework for understanding the impact of the spread of tourist habitats, using a *political ecology* approach in order to better understand relationships between tourism, communities, politics and the environment. The chapters in this collection draw on theoretical and ethnographic case studies that span Central and South America to Africa, and from South, Southeast and East Asia to Europe and the High Arctic. While geographically broad in scope, this collection of chapters illustrates how tourism-related environmental challenges are shared across vast geographical distances while also attending to the nuanced ways they materialize in local contexts and therefore demand the historically situated, place-based and multi-scalar approach of political ecology.

Political ecology has become an increasingly relevant framework from which to understand socio-environmental change. As global economic, social and environmental integration intensifies at unprecedented rates. Part of this intensified integration is the expansion of our 'knowledge of, and sensitivity to, transborder and global forms of environmental harm (ozone depletion, global climate change, toxic dumping), and the extent to which green issues are legislated through interstate agreements' (Watts and Peet, 2004, p. 4). Political ecologists' perspectives on the environment illustrate how power and structural relations at different scales have implications for local people's natural resource management and land use practices. In addition to these multiple scales of analysis, political ecology scholarship has the benefit of being diachronic in that they attend to historical factors that contribute to land use change and variability as well as to human–environment relations.

Before examining emerging themes in the political ecology of tourism, we first ask the fundamental question: What is political ecology? While never intended to be a unified theory, perspective or methodology, political ecology is an approach to enquiry that focuses on historically situated, place-based and multi-scalar perspectives of ecological change, and their deep linkages to a range of social relations. We then present a brief review of academic literature linking political ecology *with* tourism, as well as outline a starting point for developing a conceptual framework for the political ecology *of* tourism. Finally, we examine a triad of core, linked themes (communities and power; conservation and control; development and conflict) shared by political ecology and tourism studies scholars around which we have organized this collection, addressing the rationale for this tripartite division of chapters.

We intend for this collection to serve as a thoughtful point of departure for new conversations that address the overlapping agendas of political ecology and tourism studies, as well as broader academic engagements with the relationships between tourism and communities, as well as conservation and development. It is our hope that scholars from across a range of disciplinary backgrounds will find connections between their own work and the critical issues addressed in these chapters.

What is political ecology?

In this section, we examine both what political ecology is and how it is a productive – if under-utilized – analytical lens in tourism studies. Political ecology is an interdisciplinary framework with which to examine ecological matters from a broadly defined political economy perspective (Bryant and Goodman, 2006; Peet and Watts, 2004; Rocheleau et al., 2013; Sparke, 2007; Paulson et al., 2005). Yet, as Paul Robbins has notably declared, '[o]ne need not be a political ecologist to mobilize the resources, or learn from the insights, of political ecology' (2012, p. viii). Furthermore, political ecology signifies a community of practice and denotes various qualities of a text, rather than standing as a method, theory or *single* perspective.

Scholars of political ecology

> (t)rack winners and losers to understand the persistent structures of winning and losing; are narrated using human–non-human dialectics; start from, or end in, a contradiction; simultaneously make claims about the state of nature and claims about claims about the state of nature.
>
> (Robbins, 2012, p. 87)

Stories of winning and losing are 'stories of justice and injustice', where environmental actions 'have causes and consequences that are uneven between communities, classes and groups' (ibid.). These consequences must be understood as being 'non-incidental, persistent and repetitive: a structure of outcomes that produces losers at the expense of winners'. Political ecology attempts to understand the complexities of how 'winners and losers' come to be situated within 'lose-lose' outcomes. In terms of human–nonhuman dialectics, political ecology focuses not on how 'individual things . . . cause outcomes or explain other things in a straightforward way, but instead how things and relations change by becoming entangled with one another . . . things come to explain one another' (ibid., p. 94–5).

This idea of mutual entanglement is central to political ecological thinking. Building on this analytical framework, each chapter in this collection examines ways in which some of the most critical issues of our time – e.g. global economic and social inequality, human displacement, environmental degradation, natural resource competition – are linked to tourism.

Originating as a critique of an allegedly apolitical cultural ecology and ecological anthropology, political ecology illustrates the unavoidable entanglement of political economy with ecological concerns (Zimmerer, 2006). Additionally, in political ecology scholarship, we observe a shift in focus from local to cross-scale analysis. Political ecologists are often concerned with the influence of transnational political and economic processes on local, cultural-specific contexts (Clarke *et al.*, 2008, p. 28; Peet and Watts, 2004; Robbins, 2012). Additionally, the almost exclusive focus on 'Third World' political ecology has been broadened to include 'First World' political ecology scholarship as well as opened up to include the relationship between the two (McCarthy, 2002). Thus clear distinctions between so-called First and Third World political ecologies are increasingly blurred by more nuanced understandings of global and local environmental relationships (Bryant and Bailey, 1997). Despite its now widespread use by social scientists, political ecology has never occupied a single coherent theoretical position, in part because the specific definitions of politics, ecology and political economy are themselves hard to isolate and delineate (Igoe, 2010; Igoe, 2013; Watts and Peet, 2004). In its broadest sense, political ecology scholarship examines how the various articulations of politics and the economy are interconnected with ecology. Studies of neoliberalism and its entanglements with the environment, for example, are frequent topics of political ecologists' examination (Heynen *et al.*, 2007; McCarthy, 2005; McCarthy and Prudham, 2004). Building on critical theory approaches such as green Marxism and post-structuralism, political ecology scholars foreground the dialectics of nature and society and provide new ways to think about the relationship between tourism, politics, economics, culture, and the material world. Watts and Peet explain how '[i]f political ecology stood for anything from the beginning, it was not about blaming the poor or the rich for environmental change, but examining the social relations of production and the broader political economy of which actors . . . were part' (Watts and Peet, 2004, p. 41).

In their seminal text, *Land Degradation and Society* (1987), Blaikie and Brookfield define political ecology as considerations of ecology inflected by a broadly defined political economy, encompassing 'the constantly shifting dialectic between society and land-based resources, and also within classes and groups within society itself' (Blaikie and Brookfield, 1987, p. 17). Blaikie and Brookfield were instrumental in pushing forward the 'plural approach' to the meaning of ecological practices in so far as 'one person's profit was another's toxic dump' (Blaikie and Brookfield, 1987, p. 10), a point reiterated by Robbins (2004; 2012), among others. Political ecology now contains a sort of 'phenomenology of nature', building upon the notion that environmental problems are understood in sometimes disparate ways (Watts and Peet, 2004, p. 20). Thus, in political ecology analyses, the plurality, dynamism and complex interrelated nature of ecological concerns are brought the fore. Perhaps most obviously, what political ecology has done, especially for tourism studies, is to denaturalize the environment as a discursively constructed category itself and offer opportunities to examine its myriad forms in tourism contexts. This opening up to scrutiny of the category of 'environment', while readable as an admonition against reification of processes, continues to be problematic. Political ecology scholars remain instrumental in uncovering the complexities, ambiguities and productive potential of historically rooted terms such as *ecology, ecological practices, nature, environment, environmental degradation, restoration* and *production*. Characteristic of a political ecology approach to the production of knowledge is the notion of action; 'political ecology as *something people do*' (Robbins, 2012, p. 4). As such, a contextual analysis of political, economic, social, and ecological relations has the potential to provide a broader understanding of the power structures concerning people and their surroundings.

The origins of contemporary political ecology studies typically are traced to the 1970s, when anthropologist Eric Wolf, environmental scientist Grahame Beakhurst and journalist Alexander Cockburn employed the term to address the relationship between the distribution of access and control over resources and environmental degradation The emergence of political ecology as a framework coincided with the historical awakening of a global consciousness and social action aimed at addressing pressing environmental issues. In the US the natural world came to be called 'the ecology', an act of reification creating an entity perhaps best described as 'nature plus connectedness'. Barry Commoner's (1990) persistent goal of *Making Peace with the Planet* through John Bellamy Foster's (2009) *The Ecological Revolution: Making peace with the planet* in many ways find their origins in Earth Day of 1970, which remains a significant turning point in environmental activism. The *United Nations Conference on the Environment*, which took place in 1972, is also considered a defining moment that marked early calls for examining the 'environmental crisis' from a political economy perspective with 'a sensitivity to the dynamics of differing forms of, and conflicts over, accumulation, property rights, and disposition of surplus' (Watts and Peet, 2004, p. 7). These perspectives, largely informed by Marxian forms of analysis, were different from earlier accounts of environmental crises which often sought to draw links with forces such as technology, population growth or land use practice.

Such critiques, arising out of peasant studies and structural Marxism, provided important frameworks that drew attention to new kinds of explanations of environmental degradation (Shearer *et al.*, 2009), and that presented economic-based arguments to support ideas such as sustainability and biodiversity. This tradition, involving the wise use of land and economics, is often informed by the principles of the broad intellectual histories of Marxism. 'Marxian' values are predicated on the claim that there can be no universal environmental protection without concomitant social justice (e.g. Smith, 1984). Thus, poverty reduction through a reallocation of resources is a central plank of conservation strategies emanating from this tradition.

Marxists' engagement with the environment was called into being in part through the publication of Neil Smith's (1984) *Uneven Development*, Ted Benton's

(1996) *The Greening of Marxism* and Foster's *Marx's Ecology: Materialism and nature* (2000). These authors paved the way for an analytic project allowing political ecologists a soft Marxian (as opposed to a much more determined Marx*ist*) engagement with environmental protection despite their lack of a unified critical position. As such, ideas of surplus appropriation, marginalization and relations of production were applied to the analysis of environmental relationships. In addition to political economy, disciplinary approaches from geography, anthropology and environmental studies collectively contributed to the establishment of this field of inquiry.

Core elements of political ecology analyses: history, place and scale

While political ecologists may lack universal agreement on the nature of political ecology analysis, the shared objects of study among political ecologists are typically related to human–environment relations and the concomitant ways in which access to and control over land and natural resources are distributed (Watts and Peet, 2004). This focus allows political ecologists theoretical flexibility to move away from the structural Marxist focus on systems of production towards a more nuanced understanding of the significance of discourse and post-structural thought. Additionally, the politics of scale figures prominently in political ecology narratives. While political ecology focused scholars apply the framework to a variety of contexts and in sometimes disparate ways, most agree that political ecology analysis should be: (1) historically situated; (2) place-based; and (3) multi-scalar (Blaikie and Brookfield, 1987; Peet and Watts, 2004; Robbins, 2012). We will now elaborate on what these perspectives each mean and offer some thoughts on what they can lend to scholarly enquiry.

Historically situated

Political ecology frameworks are linked, at least in part, by a common acknowledgement that all ecological practices and environments require historical contextualization. In the context of tourism, this means paying attention to, for example, the legacies of practices of colonialism, migration and institutionalized discrimination. Adams (2003) traces attempts to decolonize colonial notions of nature or, as is more common, to retain vestiges of a colonial imaginary better suited to the development of contemporary ecotourism. Additionally, political ecology is concerned with writing alternative histories that often challenge previously accepted unilinear explanations of agriculture and environmental change (Watts and Peet, 2004, p. 15). In this way, the analysis of, say, local resistance to tourism development initiatives, would best be considered within the broader historical relationship between colonizers and the colonized, community members and tourism developers, and indigenous people and the state. While these kinds of ongoing power struggles have been the subject of significant tourism scholarship to date, political ecology framework necessitates historical contextualization. Nixon (2011, p. 175), for example, historically situates how 'South Africa's traumatic history of colonial conquest, land theft, racial partition and racist conservation places particular pressure on those conservation biologists, political ecologists, writers and activists committed to reimagining, during the post-apartheid era, their society's inherited culture of nature'. This inherited 'culture of nature' is where much of South Africa's inbound tourism takes place.

Place-based perspective

Political ecologists tend to conduct place-based research where they engage with local communities to experience firsthand the meanings ascribed to ecological practices. Often wielding an ethnographic approach, political ecologists are frequently familiar with local languages and create social bonds with local community members that allow them privileged access to the more nuanced understandings of nature and the environment as well as political, cultural, and environmental change. Critical to political ecology analyses that foreground the significance of 'place-based perspectives', locally based fieldwork contributes to the more subtle ways these meanings are co-constructed and negotiated within social and environmental geographies. A place-based perspective tends to depart from other forms of research that uncritically apply place-less models or theories across diverse social and geographic contexts. Similarly, empathy and sensitivity to cultural, religious, linguistic, social and economic contexts (among others) is a critical aspect of political ecological research. Feminist political ecologists, for example, have helped define the field through place-based investigations of gender, sexuality, race, class, ethnicity and disability. Sociological categories such as these play into access and control over land and natural resources and are critical to effective environmental planning and policy initiatives, including tourism-based initiatives. Significantly, feminist political ecology has contributed to advances in political ecology research around several of these concerns (Harcourt and Nelson, 2015; Mollett and Faria, 2013; Rocheleau and Edmunds, 1997; Rocheleau *et al.*, 1996; Schroeder, 1993). Feminist political ecologists are often focused on 'the ways in which environmental concerns are traced through gender roles, knowledges and practices' (Watts and Peet, 2004, p. 13) through issues such as the 'silencing' of women's environmental knowledge. In an instructive example of this kind of work, Dianne Rocheleau and David Edmunds re-examine resource use and access in the context of trees and forests through their analysis of the gendered inequalities. They illustrate how tree tenure is embedded in overlapping rights which rather than being represented on two-dimensional maps, requires an understanding of the fluidity of power relationships between different stakeholders and the flexibility of their subject positions – positions, they argue, that need to be recognized in legal and theoretical understandings of property rights. As a result, they argue that artificial dichotomies between 'haves' and 'have nots' – which they suggest tend to be respectively linked with men and women – are more negotiable than previously assumed and thus more equitable distributions of resource tenure regimes can be realized (Rocheleau and Edmunds, 1997).

Thus, they draw our attention to the fact that, within communities themselves, there are complex and ongoing struggles over the discursive and material conditions of the livelihoods of individuals that are mediated by social categories such as gender. These insights are critically relevant to tourism studies where considerations of gender, race, ethnicity, class and so on are at times overlooked. Also significant here is the way the 'community' becomes much more dynamic and hard to pin down in the same way that a 'placeless' perspective might suggest, or indeed encourage. As tourism scholars continue to embrace political ecology as a social science framework it will be important to consider the diversity of perspectives from any research community as well as the flexibility of what 'counts' as the 'community' as well as how there may still be unheard, but equally valid, voices among the non-community participants.

Multi-scalar approach

Finally, political ecology analyses foreground multi-scalar approaches that examine relationships between and across individuals, communities, the state, regions and international agencies (Painter, 1995). Political ecologists follow a mode of explanation that evaluates the influence of variables acting at a number of scales, each nested within another, with local decisions influenced by regional policies, which are in turn directed by global politics and economics (Robbins, 2012). Research pursues decision-making at many levels from the very local – where individual land managers make complex decisions about cutting trees, plowing fields, buying pesticides, and hiring labor – to international – where multilateral lending agencies shift their multi-billion dollar priorities from building dams to planting trees and farming fish (Robbins, 2012, p. 20).

A multi-scalar approach is of critical importance to understanding how power circulates through and mediates relationships and socio-environmental and economic behaviors. Duffy (2002), for example, examines ecotourists and the structures that introduce and support them in a politics of exploitation, while Brockington *et al.* (2008) links conservation and capitalism, as they are likely to influence the future, at the scale of large protected areas. More recently, Brockington and Duffy (2011) together collected illustrations of the power of capitalism to produce and reproduce itself through conservation and now biodiversity in a multi-scalar conservation effort.

Political ecology of tourism: a review

In a recent article Douglas (2014) rhetorically asks, '(w)hat's political ecology got to do with tourism?' Indeed, while there is a significant body of work in tourism studies that addresses issues relevant to political ecology analyses, (Scott *et al.*,

2012 cf.; Albrecht, 2010; Hall and Higham, 2005) with a few notable exceptions there has been relatively little work that specifically applies a political ecology framework to the study of tourism practices (Cole, 2012; Douglas, 2014; Gössling, 2003; Mostafanezhad, 2015; Stonich, 1998). In this section we explore what we believe to be promising – if not yet fully developed – collaborative opportunities between political ecology and tourism studies. We begin with a brief review of the limited-yet-emerging literature approaching tourism from a political ecology framework.

Political ecology of tourism has, to date, been most clearly spearheaded by work that highlights the familiar dilemma that exists between the simultaneous desire for much-needed foreign exchange from tourism development, and environmental sustainability. In one such groundbreaking article, 'Political Ecology of Tourism' (1998), Susan Stonich examined how the poorest community residents in Bay Islands, Honduras, were adversely affected by unchecked tourism development while those who benefited the most were the more powerful national and international stakeholders. Stroma Cole and Nigel Morgan (2010) build on Stonich's work through Cole's examination of water equity and tourism in Bali (Cole, 2012; Cole and Browne, 2015; Cole and Ferguson, 2015). Cole highlights the relationship between social power and ecology in Bali where 80 percent of the economy depends on tourism. In pointing out how developing countries are often more vulnerable to some of the issues related to water inequity as a result of power differences between stakeholders, she paved the way for future tourism scholars to examine relationships between tourism development and unequal access to land and natural resources. Stonich's and Cole's work is indicative of how political ecology can contribute to a more nuanced understanding of the ways tourism is implicated in broader, historically rooted social and economic inequalities at local, national and international scales. Other scalar approaches involving tourism include Stonich's (2000), *The Other Side of Paradise*, where she calls for the broadening of focus on tourism, conservation and development to include consideration of unintended negative consequences for a group of islands. Gössling has been particularly prolific in writing on the political ecology of tourism, especially in relation to climate change and small island states (Gössling, 2003; Gössling *et al.*, 2008; Patterson and Rodriguez, 2003; Scott *et al.*, 2012; Stonich, 2003). Indeed, islands as well as other topographically isolated regions have proven solid fodder for calling into relief issues of concern via political ecology frameworks. Gezon's (2006) edited volume *Global visions, local landscapes*, for instance, provides a political ecological analysis of conservation, conflict and control in one section of a large island, northern Madagascar – an analysis demonstrating simultaneous global implications. Kütting's work (2010) examines the relationship between the environment and tourism development in Greece while Sharma, Manandhar and Khadka (2011) examine the political ecology of tourism on Mount Everest in Nepal. In contrast to Smith's (1984) earlier focus on the production of space, these authors attend to particularized place, either in separate countries or in collective insights into tourists experiencing 'poverty, power and ethics' within slums (Frenzel and Koens, 2012). In a related vein, Brockington and Duffy (2010) as well as Brockington *et al.* (2008) examine the political economy of biodiversity conservation and protected areas, and while not their explicit focus, this collection addresses myriad issues relevant to the political ecology of tourism.

Despite notable exceptions such as those addressed above, political ecology of tourism is still a fledgling, yet provocative, framework with the potential to radically shape the way we think about the relationship between tourism and environmental change. Indicative of this lack of attention, the index of the first edition of Robbins's (2004) wide-reaching book *Political Ecology: A Critical Introduction* has no entry for *tourism*, while the second edition has just two. This is perhaps not surprising, since tourism scholars themselves have little engaged with or illustrated the relevance of political ecology to tourism. Additionally, there are numerous points of intersection between current work in tourism studies and political ecology that are rarely brought to their fullest potential.

Political ecology of tourism: towards a conceptual framework

In this section we offer a point of departure for a conceptual framework for analyzing the political ecology of tourism. Specifically, we illustrate how the diversity of tourism forms, as well as the theoretical and topical foci of analysis in tourism studies, could be productively reconsidered using political ecology as an analytical lens. While it is beyond the scope of this introduction to address all potential topics of relevance to political ecology of tourism, we do however offer some initial (and, we believe, critical) considerations from which to examine several core concerns in tourism studies.

Drawing extensively from post-structuralism, political ecology scholars emphasize the role of power, knowledge and discourse in the construction of what passes as the 'environment'. Since the 1990s, political ecologists have foregrounded the role of a number of key issues summarized by Watts and Peet as 'knowledge, power and practice, and justice governance and ecological democracy' (Watts and Peet, 2004, p. 20). Environmental knowledge is plural and contested. Watts and Peet point out how 'most knowledges are not simply local but complex hybrids drawing upon all manner of knowledges', and identify three core issues in regards to these knowledges: '[f]irst a recognition that environmental knowledge is unevenly distributed within local societies; second that it is not necessarily right or best just because it exists . . . ; and third, that traditional or indigenous knowledge may often be of relatively recent invention' (Watts and Peet, 2004, p. 20). In this way, environmental knowledge itself is always, already political and the ways in which some knowledges are privileged while others are marginalized are a significant focus of research among political ecology scholars.

Since political ecology is something people 'do' rather than something people 'are', many people do political ecology who do not necessarily call themselves political ecologists (Robbins, 2012, p. 20). Three common assumptions link political ecology analysis including:

> the idea that costs and benefits associated with environmental change are for the most part distributed among actors unequally ... [which inevitably] reinforces or reduces existing social and economic inequalities ... [which holds] political implications in terms of altered power relations to other actors.
> (Bryant and Bailey, 1997, p. 289 cited in Robbins, 2012, p. 20)

Core considerations for political ecology scholars include questions such as: What causes regional forest loss? Who benefits from wildlife conservation efforts and who loses? What political movements have grown from local land use transitions? These questions attend to the five dominant narratives in political ecology including: (1) degradation and marginalization; (2) conservation and control; (3) environmental conflict and exclusion; (4) environmental subjects and identity; and (5) political objects and actors (Robbins, 2012, p. 20).

The first narrative – degradation and marginalization – is based on the observation that land degradation is often blamed on marginalized people rather than the political and economic forces that compel environmentally unsustainable practices (Robbins, 2012). The second theme – conservation and control – highlights how conservation efforts can sometimes (albeit inadvertently) contribute to political and economic marginalization. Third, conflict and exclusion point to how environmental conflicts are embedded in broader power struggles that are often entangled in race, class, gender and ethnic politics around access to land and natural resources. The fourth theme – environmental subjects and identity – addresses how identities (including new and emerging identities) are linked to access to control and struggles over livelihood and environmental resources. The fifth and final theme – political objects and actors – foregrounds the role of nonhuman actors (e.g. trees, cattle, climate, volcanoes, palm tree oil etc.) in political and economic struggles for control over them. Thus, Robbins views political ecology as offering both 'a "hatchet" to take apart flawed, dangerous, and politically problematic accounts, and a "seed," to grow into new socio-ecologies' (2012, p. 20) thereby characterizing a new kind of narrative.

These themes serve as a framework upon which political ecology scholars often situate their own work and also provide useful analytical starting points from which to revaluate the framework for the analysis of tourism practices. As mentioned, political ecology relates to a variety of different issues that are of relevance to tourism practices. Below we elaborate on several themes that tourism scholars have traditionally been concerned with and how they might be re-examined from a political ecology perspective.

Douglas has written that '[b]roadly speaking, political ecology scholars seek to understand how the human–environment relationship is produced, reproduced, and altered through discursive and material articulations of nature and society' (Douglas, 2014, p. 9). From this perspective, the extensive literature on tourism discourse and the environment (cf. Norton, 1996) could aptly engage with what Watts and Peet (2004) refer to as 'discursive ecological formations'. Discursive ecological formations can be used to analyze the role of the media and popular culture in creating environmental imaginaries of a region, people and/or place as well as how tourism imaginaries are mediated by history, politics, economics and culture (West, 2012). Cultural and historical representations intersect with colonial legacies and postcolonial rule that plays out in unique and place-based ways in tourism practice. While there is a significant literature on postcolonialism in tourism (cf. Hall and Tucker, 2004), the relationship between tourism, power and the environment is ripe for a political ecology of tourism critique. Concepts such as 'green governmentality' (cf. Chapters 6 and 7 in this volume) as well as the role of what Michel Foucault referred to as the 'conduct of conduct' – by which he means the ways thought and behavior are managed through multi-scalar power relations – are deeply relevant to tourism studies.

Sustainable tourism too, is an aspect well worth consideration with a political ecology lens. Sustainable tourism advocates have sought to distinguish its practices from mainstream mass tourism through their facilitation of tourism, which they often intend to be sensitive to ecological, cultural and social contexts as well as help to generate income for local communities while contributing to minimal or positive impacts on the destination. Douglas highlights how 'sustainable tourism and tourism more broadly are not merely rooted in developmentalism, but are fundamentally political, economic, social, and ecological' (Douglas, 2014, p. 11). As a result, discursive imaginaries of place are core elements of understanding possibilities for sustainable development in the context of people and nature (Escobar, 1995; 1999; West, 2008; West and Carrier, 2004). Sustainable and community-based tourism research has drawn attention to 'the panoply of political forms . . . contestations within state bureaucracies – and the ways in which claims are made, negotiated and contested' (Blaikie and Brookfield, 1987, p. 12). The focus on community and justice approaches in political ecology creates a number of critical implications. Watts and Peet (2004, p. 25), for instance, highlight the cultural politics of the 'community'. They explain how the community is a political unit and as such, there are methodological implications for social science research on communities. They note the importance of understanding the multiple and often competing voices within communities and community-based development initiatives. This perspective politicizes the role of the 'stakeholder' in the sense that it draws attention to their sometimes disparate access to power and authority:

> First, and most obviously, the forms of community relations and access to resources are invariably wrapped up with questions of identity. Second, these forms of identity (articulated in the name of custom and tradition) are not stable (their histories are often shallow), and may be put to use (they are interpreted and contested) by particular constituencies with particular interests. Third, images of the community, whether articulated locally or nationally, can be

put into service as a way of talking about, debating and contesting various forms of property (and therefore of claims over control and access). Fourth, to the extent that communities can be understood as differing fields of power – communities are internally differentiated in complex political, social and economic ways – we need to be sensitive to the internal political forms of resource use or conservation . . .

The authors further explain situations in which power is non-local, such as when ecotourism – clearly a frequent foci for tourism researchers – works through local chiefs, which are often directed by local elites who may be directed by local, national and/or transnational corporations. Ecotourism, as defined by the International Ecotourism Society (TIES) is 'responsible travel to natural areas that conserves the environment, sustains the well-being of the local people and involves interpretation and education' (TIES, 2015). Ecotourism is particularly apt for political ecology analysis in that many of the laudable and perhaps benign aims of ecotourism inevitably set up, in Robbins's terms, winners and losers.

In contrast (to political economy's focus on commodity chains and globalization), post-structuralism and neo-Marxism have come to the fore in an analysis of how people remake nature through their everyday interactions and broader societal understanding of the relationship between people and nature. Ecotourism may happen on a local level but the conservation implications of the activity has international implications that can be analyzed using political ecology theory as demonstrated in Campbell's critique of global discourse and sea turtle conservation in Costa Rica (2007). As such, the conceptual framework of political ecology provides a contextual lens for analyzing the problems and potentials of tourism in the context of people, nature and power by examining ecological issues from a place-based, multi-scalar and historically situated perspective (Douglas, 2014, p. 12).

The role of late capitalism and inequality in tourism practices is a frequent topic of consideration in tourism studies (cf. Cole and Morgan, 2010) that could, in the context of human–environment relations, offer new insights when analyzed from a political ecology framework. Adopting this approach, Fletcher (2010, p. 172) highlights:

1) The creation of capitalist markets for natural resource exchange and consumption; 2) privatisation of resource control within these markets; 3) commodification of resources so that they can be traded within markets; 4) withdrawal of direct government intervention from direct market transactions; and 5) decentralisation of resource governance to local authorities and non-state actors such as non-governmental organisations (NGOs).

In a similar vein, the introduction in 2009 of *the conservation economy* to New Zealand 'signals a move from intrinsic valuation of the (conservation) estate to extrinsic valuation: the question being, what are the ecosystem services delivered and how is tourism serviced?' (Shelton, 2013, p. 184). Duffy's groundbreaking work on the politics of ecotourism as it intersects with broader trends in sustainable development, transnational capital and economic globalization in the Global South is also instructive here in that it lays the groundwork for later work in political ecology of tourism.

Political ecology themes of development and marginalization are deeply relevant to tourism studies. For example, what is often referred to as 'pro-poor' tourism is of particular relevance to political ecology analysis in its aim to facilitate opportunities for marginalized people to benefit economically and politically from tourism development (cf. http://www.propoortourism.info). Pro-poor tourism advocates highlight the ways in which mainstream mass tourism development often leads to benefits for some groups while further marginalizing the poor (Scheyvens, 2010; 2012). On the other hand, cultural, political, economic and environmental implications of mass tourism similarly call out for political ecology analysis. Political ecology frameworks for understanding relationships between development and marginalization in the context of tourism and the claims, evaluation and outcomes of pro-poor and mass tourism analysis would be welcome and much needed contributions to tourism studies.

Non-representational theory urges us to conduct geographical research that goes beyond representation to focus on 'embodied' experience (Thrift, 2004; 2008). In a related vein, relationships between human and nonhuman actors are of critical importance to political ecology and tourism studies scholars. Tsing's work is particularly instructive for tourism scholars in that she draws attention to the unsatisfactory way in which researchers approach or construct relationships between human and nonhuman animals, since in her view, '[c]onservation biologists segregate nonhumans; political ecologists too often take them for granted as resources for human use' (2005). Thus, Tsing's critique presents an opportunity for political ecologists of tourism to engage further with research questions about the often integral role of animals in tourism, engaging with issues of e.g. authenticity, captivity, representation and/or animal rights.

Themes and structure of the book

Situated at the intersection of politics, ecology and tourism, the chapters in *Political Ecology of Tourism* offer theoretically rich case studies that span the globe. From Africa to North and South America and the Asia-Pacific region, these chapters are illustrative of how a range of political, economic and cultural relations invariably mediate tourism practices. As a result of these relations, some ecological concerns are privileged while others are marginalized. *Political Ecology of Tourism* is organized into three sections: (I) Communities and power; (II) Conservation and control; and (III) Development and conflict. The chapters in these sections provide theoretically situated, empirically grounded illustrations of how several core themes in political ecology research can be productively applied to the analysis of tourism practices. *Political Ecology of Tourism* concludes

Introduction 15

with an overview of the new interdisciplinary terrain that is carved out by these preceding chapters and possible routes for new research on the intersection of political ecology and tourism studies followed by an afterword (authored by James Igoe), which addresses ways forward for political ecology-inspired and tourism-focused scholars. To be clear, we recognize that many of the chapters in each section overlap theoretically and topically with chapters in other sections. Indeed, these overlaps, articulations and convergences are both unavoidable and productive in that they draw attention to the ways politics, economics, culture and the environment are so tightly entangled that any attempt to disengage one from the other inevitably leads to a partial perspective at best.

Part I: Communities and power

Communities and tourism stakeholders are core foci of analysis in tourism studies. The community itself has been thoroughly deconstructed and political ecology focused scholars now regularly ask: Whose community? (cf. Watts, 2000; Young *et al.*, 2001), Whose knowledge? (cf. Escobar, 1998; Peluso, 1993), Whose nature? (cf. Escobar, 1998; Peluso, 1995; Zimmerer and Bassett, 2003) and Whose landscape? (cf. Walker and Fortmann, 2003). Stronza, for example, examines the sometimes disparate implications of tourism for community members and between communities (Stronza, 2001; Stronza and Gordillo, 2008). The chapters in Part I examine complex relationships between communities, the landscapes or settings within which they live or have attachments, and tourism practices. Theoretical concepts such as *sense of place*, *traditional ecological knowledge* and *alternative ways of knowing* are used to investigate issues facing communities – including indigenous communities – which are too frequently disenfranchised from their traditional lands as a result of tourism development (see Chapters 2 and 3). The involvement of local communities within specific ecological and environmental management issues provides both theoretical and practical insights into the management of tourism at contentious landscapes and sites where traditional land occupiers are being recognized within political structures. The roles of race, class, gender and religion within communities are investigated in the context of tourism practices (see Chapters 1 and 4). Additionally, the boundaries of what constitutes a community in urban spaces are examined (see Chapter 5). Each chapter emphasizes the discursive construction as well as the relationship between communities and the management of ecosystems, including protected areas and/or species. Finally, the politics of community 'voice' regarding research as well as the management of places at international, national, regional and local levels are examined throughout the section.

Part II: Conservation and control

The chapters in Part II engage with critical approaches to the study of the deliberate production and sustainable consumption of 'natural' environments (Peluso,

1993; 1995). The political ecology approaches put forth here illustrate how the production of tourism in protected areas is never purely a question of conservation or sustainability (Duffy, 2002; Duffy and Smith, 2003). Rather, nature-based tourism is deeply embedded in broader questions of conservation for whom, and concerns over the control of land and natural resources, and the power of discourse to produce environments (Walker, 2005; West, 2006). For example, Watts and Peet explain how practices of 'green governmentality' refer to both relationships between people and resources/environment as well as governable spaces (Watts and Peet, 2004, p. 32). In this way, '[t]he scales at which government is "territorialized" – territory is derived from terra, land, but also terror, to frighten – are myriad: the factory, the neighborhood, the commune, the region, the nation' (Watts and Peet, 2004, p. 28). Contemporary trends in ecotourism studies often stray from Ceballos-Lascuráin's postulation that central to ecotourism is the observation and protection of nature (1996). The Brundtland Report, invoking natural environments as a set of natural resources, drew ecotourism to a position within political economy. By empirically and theoretically highlighting the existence of social processes such as the distribution of power, disenfranchisement and marginalization within tourism more broadly, the chapters in this section move critical nature-based tourism research from political economy to political ecology. Thus, through the lens of political ecology, the so-called 'winners' and 'losers' of conservation are identified and historically contextualized within a place-based and multi-scalar analysis. Additionally, environmental discourse (see Chapter 9) as well as 'eco-governmentality' (Goldman, 2001; 2004; 2005) and 'green governmentality' (Bäckstrand and Lövbrand, 2006; Dressler, 2014; Rutherford, 2007; Watts and Peet, 2004) are examined by the authors in this section from multiple geographical locations (see Chapters 6, 7 and 8).

Part III: Development and conflict

Some of the most critical issues of our time such as climate change, indigenous rights, urban pollution and human displacement are addressed in Part III. The chapters here examine core points of convergence between political ecology and tourism studies including how tourism development often leads to both 'winners' and 'losers' as well as how this relationship is often embedded in, or can lead to, long-standing environmental conflict. Initiatives to facilitate tourism to new areas such as 'cleaning up the streets' (see Chapter 11) or the development of infrastructure and nature viewing areas are examined (see Chapters 10, 12 and 14), as is the discursive construction and negotiation of development through embodied tourism experiences (see Chapter 13). The examination of climate change debates through the context of tourism development can highlight the political, socio-economic and cultural (e.g. lifestyle) differences between global consumers and local citizens and producers, which are reflected in differential access to tourism experiences as well as disparate consequences from the effects of climate or ecological change. For example, they illustrate how climate politics and cultural politics intersect with tourism in ways that beckon political ecology analyses. We can think of, for instance, how climate change discourse in popular culture is a mediating force in tourists' experience (Boykoff, 2008). Tourists are appealed to by reports of how to reduce their carbon footprints through washing their towels less frequently and buying off their carbon footprints. In this way, the media plays a key role in addressing climate questions, but in sometimes unproductive or even pernicious ways. Like celebrity cultures, tourism cultures have become 'a currency that spends (overly) well in the neoliberal spaces carved out by the increasingly marketized, privatized, voluntary and individualized ways of addressing climate questions' (Boykoff and Goodman, 2009, p. 148). As a result, popular representations of individualizing and neoliberal responses to climate change are perpetuated (Prudham, 2009). Tourism practices are similarly represented through, for example, green consumption and carbon capitalism which are also embedded in market-based development approaches and 'actually existing neoliberalisms' (Brenner and Theodore, 2002; Ferguson, 2010; Peck and Tickell, 2002).

Together, these three sections comprise the major themes around which this book is structured as well as topics representing critical spaces of analytical overlap between political ecology and tourism studies. While this collection does not engage in 'pro' versus 'anti' tourism debates, or address all of the potential examples of how these impacts play out in local contexts, it does consider a captivating range of tourism practices through the lens of political ecology in the anticipation of future research that informs practice and theory. We hope that the contributions in this collection will spark meaningful discussions amongst critically engaged tourism scholars and industry practitioners and planners, thus addressing the numerous ecological problems affecting communities and the ecosystems they inhabit.

References

Adams, W. M. (2003) Nature and the colonial mind. In: Adams, W. M. and Mulligan, M. (eds), *Decolonizing Nature: Strategies for conservation in a post-colonial era*. London: Earthscan Publications, pp. 16–50.

Albrecht, J. N. (2010) Challenges in tourism strategy implementation in peripheral destinations – the case of Stewart Island, New Zealand. *Tourism and Hospitality Planning and Development*, 7, pp. 91–110.

Bäckstrand, K. and Lövbrand, E. (2006) Planting trees to mitigate climate change: Contested discourses of ecological modernization, green governmentality and civic environmentalism. *Global Environmental Politics*, 6, pp. 50–75.

Benton, T. (1996) *The greening of Marxism*. New York: Guilford Press.

Blaikie, P. and Brookfield, H. (1987) *Land Degradation and Society*. New York: Methuen.

Boykoff, M. T. (2008) The cultural politics of climate change discourse in UK tabloids. *Political Geography*, 27, pp. 549–569.

Boykoff, M. T. and Goodman, M. (2009) Cultural politics of climate change: interactions in everyday spaces. In: Boykoff, M. T. (ed.), *The Politics of Climate Change: A survey*. London: Routledge.

Brenner, N. and Theodore, N. (2002) Cities and the geographies of 'actually existing neoliberalism'. *Antipode*, 34, pp. 349–379.

Brockington, D. and Duffy, R. (2010) Capitalism and conservation: the production and reproduction of biodiversity conservation. *Antipode,* 42, pp. 469–484.

Brockington, D. and Duffy, R. (2011) *Capitalism and conservation.* Chichester: Wiley.

Brockington, D., Duffy, R. and Igoe, J. (2008) *Nature unbound: Conservation, capitalism and the future of protected areas.* London: Earthscan Publications.

Bruner, E. M. (2001) The Maasai and the Lion King: authenticity, nationalism, and globalization in African tourism. *American Ethnologist,* 28, pp. 881–908.

Bryant, R. and Michael, G. (2006) A pioneering reputation: assessing Piers Blaikie's contributions to political ecology. *Geoforum,* 39, pp. 344–366.

Bryant, R. L. and Bailey, S. (1997) *Third world political ecology.* London: Psychology Press.

Campbell, L. M. (2007) Local conservation practice and global discourse: a political ecology of sea turtle conservation. *Annals of the Association of American Geographers,* 97, pp. 313–334.

Ceballos- Lascuráin, H. (1996) *Tourism, Ecotourism, and Protected Areas: The state of nature-based tourism around the world and guidelines for its development.* Cambridge: Cabi Direct.

Clarke, N., Cloke, P., Barnett, C. and Malpass, A. (2008) The spaces and ethics of organic food. *Journal of Rural Studies,* 24, pp. 219–230.

Cole, S. (2012) A political ecology of water equity and tourism: a case study from Bali. *Annals of Tourism Research,* 39, pp. 1221–1241.

Cole, S. and Browne, M. (2015) Tourism and water inequity in Bali: a social-ecological systems analysis. *Human Ecology,* pp. 1–12.

Cole, S. and Ferguson, L. (2015) Towards a gendered political economy of water and tourism. *Tourism Geographies,* pp. 1–18.

Cole, S. and Morgan, N. (2010) *Tourism and inequality: Problems and prospects.* Wallingford: CABI Publishing.

Commoner, B. (1990) *Making Peace with the Planet.* New York, Pantheon.

Douglas, J. A. (2014) What's political ecology got to do with tourism? *Tourism Geographies,* 16, pp. 8–13.

Dressler, W. (2014) Green governmentality and swidden decline on Palawan Island. *Transactions of the Institute of British Geographers,* 39, pp. 250–264.

Duffy, R. (2002) *A Trip Too Far: Ecotourism, politics and exploitation.* Sterling, VA: Earthscan Publications.

Duffy, R. (2008) Neoliberalising nature: global networks and ecotourism development in Madagasgar. *Journal of Sustainable Tourism,* 16, pp. 327–344.

Duffy, R. and Smith, M. (2003) *The Ethics of Tourism Development.* London and New York: Routledge.

Eijgelaar, E., Thaper, C. and Peeters, P. (2010) Antarctic cruise tourism: the paradoxes of ambassadorship,'last chance tourism' and greenhouse gas emissions. *Journal of Sustainable Tourism,* 18, pp. 337–354.

Escobar, A. (1995) *Encountering Development: The making and unmaking of the Third World.* Princeton, NJ: Princeton University Press.

Escobar, A. (1998) Whose knowledge, whose nature? Biodiversity, conservation, and the political ecology of social movements. *Journal of political ecology,* 5, pp. 53–82.

Escobar, A. (1999) After nature: steps to an antiessentialist political ecology. *Current Anthropology,* 40 (1), pp. 1–30.

Ferguson, J. (2010) The uses of neoliberalism. *Antipode,* 41, pp. 166–184.

Fletcher, R. (2010) Neoliberal environmentality: towards a poststructuralist political ecology of the conservation debate. *Conservation and Society,* 8, pp. 171.

Foster, J. B. (2000) *Marx's Ecology: Materialism and nature*. New York: NYU Press.
Foster, J. B. (2009) *The Ecological Revolution: Making peace with the planet*. New York: Monthly Review Press.
Frenzel, F. and Koens, K. (2012) Slum tourism: developments in a young field of interdisciplinary tourism research. *Tourism Geographies*, 14 (2), pp. 195–212.
Gezon, L. L. (2006) *Global Visions, Local Landscapes: A political ecology of conservation, conflict, and control in northern Madagascar*. Lanham, MD: Rowman Altamira.
Goldman, M. (2001) Constructing an environmental state: eco-governmentality and other transnational practices of a 'green' World Bank. *Social problems*, 48, pp. 499–523.
Goldman, M. (2004) Eco-Governmentality and Other Transnational Practices of a 'Green' World Bank. In: Peet, R. and Watts, M. (eds) *Liberation Ecologies: Environment, development and social movements*. New York: Routledge.
Goldman, M. (2005) *Imperial Nature: The World Bank and struggles for social justice in the age of globalization*, New Haven, CT: Yale University Press.
Gössling, S. (2002) Human–environmental relations with tourism. *Annals of Tourism Research*, 29, pp. 539–556.
Gössling, S. (2003) *Tourism and Development in Tropical Islands: Political ecology perspectives*. Cheltenham: Cabi Direct.
Gössling, S., Peeters, P. and Scott, D. (2008) Consequences of climate policy for international tourist arrivals in developing countries. *Third World Quarterly*, 29, pp. 873–901.
Hall, C. M. and Higham, J. E. (2005) *Tourism, Recreation, and Climate Change*. Tonawanda, NY: Channel View Publications.
Hall, C. M., Saarinen, J., Hall, C. and Saarinen, J. (2010) Last chance to see? Future issues for polar tourism and change. In: Hall, C. M. and Saarinen, J. (eds), *Tourism and Change in Polar Regions: Climate, environments and experiences*. New York: Routledge, pp. 301–310.
Hall, C. M. and Tucker, H. (2004) *Tourism and Postcolonialism: Contested discourses, identities and representations*. New York: Routledge.
Harcourt, W. and Nelson, I. L. (eds) (2015) *Practising Feminist Political Ecologies*. New York: Zed Books.
Heynen, N., McCarthy, J., Prudham, S. and Robbins, P. (2007) *Neoliberal Environments: False promises and unnatural consequences*. New York: Routledge.
Igoe, J. (2010) The spectacle of nature in the global economy of appearances: anthropological engagements with the spectacular mediations of transnational conservation. *Critique of Anthropology*, 30, pp. 375–397.
Igoe, J. (2013) Consume, connect, conserve: consumer spectacle and the technical mediation of neoliberal conservation's aesthetic of redemption and repair. *Cultural Geography*, 6, pp. 16–28.
Kütting, G. (2010) *The Global Political Economy of the Environment and Tourism*. New York: Palgrave Macmillan.
Lemelin, H., Dawson, J. and Stewart, E. J. (2013) *Last Chance Tourism: Adapting tourism opportunities in a changing world*. Abingdon: Routledge.
McCarthy, J. (2002) First World political ecology: lessons from the Wise Use movement. *Environment and Planning A*, 34, pp. 1281–1302.
McCarthy, J. (2005) Devolution in the woods: community forestry as hybrid neoliberalism. *Environment and Planning A*, 37, pp. 995–1014.
McCarthy, J. and Prudham, S. (2004) Neoliberal nature and the nature of neoliberalism. *Geoforum*, 35, pp. 275–283.
Mollett, S. and Faria, C. (2013) Messing with gender in feminist political ecology. *Geoforum*, 45, pp. 116–125.

Mostafanezhad, M. (2015) 'They Came for Nature': A political ecology of volunteer tourism development in Northern Thailand. In: Finney, S., Mostafanezhad, M., Pigliasco, G. and Young, F. (eds) *At Home and in the Field: Ethnographic encounters in Asia and the Pacific Islands.* Honolulu, HI: University of Hawai'i Press.

Nixon, R. (2011) *Slow Violence and the Environmentalism of the Poor.* Boston, MA: Harvard University Press.

Norton, A. (1996) Experiencing nature: the reproduction of environmental discourse through safari tourism in East Africa. *Geoforum,* 27, pp. 355–373.

Painter, M. (1995) *The Social Causes of Environmental Destruction in Latin America.* Ann Arbor, MI: University of Michigan Press.

Patterson, T. and Rodriguez, L. (2003) The political ecology of tourism in the commonwealth of Dominica. In: Gössling, S. (ed.), *Tourism and Development in Tropical Islands: Political ecology perspectives.* Cheltenham: Cabi Direct, pp. 60–87.

Paulson, S., Gezon, L. and Watts, M. (2005) Politics, ecologies, genealogies. In: Paulson, S., Gezon, L. L. (eds), *Political ecology across spaces, scales, and social groups.* New Brunswick, NJ: Rutgers University Press, pp. 17–37.

Peck, J. and Tickell, A. (2002) Neoliberalizing space. *Antipode,* 34, pp. 380–404.

Peet, R. and Watts, M. (eds) (2004) *Liberation Ecologies: Environment, development, social movements.* New York: Routledge.

Peluso, N. L. (1993) Coercing conservation? The politics of state resource control. *Global Environmental Change,* 3 (2) pp. 199–217

Peluso, N. L. (1995) Whose woods are these? Counter-mapping forest territories in Kalimantan, Indonesia. *Antipode,* 27, pp. 383–406.

Prudham, S. (2009) Pimping climate change: Richard Branson, global warming, and the performance of green capitalism. *Environment and Planning A,* 41 (7), p. 1594.

Robbins, P. (2004) Political ecology: a critical introduction. *Book Reviews,* XXI: p. 242.

Robbins, P. (2012) *Political Ecology: A critical introduction.* New York: Blackwell.

Rocheleae, D. E., Thompson-Slayter, B. and Wangari, E. (1996) Gender and Environment: A feminist political ecology perspective. In: Rocheleae, D. E., Thompson-Slayter, Barbara and Wangari, Esther (eds) *Feminist Political Ecology: Global issues and local experience.* New York: Routledge.

Rocheleau, D. and Edmunds, D. (1997) Women, men and trees: gender, power and property in forest and agrarian landscapes. *World Development,* 25, pp. 1351–1371.

Rocheleau, D., Thomas-Slayter, B. and Wangari, E. (2013) *Feminist political ecology: Global issues and local experience.* New York: Routledge.

Rutherford, S. (2007) Green governmentality: insights and opportunities in the study of nature's rule. *Progress in Human Geography,* 31, pp. 291–307.

Scheyvens, R. (2010) Tourism and development in the developing world. *Tourism Management,* 31, pp. 292–293.

Scheyvens, R. (2012) *Tourism and Poverty.* New York: Routledge.

Schroeder, R. A. (1993) Shady practice: gender and the political ecology of resource stabilization in Gambian garden/orchards. *Economic Geography,* pp. 349–365.

Scott, D., Hall, C. M. and Gössling, S. (2012) *Tourism and Climate Change: Impacts, adaptation and mitigation.* New York: Routledge.

Sharma, S. K., Manandhar, P. and Khadka, S. R. (2011) Everest tourism: forging links to sustainable mountain development-a critical discourse on politics of places and peoples. *European Journal of Tourism, Hospitality and Recreation,* 2, pp. 31–51.

Shelton, E. J. (2013) Narrative frameworks of consideration: Horizontal and vertical approaches to conceptualizing the sub-Antarctic. In: Müller, D. K., Lundmark, L. and Lemelin, R. H. (eds), *New Issues in Polar Tourism.* Dordrecht: Springer.

Smith, N. (1984) *Uneven Development: Nature, capital, and the production of space*. Oxford: Blackwell.

Sparke, M. (2007) Geopolitical fears, geoeconomic hopes, and the responsibilities of geography. *Annals of the Association of American Geographers,* 97, pp. 338–349.

Stonich, S. C. (1998) Political ecology of tourism. *Annals of Tourism Research,* 25, pp. 25–54.

Stonich, S. C. (2000) *The Other Side of Paradise: Tourism, conservation and development in the Bay Islands*. New York: Cognizant Communication Corporation.

Stonich, S. C. (2003) The political ecology of marine protected areas: the case of the Bay Islands. In: Gössling, S. (ed.), *Tourism and Development in Tropical Islands: Political ecology perspectives*. Cheltenham: Cabi Direct, pp. 121–147.

Stronza, A. (2001) Anthropology of tourism: forging new ground for ecotourism and other alternatives. *Annual Review of Anthropology,* 30, pp. 261–283.

Stronza, A. and Gordillo, J. (2008) Community views of ecotourism. *Annals of Tourism Research,* 35, pp. 448–468.

Thrift, N. (2004) Intensities of feeling: towards a spatial politics of affect. *Geographical Ann.,* 86, pp. 57–78.

Thrift, N. (2008) *Non-Representational Theory: Space, politics, affect*. New York: Routledge.

TIES (2015) TIES announces ecotourism principles revision. Available from: https://www.ecotourism.org/news/ties-announces-ecotourism-principles-revision [Accessed 30 August 2015].

Tsing, A. (2005) *Friction: An ethnography of global connection*. Princeton, NJ: Princeton University Press.

Tucker, H. and Shelton, E. (2014) Traveling through the end times: the tourist as apocalyptic subject. *Tourism Analysis,* 19, pp. 645–654.

Walker, P. (2005) Political ecology: where is the ecology. *Progress in Human Geography,* 29, pp. 73–82.

Walker, P. and Fortmann, L. (2003) Whose landscape? A political ecology of the 'exurban' Sierra. *Cultural Geographies,* 10, pp. 469–491.

Watts, M. (2000) Political ecology. In: Barnes, T. and Sheppard, E. (eds), *A Companion To Economic Geography*. Oxford: Blackwell, pp. 257–275.

Watts, M. and Peet, R. (2004) Liberating political ecology. In: Peet, R. and Watts, M. (eds) *Liberation Ecologies: Environment, development, social movements*. New York: Routledge.

West, P. (2006) *Conservation is Our Government Now: The politics of ecology in Papua New Guinea*. Durham, NC: Duke University Press.

West, P. (2008) Tourism as science and science as tourism. *Current Anthropology,* 49, pp. 597–626.

West, P. (2012) *From Modern Production to Imagined Primitive: The social world of coffee from Papua New Guinea*. Durham, NC: Duke University Press.

West, P. and Carrier, J. (2004) Ecotourism and authenticity: getting away from it all? *Current Anthropology,* 45, pp. 483–498.

Young, Z., Makoni, G. and Boehmer-Christiansen, S. (2001) Green aid in India and Zimbabwe–conserving whose community? *Geoforum,* 32, pp. 299–318.

Zimmerer, K. S. (2006) Cultural ecology: at the interface with political ecology – the new geographies of environmental conservation and globalization. *Progress in Human Geography,* 30, p. 63.

Zimmerer, K. S. and Bassett, T. J. (2003) *Political Ecology: An integrative approach to geography and environment-development studies*. New York: Guilford Press.

Part I
Communities and power

Introduction to Communities and power

Anna Thompson-Carr

Communities are messy places. Watts and Peet point to how the term 'community' 'is often invoked as a unity, as an undifferentiated entity with intrinsic powers, which speaks with a single voice to the state, to transnational NGOs or the World Court. Communities, of course, are nothing of the sort' (2014, p. 24). Communities are collective in nature; the social cohesion that results in a community relies not on the individual but a group of individuals who may share characteristics such as ethnicity, heritage, religious, spiritual, cultural beliefs and values as well as geographical spaces. Watts and Peet note how communities are regarded as:

> a locus of knowledge, a site of regulation and management, a source of identity and a repository of 'tradition', . . . which necessarily turn on questions of representation, power, authority, governance and accountability, an object of state control, and a theatre resistance and struggle.
>
> (2014, p. 24)

Thus, even members of environmental organisations that seek to protect ecological systems through acts of resistance or protesting (their values manifesting in 'political ecology in action') can consider themselves to be united as a community through their shared values (c.f. Lovelock, 2005).

In the past, communities have often been linked by their geographical proximity. Physically, tourism destinations consist of individual or networked communities that are the core of tourism spaces. Thus, today community members engage with tourism from geographically distant spaces, not only as providers of tourism experiences but also through their involvement in planning tourism or supplying tourism businesses. The people living and working within communities include the personalities that are the human face of tourism interactions – providers of positive tourism experiences. Critically, community members are also tourists themselves and travel domestically or internationally thus bringing back new forms of knowledge (e.g. ecological knowledge) that can be rejected or adapted and utilised in the sustainable management of, or involvement in, local tourism. Newcomers to communities, including tourists and migrants, non-resident business and property owners, politicians, government agencies and NGOs significantly influence how communities act. Community activists may be involved in environmental non-government organisations (ENGOs), yet live alongside other community members who are working for businesses or government agencies that could be perceived as being at the core of local environmental problems. Thus community members directly influence political processes as insiders. Community members may be politicians, business leaders or activists. Responsible tourism research requires an awareness of the diverse makeup of the communities in which researchers work. Communities have the right to participate in decision-making processes that influence how tourism activities or infrastructure are planned. Positive outcomes for sustainable tourism development can hinge on the ability of communities to protect, manage and, if necessary, reclaim their traditional resources as a step towards self-determination. Community conflicts may still arise if the need to provide for visitors results in environmental degradation from tourism, with or without community consent. The loss of community access to and control of natural resources can affect their ability to participate in the tourism sector as a means of sustainable development. Decades have passed since Garrett Hardin's seminal paper, *The Tragedy of the Commons*. Hardin called attention to the impacts of over-population and unceasing human demands for using natural resources (Hardin, 1968). Since then, pro-environmental endeavours within communities have dramatically increased. Sustainability development initiatives, such as those advocated in ecotourism or voluntourism, have sometimes been viewed as futile - James Lovelock observes such efforts as carbon off-setting may simply be a 'rearrangement of Titanic deck chairs' (Aitkenhead, 2008) – but communities can influence positive change. Communities in countries with weak (or non-existent) environmental legislation or immature natural area planning systems are at a disadvantage when it comes to the management or governance of their natural resources. Unfettered tourism development may not be as harmful as other industries that are more resource intensive. However, communities also face increasing battles when 'growth at all costs' is being pursued.

Within tourism studies there has been considerable interest in researching and understanding the impacts of tourism development and tourists' activities on communities. However, this attention rarely draws explicitly from political ecology frameworks. For example, Takeda (2015) wrote about the Haida Gwaii First Nation people's management of forestry and fishing industries on their islands, but tourism is barely mentioned as it is seen as a recent development option under negotiation by the community. Adams and Hutton (2007) provide a notable contribution in their book exploring communities alienated from their lands as a result of protected area designation. Stonich's (1998) seminal article details one of the first studies of tourism development using a political ecology approach to research the impact on Honduran Ladino and Afro-Antillean communities' members access to and use of freshwater, land and marine resources. Fifteen years after Stonich's paper was published, Cole (2012) used a mixed-methods approach to explore the issues surrounding water inequality in the Canggu community in Bali, where local and outsider activities in the tourism, agriculture, mining and forestry sectors impacted on freshwater resources to the detriment of the local community's future sustainable viability. For indigenous communities the processes of colonialism have often resulted in their being denied access to natural resources. However, post-colonial experiences have seen opportunities for improving livelihoods if lands and waterways are returned or reclaimed through collective action. According to the Secretariat of the Permanent Forum on Indigenous Issues of the United Nations (2004, p. 2):

> Indigenous communities, peoples and nations are those which, having a historical continuity with pre-invasion and pre-colonial societies that developed on their territories, consider themselves distinct from other sectors of the societies now prevailing on those territories, or parts of them. They form at present non-dominant sectors of society and are determined to preserve, develop and transmit to future generations their ancestral territories, and their ethnic identity, as the basis of their continued existence as peoples, in accordance with their own cultural patterns, social institutions and legal system.

Intergenerational empowerment of indigenous people through their ownership or management of the natural resources that are utilised for tourism purposes by or within their communities is an important goal of sustainable tourism development. Political management of resources has implications for how tourism evolves in a community. For example, in New Zealand the Department of Conservation granted local Kai Tahu Māori the sole concession for whale watching in Kaikoura resulting in an internationally recognised ecotourism venture controlled by local indigenous peoples (Thompson, 2013). In an indigenous tourism context, a political ecology approach can include topics such as 'indigenous politicisation and empowerment, indigenous social movements and social movements for alternative and anti-globalisation tourism' (Pereiro, 2013, pp. 214–215). Thus, indigenous community development at local levels is not immune from the influence of international politics and the United Nations implores that governments support the maintenance of traditional relationships with natural resources in the Universal Declaration on the Rights of Indigenous Peoples, Part Six, Article 25 (United Nations, 2007).

The theories of Foucault are significant when one considers community actions involving discourse and resistance, particularly with respect to Robbin's five theses of political ecology (Robbins, 2004, p. 22). When considering the applicability of Foucault to political ecology studies, Robbins noted the relevance to communities when reflecting on Foucault's notion of governmentality which is 'the way governmental management and governance become normalised within communities and individuals themselves' (Robbins, 2004, p. 75). Such normalisation can result in individuals and societies going about daily actions (without questioning) as subjects of the state. A paradox can arise here. If people are 'normalised' as governed beings then will they become politically active and free thinking with a genuine concern for their environmental and ecological wellbeing? Or are they perhaps engaging in ways that those in governance roles may have already anticipated and can thus pre-empt? Fortunately Foucault posits that power is not confined to those in government – power permeates human relationships throughout all of society therefore power can be wrested back from those in governance as an outcome of political action. This can occur when community groups, asserting their environmental concerns, engage in knowledgeable discourse (in particular discourses of local ecological knowledge or traditional ecological knowledge). It has distinct applicability for researchers within a political ecology of tourism context, for instance applying Foucault's understandings of 'power' within tourism studies enables researchers to identify and explore individual and community discourses as a force for social change (whatever the outcome).

Hollinshead (1999) was one of the first tourism/cultural studies scholars to argue the benefits of a Foucauldian approach for enriching our understandings of tourism studies. Communities and individuals, both indigenous and non-indigenous, may present a diversity of discourses and acts of grass roots resistance. Such discourses or acts can challenge those in power (i.e. Robbins's 'corporate, state and international authorities') and assert a community sense of belonging to place (of place identity). Power dynamics within tourism and communities may therefore be understood when local histories and knowledge are examined for insights into socio-political conditions through adopting the Foucauldian methodology of discourse analysis.

Readers may or may not agree with the use of Foucauldian approaches in contributing to our understandings of communities and power dynamics within a political ecology framework. While the researchers in this section make no mention of Foucault there is nevertheless evidence within their chapters that asserts the truth(s) that power relationships are inescapable when exploring the political ecology of communities. The chapter discourses presented suggest that socio-ecological changes occur for the betterment of society and the natural world, as individuals and communities seek to assert the right to have meaningful relationships with the environments within which they live.

Drawing on emerging themes in political ecology of tourism, the five chapters in this section examine communities' experiences with tourism. Examples of both 'First World' and 'Third World' political ecology are represented. In Chapter 1, Stroma Cole explores the gritty implications of tourism developments diverting water resources from local communities in Costa Rica and Bali. Cole uses a gendered approach to analyse her interviews and identifies how tourism development can have different implications for men and women. The following two chapters in this section explore communities whose traditional lands are designated as protected areas and therefore are renegotiating their involvement in national park management systems or with government agencies as they seek to assert traditional land rights. Chapter 2 explores the outcome of community collaboration where a community member (Rigney) was a vital member of the research team throughout the project. The chapter examines the cultural significance of water to a community by recounting the Ngarrindjeri Aboriginal community of South Australia's struggle against a legacy of colonialism. In Chapter 3, Chaderopa reports on his in-depth study of the Makuleke Community (South Africa) and the complexities surrounding their co-management (with SANParks) of an area in Kruger National Park. The remaining two chapters explore urban communities. In Chapter 4 on Vrindavan (India), Luthy explores adaptive management and the ability of devotees on pilgrimage to unite because of their spiritual beliefs in order to conserve a sacred forested landscape. Finally, Chapter 5 on Denmark's Roskilde Festival investigates the social phenomenon of festival eco-voluntourism, showing the implications of new (albeit, transient) communities being forged around shared ideals. The five chapters present a diversity of experiences within equally diverse communities but are united by an emphasis on the significance of active community participation and community-led decision making in planning and development processes.

References

Adams, W. and Hutton, J. (2007) People, parks and poverty: Political ecology and biodiversity conservation. *Conservation and Society,* 5 (2), pp. 147–183.

Aitkenhead, D. (2008) James Lovelock: 'enjoy life while you can: in 20 years global warming will hit the fan', *The Guardian*, 1 March 2008. Available from: http://www.theguardian.com/theguardian/2008/mar/01/scienceofclimatechange.climatechange [Accessed 10 June 2015].

Cole, S. (2012) A Political ecology of water equity and tourism: A case study from Bali. *Annals of Tourism Research,* 39 (2), pp. 1221–1241.

Hardin, G. (1968) The tragedy of the commons. *Science*, 162, December 1968, pp. 1243–1248.

Hollinshead, K. (1999) Surveillance of the worlds of tourism: Foucault and the eye-of-power. *Tourism Management*, 20(1), pp. 7–23.

Lovelock, B. (2005) Tea-sippers or arsonists? Environmental NGOs and their responses to protected area tourism: a study of the royal forest and bird protection society of New Zealand. *Journal of Sustainable Tourism*, 13 (6), pp. 529–545.

Pereiro, X. (2013) Understanding indigenous tourism. In: Smith, M. and Richards, G. (eds), *Handbook of Cultural Tourism*. Abingdon: Routledge, pp. 214–219.

Robbins, P. (2004) *Political Ecology: A Critical Introduction*. New York: Blackwell.

Secretariat of the Permanent Forum on Indigenous Issues of the United Nations (2004) *The Concept of Indigenous Peoples.* Paper presented at the Workshop on Data Collection and Disaggregation for Indigenous Peoples, New York, United Nations.

Stonich, S. (1998) Political ecology of tourism. *Annals of Tourism Research*, 25 (4) 979–983.

Takeda, L. (2015) *Islands' Spirit Rising: Reclaiming the Forests of Haida Gwaii*. University of Vancouver: British Columbia Press.

Thompson, A. (2013) Māori tourism: A case study of managing indigenous cultural values. In: Smith, M. and Richards, G. (eds), *Handbook of Cultural Tourism*. Abingdon: Routledge, pp. 227–235.

United Nations (2007) *Universal Declaration on the Rights of Indigenous Peoples*. New York: United Nations.

Watts, M. and R. Peet (2004) Liberating political ecology. In: R. Peet and Watts, M. (eds), *Liberation Ecologies: Environment, development, social movements*. New York: Routledge, pp. 3–43.

1 A gendered political ecology of tourism and water

Stroma Cole

This chapter sets the context of a gendered political ecology of tourism and water by tracing the genealogy of political ecology from political economy, both in tourism and water as well as examining the gendered components to both. In doing so, I provide an overview of a gendered political ecology of tourism and water. This overview is followed by two case studies which examine the gendered political ecology of tourism and water in Bali, Indonesia and Tamarindo, Costa Rica. In each case study I detail the diverse social and environmental contexts, before examining the salient aspects of a framework for a gendered political ecology of tourism and water.

Before examining the linkages between gender and political ecology, it is useful to map out the context by exploring its genealogy from political economy to political ecology and of political economy to gendered political economy. It is in this transition that the foundations for a gendered political ecology of tourism are located. Political economy is itself not a unified approach, but rather a range of approaches to study the relationship between the economy and the non-economic aspects of society; the common thread being that the political and the economic are irrevocably linked (Mosedale, 2011). From its origins, and following the 'cultural turn' over the past few decades, social scientists increasingly began to move away from a positivist epistemology and recognise the neglect of cultural factors. Political economists started to take cultural meanings and discourses into consideration. Political ecology is a branch that considers the environment as critical to our understanding of the relationship between the social and environmental disparities that are experienced unequally. Thus, with a strong emphasis on understanding conflicts and power dynamics, it provides new insights into how some actors are privileged, while others remain marginalised in their access to and control over natural resources.

The purpose of political ecology is to understand the complex relations between nature and society through a careful analysis of access and control over resources and their implications for the environment (Watts, 2000). There are a number of emerging disciplines with similar approaches and goals, such as environmental justice and socio-ecological systems analysis (Ostrom, 2009). However political ecology and its related fields have been given relatively little attention by tourism studies scholars, with a few notable exceptions including Stonich (1998),

Gössling (2003), Cole (2012) and Cole and Ferguson (2015). In relation to the political ecology of water, the work of Swyngedouw (2009) is most notable; he examines how the mobilisation of water for different uses in different places is a conflict-ridden process and the organisation of the flow of water shows how social power is distributed in a given society. Thus, he illustrates how 'the political ecology approach traces the fundamentally socially produced character of inequitable hydro-social configurations' (Swyngedouw, 2009, p. 58).

Adding gender to political economy studies has been particularly instructive, as so clearly articulated by Peterson:

> feminists have exposed how men dominate the practice of and knowledge production about (what men define as) 'economics'; how women's domestic, reproductive and caring labour is deemed marginal to (male-defined) production and analyses of it; how orthodox models and methods presuppose male-dominated activities (paid work, the formal economy) and masculinised characteristics (autonomous, objective, rational, instrumental, competitive). As a corollary, 'women's work' and feminised qualities – in whatever sphere – are devalued: deemed 'economically' irrelevant, characterised as subjective, 'natural' and 'unskilled', and typically unpaid. For most economists, social reproduction through heterosexual families and non-conflictual intra-household dynamics are simply taken for granted; alternative household forms and the rising percentage of female-headed and otherwise 'unconventional' households are rendered deviant or invisible.
>
> (Peterson, 2005, p. 501)

Tourism development is part of the same development narrative. As Tribe *et al.* (2015) argue, tourism research is undertaken within an overarching neoliberal paradigm, despite the obvious paradoxes. Neoliberal policies aimed at economic development in the world's least developed countries have often promoted tourism growth at any cost, with little space for alternative discourses. Even alternative tourisms have been criticised for pursuing the dispersal of tourism to increase industry profitability rather than to empower host communities, improve their livelihoods and/or preserve environments (Blackstock, 2005; Harrison, 2008; Wheeler, 2003).

While tourism as part of economic growth, guided by neoliberal policies, has been an objective realised in some areas and by some sectors, this process has brought about increased inequalities, not least of all in gender distribution. For example, it is noted how 'Men, especially those who are economically, ethnically and racially privileged, continue to dominate institutions of authority and power worldwide' (Peterson, 2005). If we look at the mission of the World Travel and Tourism Council, for example, whose mission is to promote freedom to travel, policies for growth and tourism for tomorrow, only two out of twenty of the board members are women.

A second crucial feminist addition to the understanding of political economy is the concept of intersectionality; that is, that gender and race are mutually constituted. This is credited to critical race theorist Kimberle Crenshaw who worked to understand race, gender, class and ethnicity as interdependent and interlocking rather than disparate and exclusive social categories (Crenshaw, 1989). Not only have our understandings of economics, politics and development been shaped by men, but by white men. Indeed, as Mollett and Faria (2013) discuss, development thought and practice are deeply racialised.

Rocheleau *et al.* (1996) produced the first feminist political ecology edited collection as a conceptual framework for critiquing international development practice, treating gender as 'a critical variable in shaping resource access and control, interacting with class, caste, race, culture, and ethnicity to shape processes of ecological change' (Rocheleau *et al.*, 1996, p. 4). Gendered political ecology considers a range of environmental rights and responsibilities including property, resources and use of space. People who are socially, economically, politically, culturally, institutionally or otherwise marginalised are highly vulnerable to the impacts of environmental change (Hanson and Buechler, 2015). Feminist political ecologists' privileged knowledge of those most affected or marginalised by neoliberal, colonial or patriarchal systems in which tourism and water policy and practice are carried out. They consider gendered issues of resource access and control as well as gendered collective action or social movements.

As care givers, food providers and health care suppliers in many societies, women are responsible for domestic water provision and management. These roles are often 'naturalised', unpaid and unrecognised but mean that women live with issues of water scarcity and contamination on a daily basis. This is not about a view of women as more natural or closer to nature, but rather that women have different spaces in which they operate. These spaces vary across cultures, but frequently are in reproductive labour (and therefore un-counted by male economists) and are less visible, and the women's voices less heard. A lack of water access is frequently linked to lack of land ownership and women are frequently excluded from water distribution policy and decision-making.

Despite gender being a significant dimension of environmental sustainability and development, gendered analyses of socio-environmental issues are still niche, no more so than in tourism. Plenty has been written on the unequal gendered power relations embedded in the tourism sector (Ferguson, 2011; Gentry, 2007; Schellhorn, 2010; Tucker and Boonabaana, 2012; Vandegrift, 2008). As early as 1996, Kinnaird and Hall set out the highly gendered nature of the tourism industry, arguing that: 'Unless we understand the gendered complexities of tourism, and the power relations they involve, then we fail to recognise the reinforcement and construction of new power relations that are emerging out of tourism processes' (1996, p. 100). There has been significant work on gendered aspects of tourism employment (Sinclair, 1997; Vandegrift, 2008), and recently Baum and Cheung's (2015) study undertook to examine the barriers to women's equality in employment, increase their positions in leadership and derive economic benefits from tourism development. Collectively, this work lies in the dominant discourse that increased tourism is good, and fails to consider differential impacts or consequences of such developments on men and women. Further topic areas within the gender and tourism nexus have included gendered tourism imagery (Marshment, 1997; Pritchard and Morgan, 2000); sex tourism (Dahles and Bras, 1999; Enloe, 1989; Sanchez Taylor, 2001, 2006, 2010; Truong, 1990) and female consumers (Frew and Shaw, 1999; Kim *et al.*, 2007). A number of studies (Harrill and Potts, 2003; Mason and Cheyne, 2000; Ritzdorf, 1995) have shown that women tend to view tourism development more negatively than men. However, these insights are not linked to inequalities of resource allocation but to increased noise, traffic and crime and decreased safety. Nunkoo and Ramkissoon (2010) postulate that women's lack of support for tourism comes from their lack of control and power due to their reduced ownership of resources. More recent studies have considered the additional burdens tourism places on women (Ferguson, 2010), the feminisation of poverty through tourism and the need to both 'move beyond simplistic and fixed thoughts of women's economic empowerment' and to 'consider more fully the cultural complexity and the shifting dynamics of how gender norms, roles and inequalities affect, and are affected by, development and poverty reduction outcomes' (Tucker and Boonabaana, 2012, p. 438). Nevertheless, scholarly work on the environmental impacts of tourism have been gender blind, failing to acknowledge the differences between men and women and frequently reinforcing gender stereotypes. Inequalities in terms of access to resources, greater vulnerabilities and disproportionate negative impacts have not yet been subject to systematic gender analysis. This chapter is, in part, an attempt to rectify this lacuna in social scientific research on these issues.

A brief reflection on the contributions of political economy and political ecology to the study of gender and water provides further critical context for the development of a gendered political ecology of water and tourism. Scholars tend to agree that the social relations of water are poorly understood (Crow and Sultana, 2010; Tortajada, 1998; Truelove, 2011). Yet, it is also widely accepted that historically women have been fundamental in the provision and management of water at household and community levels and that when daily practice is closely examined, it becomes apparent how gender relations interact with class, material inequalities and other social power relations resulting in unequal water security and access. While adult women and young girls frequently take responsibility for providing water, men commonly own the productive assets and make decisions in government offices and communal institutions. The growing field of research on women and water has emerged illustrating how gender has become an important feature of water campaigning, with women taking centre stage in struggles over water management issues (Laurie, 2010). Yet, despite this expanding awareness of the role of women in water resource control, 'most water sector decisions continue to be made based on the false assumption that they are gender neutral, that the population is a homogenous whole, and that benefits reach everyone equally' (Dávila-Poblete and Nieves Rico, 2005, p. 49).

While there is substantive research on women and water, on the whole, gender issues remain 'under-theorized and marginal' (Laurie, 2010, p. 172) contributing to a number of key pitfalls. Through a narrow focus on the involvement of women – as opposed to a larger-scale, more holistic look at gendered power relations – such work tends to obscure the broader power dynamics at work in processes of the neoliberalisation of water (Harris, 2009, p. 391). There are some exceptions, however. Ahlers and Zweeten's work on gender and water demonstrates 'the power dynamics underlying resource allocation with gender inequality being a critical structuring force' (Ahlers and Zweeten, 2009, p. 411) as well as how feminist strategies in water politics should not just seek to 'equalize', but rather to construct possibilities for seeing beyond 'women' (Ahlers and Zweeten, 2009, p. 417). A recent collection of specifically political ecology studies of women and water edited by Buechler and Hanson (2015) highlights a number of important considerations: First, they reiterate women's proactive roles and the politics of emerging social movements and collective action. Second, they draw links between different places and different scales, problematising the urban-rural divides and emphasising the relationship between the two and the importance of the watershed scale of analysis. Third, they examine how climate change impacts on water resource-related market dynamics, adding to the body of knowledge that demonstrates how privatisation of water creates deeper gender and other inequalities (Bakker, 2003; Swyngedouw, 2009; Truelove, 2011). Finally, the chapters stress the integration of multiple scales of analysis from the intimate to the international and remind us of the importance of intersectionality and how other forms of social difference interact with gender to determine responsibilities, vulnerabilities, and governance of water resources. They point out that age has largely been ignored from water resource accounts and analysis.

Before examining the case studies I will briefly reflect on literature that has considered the links between water and tourism. As this has been done in detail elsewhere (Cole, 2012, 2014) a summary here will suffice. Most research on the direct consumptive use of water for tourism that has taken place in the dry land regions of Australia (Crase *et al.*, 2010; Lehmann, 2009; Pigram, 2001) or in relation to the Mediterranean (De Stefano, 2004; Essex *et al.*, 2004; Garcia and Servera, 2003; Kent *et al.*, 2002; Rico-Amoros *et al.*, 2009; Tortella and Tirado, 2011). Elsewhere, I note how:

> despite access to water being a key indicator of progress towards achieving the Millennium Development Goals, the intensification of global concerns over water access and availability and the increasing importance of tourism in developing countries, there has been remarkably little academic research into the link between tourism and the impact of water scarcity on destination populations.
>
> (Cole, 2012, p. 1223)

The exceptions include work by Stonich (1998) in Honduras and Gössling (2001) in Zanzibar. Studies have shown that the per capita use of water by tourists far exceeds that of locals (Crase *et al.*, 2010; De Stefano, 2004). Overall, high-end and luxury tourism establishments tend to consume greater volumes of water than smaller guesthouses (Deng and Burnett, 2002; Gössling, 2001; Gössling *et al.*, 2012). While Gössling *et al.*'s (2015) recent publication highlights the importance of embedded or virtual water, it is the disproportionate water consumption by tourism establishments as compared to local households that has directly impacted on destination communities – both environmentally and socio-economically. Pressure on water resources is 'directly contributing to water scarcity and inequity' (Tourism Concern, 2012, p. 4), posing a direct threat to people's right to health while exacerbating existing poverty and generating conflict and societal instability (Cole, 2012, 2014; King, 2005; OECD/UNEP, 2011) and affecting gender relations. Given the central importance of water to the tourism industry as well as to women's daily lives and the climate induced environmental changes taking place in coastal destinations, this chapter unpicks some of the critical aspects where these topics coalesce.

In what follows I provide outline sketches of the two distinct case studies before going on to the comparative gendered political ecology analysis of their tourism and water. Both case studies are outlined here as detailed examinations have been covered elsewhere (Cole, 2012, 2014, 2015). Research in Bali was conducted in 2010 and in Costa Rica in 2012. In both cases the research conducted made use of a range of ethnographic methods including participant-observation, structured and semi-structured interviews and focus groups. In both destinations a local research assistant was engaged. In Costa Rica 44 women and men took part; in Bali 39 participants were involved. Informants were located by a combination of serendipitous meetings, strategic place-based conversations and snowball sampling. The data was analysed through discussions in the field, saturation, triangulation and the identification of themes for the process of conveying my findings.

Case study 1: Bali, Indonesia

Bali is a small, rugged, tropical island in the centre of the Indonesian archipelago. Measuring 140km by 90km the island has an area of 5,632km^2. As a tropical island it has a warm, humid climate with two seasons: the wet season from October to March and the dry season from April to September. Mean annual rainfall ranges from less than 500mm to as much as 3,500mm in the mountains, which reach 3,142m (McTaggart, 1988). There are three categories of water resources in Bali: crater lakes (which make an important contribution to underground reserves), rivers and groundwater.

The population of Bali is over 4 million. About 15 percent of the population are non-Balinese, the majority of these from the neighbouring islands of Java and Madura and Lombok. While tourism has lured people from across the archipelago, there are additionally around 30,000 foreign migrants living in Bali (Cohen, 2008). The island is a single Indonesian Province, but is divided into nine Regencies (*kabupaten*), each with a Regent (*Bupati*) or area head. In 1999, following 30 years of highly centralised dictatorship, Indonesia gave considerable autonomy to each of the Regencies; new laws invited intense competition over local resources and political power. As Usman (2001), Benda-Beckmann and Benda-Beckmann (2001), and Antlov (2001) have all noted, with Regency autonomy comes the power, obligation and responsibility to raise local revenues.

At the local level Bali is divided into *Banjars*. These traditional neighbourhoods are territorial, social and cultural units (Hussey, 1989). *Banjars* have a significant impact on local-level decision-making. The head of a *Banjar* is democratically elected and decisions are made democratically, but only by male heads of households. Balinese fall into one of four castes; the first three castes, collectively referred to as gentry or nobility, make up about 10 percent of the population and disproportionately fill high political office and own large tracts of land (Howe, 2005). The *Subak*, headed by a *pekaseh*, is a third dimension of social organisation, of particular importance to water management. These are self-governing, democratic associations of farmers who have managed sharing Bali's water for centuries (Lansing, 2007). While neither homogenous nor harmonious, with internal workings that are complex and contested (MacRae and Arthawiguna, 2011), the *Subak* is a religious as well as administrative community. The *Subak* carries out the necessary rituals and ceremonies related to the capture and use of water, itself a sacred substance in Balinese society (Hauser-Schäublin, 2011; McTaggart, 1988). The water temples of Bali are still actively used and maintained by local populations, but the *Subak* system is endangered (Lorenzen and Lorenzen, 2011).

Balinese people live in a highly patriarchal society where land is handed down through male lines. Women play triple roles in society, responsible for not only the domestic but also the economic and customary/ritual/spiritual spheres in life. This three-sided role has led to commentators calling them 'wonder women' (Nakatani, 2004). Women leave their father's home to join their husbands; in the case of divorce children remain part of the husband's family, making divorce rare. Regional autonomy since the fall of the New Order has been linked with a rise in concern about regional cultural identities. The re-conservatism and the reinforcement of patriarchal values that stem from this new focus on regional issues have already had considerable repercussions for women in Bali where issues of gender intersect closely with calls to regional identity (Creese, 2004). Ritual is an important aspect of Balinese life, and while 'constant ritual performance might sometimes be felt as a burden, it appears to be a necessary one' (Howe, 2005, p. 63). Some people are concerned about the effect that lengthy ritual procedures have on their jobs and some employers have become explicit about their irritation due to many requests for time off for ceremonial matters, and thus hire migrant labour considered to be more 'reliable'. According to Howe (2005), this exacerbates already hostile ethnic relations between Balinese and migrant workers.

Bali has been promoted as a tourism destination since Dutch colonial times in the early 1900s. By the end of the 1930s tourists were arriving in the thousands (Picard, 1997). This escalated to tens of thousands in the 1960s and hundreds of thousands by the 1980s. Bali was opened to mass tourism during Indonesia's New Order era (1966–1998), with growth coming at any cost (Lewis and Lewis, 2009; Warren, 1998). From 5,000 rooms across the island in 1987, there were 13,000 in 1992, over 50,000 in 2010 and some 77,500 in 2014 (PHRI, 2015). Despite disquiet among islanders over the exploitation of their island, outside investors and powerful government officials with links to Regents have continued to gain concessions. Tourism has become an integral part of Balinese culture (Picard, 1997) and economy, providing 481,000 direct jobs – equating to 25 percent of the work force – and supporting a further 55 percent, thereby contributing 30 percent of Bali's GDP (BPS,

2010). While many Balinese have certainly benefited from tourism, it is estimated that 85 percent of the tourism economy is in the hands of non-Balinese (MacRae, 2010), while tourism accounts for 65 percent of the island's water consumption (Merit, 2010). However, all is not well in this tourist paradise. The tourism industry has reached saturation, and may even be in decline (Kuntjoro, 2009), with income from tourism having continually dropped since 2000 despite continuing development and a steady increase in hotel rooms (Kuntjoro, 2009).

The rapid and unchecked development of tourism has led to extensive mismanagement of water resources and negative consequences for the local communities, infringing on their right to water. As discussed elsewhere (Cole, 2012) Bali's water crisis is due to a complex of interrelated factors including: inadequate regulation, social and cultural factors (discussed below), deforestation and a lack of awareness. However, groundwater has been overused by the tourism industry and as a consequence there has been a dramatic fall in the water table, salt water has intruded up to 4km in South Bali (Pdam, 2015), underground supplies have been polluted, and are 'unfit for human consumption' (Sundra, 2007). Hand-dug wells have run dry, there is an increased use of bore wells – for those who can afford them – and there has been increasing privatisation of drinking water, much of which is unregulated and of dubious quality.

Case study 2: Tamarindo, Costa Rica

Tamarindo is a beach destination on the North Pacific Guanacaste Coast of Costa Rica, which receives approximately 1,700mm of rain per year. The district of Tamarindo covers 123 square kilometres and includes the neighbouring settlements of Villa Real and Santa Rosa. It has a population of 6,375, a quarter of which comprises people born outside Costa Rica. Nearly 26 percent have no national insurance, approximately 5 percent because they are unemployed and just over 20 percent because they are illegally employed (INEC, 2012); however these figures obscure the presence of unemployed illegal migrants from Nicaragua and other countries. Costa Rica has the highest levels of migration of any Central American country; 75 percent of migrants arrive from Nicaragua (OECD, 2009). Most of Tamarindo's original population, and those who work in Tamarindo, live in the neighbouring villages of Villa Real and Santa Rosa. Before the late 1970s, Tamarindo was a remote village, only accessible on foot or by horse. In the 1980s a road was built and development began, reaching a frenzy between 2005 and 2008. A great deal of the later development was in real estate and residential tourism, resulting in a large number of condominiums (van Noorloos, 2011). There are now 64 registered hotels with a total of over 1,600 rooms, including three large hotels with 200+ rooms each, eight medium-sized establishments with between 30 and 100 rooms, and the rest comprising small lodges (Cole and Ferguson, in press).

In the 1970s, water for the area's sparse agricultural population came from artesian wells. The Costa Rican state owns all the country's underground water resources and grants access and withdrawal rights. The autonomous state institution for water and sanitation (AYA) supplies water to half of Tamarindo. A voluntary, community-based drinking water organisation (ASADA) provides the other half. Both AYA and ASADA obtain their water from the same aquifer, which is limited in capacity and is starting to show signs of salt water intrusion.

Tourism in Tamarindo has followed a similar unmanaged pattern, involving unsustainable resource use and unsustainable tourism and lack of regulation (Duffy, 2002; Honey, 2008). It was described by an official at the Costa Rican Tourism Board as 'out of control . . . exactly how tourism shouldn't be'. There are no locally-owned hotels and only three establishments are owned and managed by Costa Ricans. Tamarindo experiences extreme water stress during the dry season. At such times the state water department, AYA, 'cannot deal with peak demand, the supply can be interrupted for 6–7 or 10–11 hours per day, we use trucks to deliver water to hotels', but residents are left no choice but to wait until their water flows again. There are over 100 illegal wells and a backlog queue of over 1,000 requests for wells and reports about illegal wells according to Department of the Environment (MINEA). Poor water pressure was a common complaint from women residents of Tamarindo, as was the unpleasant taste resulting from high mineral or chlorine content. For those able to afford it, the privatised solution of bottled or filtered water was the answer.

Traditionally, men in Guanacaste have migrated to find seasonal work. Coupled with a culture of multiple partners and alcohol consumption, the concept of the family in Guanacaste 'has always been a rather fragile entity and a source of struggle for women and children' (Chant, 2000, p. 204). Despite being a Catholic country, the region has a high number of births 'out of wedlock' (approximately 60 percent) and serial consensual unions and female-headed households are the norm. Females are the head of 31 percent of households. While tourism has brought employment and autonomy for many women, much of the work in the region is seasonal and informal (approximately 50 percent), and women's mean earnings in hotels and restaurants make up only 64 percent of male earnings in the sector (Chant, 2008). As Chant has argued, the 'feminisation of poverty' in the region 'is about increased unevenness of inputs to household survival between men and women' (2008, p. 81). Since construction halted in Tamarindo in 2008, job opportunities for men have shrunk; the only fields of work that remain are security, gardening and taxi driving. Many front-line service and management jobs in tourism are held by women from North America and Western Europe. In contrast, the lower paid and menial tasks such as cleaning and other domestic work are performed by Nicaraguan migrant women, who make up the bulk of the Tamarindo tourism workforce. Local Costa Rican women tend to work in more professional or clerical positions in the public and community sector.

Salient issues in a gendered political ecology of tourism and water

As discussed earlier, political ecology scholars attempt 'to understand how environmental and political forces interact to affect social and environmental changes through the actions of various social actors at different scales' (Stonich, 1998,

p. 28). A gendered political ecology ensures that the voices of women and other marginalised sections of the community are heard, their issues of access and control are considered, and their conflicts and activism understood. While it is accepted that the complex interactions between changing environment and society lead to culturally and historically contextualised conclusions (Derman and Ferguson 2000), by making a comparison between these two destinations I aim to show that while there are differences, there are far more similarities. These will be considered from the perspectives of unsustainable development; access and control over water resources; intersectionality; conflict and activism; and issues of scale from intimate to international. In the sections that follow, I will attempt to draw on relevant aspects of both cases in order to demonstrate the common themes and most salient aspects of a gendered political ecology of tourism and water.

Unsustainable development

Both Bali and Tamarindo are destinations suffering from unsustainable overdevelopment. In both cases there has been rapid and unchecked growth of tourism at the expense of the environment. A lack of governance resulting from a lack of law enforcement, overlapping mandates between government departments and the deliberate misuse of terms to get around water laws was evident in both destinations. In Tamarindo the 'Hacienda Panilla' was just one of a number of properties masquerading as a farm in order to drill wells without permits. The 4,500-acre property, which has 25 luxury villas, a golf course, beach club, five restaurants and a 316-room Marriott hotel, possesses only household well permits. By calling themselves a 'Hacienda' or farm, they were enabled to bore with impunity. In Bali, where the term 'villa' is (mis)used, the situation was similar, since 'a villa originally assumed as a private house that is sometimes rented out, pays a lower premium for water than a star rated hotel or even a non-star rated hotel' (Cole, 2012, p. 1231). But 'anything can be a villa', as the head of the Bali Villa Association stated, meaning that 'villa' could denote a tourism accommodation structure with 70 rooms, each having its own swimming pool, for which up to US$ 750 per room per night could be charged (ibid.).

In both destinations there has been a rapid change from rural to urban development and as a consequence the infrastructure has been unable to keep up. While in Tamarindo over 95 percent of residents are connected to a piped water supply, in Bali it is only 64 percent. However, sewage systems had not been developed in either area, although they were still reliant on septic tanks in urban areas. This, combined with an over-reliance on additional (and illegal) wells, is a cause of concern. A combination of saline intrusion and proximity to septic tanks means that groundwater in southern Bali is unpotable. In Tamarindo the ground rock is hard and the septic tanks very small so they need to be emptied regularly (at least every six months). This led to conflict: trucks that were used to empty the septic tanks emptied the contents into the river or onto waste land. Reportedly, the same trucks were used to distribute water to hotels during the dry season.

Access and control

In both destinations women were responsible for water for domestic purposes. It was the poor women who suffered when hand-dug wells ran dry and it was women who worried about collecting drips all day or filling pails to ensure there was back up when sources ran dry. The patriarchal nature of Balinese society places women below men. The effect of this is that while women from lower ranks are the most likely to be impacted by water scarcity – since they must feed, wash and prepare food for their families – these women will have the least opportunity to voice their concerns for fear of bringing disrepute or not showing respect. Whereas in Guanacaste, one Costa Rican respondent explained:

> It is the woman who is at home on a daily basis, we are the ones that know, we have to wash, to cook . . . We always have that culture that we use water all the time; and also the woman is feeling it more, because at the moment they have no water, if the man goes to work, only the woman knows, the pressure is on women, because we spent more time stuck at home.

By contrast, in both destinations we conducted interviews in all the relevant department offices connected to water and its supply, the environment and tourism, In every case a man was in charge, and – apart from in the health department in Tamarindo and administrators – women were absent from the offices visited. In each case it was a man who gave the reasons and justifications, who chose the narratives and who steered the debate. It was men who discussed the pricing of water, who planned supplies, who made decisions to supply the tourism sector over the community and who had control. Notably, in Costa Rica the rural water supply is organised by community associations (or ASADA), the boards of which are made up of a 50–50 ratio of men and women. In Tamarindo the people supplied by the ASADA made far fewer complaints than those supplied by the semi-autonomous state supplier (AYA).

Intersectionality

Struggles with water supply were not felt equally by all women in both destinations. The worst impacts affect the poorest and most marginalised members of society first. In Bali it is those from the commoners' caste or migrants from other islands whose hand-dug wells have run dry and could not afford to be connected to state water supplies as the standard fees are unaffordable to many households (Straus, 2011). Additionally, this same population cannot afford to buy sealed and treated bottle water (Aqua) by the gallon. The price of this relatively new but increasingly dominant supply of drinking water for the middle classes had increased by 25 percent between 2007 and 2010. Residents of Bali have to make do with unregulated 'refill' drinking water, increasing their risk of disease. The villagers who buy refill gallons for drinking purposes spoke of increased cases of diarrhoea. One respondent explained how, 'sometimes there was mud or worms/

larvae in the water'. A similar privatisation of drinking water has taken place in Tamarindo, where middle-class Costa Ricans and migrants from the Global North tended to drink bottled or filtered water. As one guesthouse owner from the capital told me 'I don't drink the water . . . a lot of people have kidney problems in old age . . . I do not trust it. We filter water for ourselves and even our dogs'. Low-income Nicaraguan migrants were unable to afford this privatised solution.

In both destinations tourism work was divided along both gender and ethnic lines. As mentioned above, in Tamarindo most of the cleaning was done by Nicaraguan women. Gender stereotypes depict women as suitable for hospitality work and these women were frequently so desperate to access income for their families that they took whatever work was available. Undemanding and unorganised because of their frequently illegal status (despite decades living in Costa Rica) they provide the perfect flexible labour force that characterises so much of tourism employment. Juana's experience was typical: she cleaned houses, condos and offices in Tamarindo between 7 a.m. and 5 p.m. every weekday. None of her employers has given her a contract, so if she gets sick, she explained how: 'there's no way . . . no one pays me insurance . . . nobody, nobody'. Relatives look after her children while she is at work. As Ferguson discusses, Nicaraguan migrants represent a 'new "tier" of workers, who tend to sleep in the lowest quality accommodation and work in hyper-flexibilised tourism activities' (2010, p. 869).

In Bali the starkest intersectionality exists between the rural and urban populations and between the island's southern and north-eastern regions. Despite the US$ 3–5 billion generated in Bali from tourism, many rural people can neither live from their land nor afford the increased prices of essential goods. In the resort towns, rural migrant workers still earn very little: the minimum wage (which in 2015 was raised to just US$ 160 per month) is far below the living wage. Bali's governor, I Made Mangku Pastika, previously condemned tourism, calling it 'a disaster for the poor' (Auskar Surbakti, 2012).

Conflict and activism

An understanding of power dynamics and conflicts lies at the centre of political ecology. Threats to water security bring struggles, conflict and also activism as a form of resistance to the inequity and injustice people feel. The differences between the two case studies in how women are able or not to express the injustice they experience, depends on the cultural context. In Costa Rica women's participation in activism appears to be increasing. Of the six people we interviewed who took on strong activist roles, five were women. The groups that have protested about water shortages due to tourism have been largely shaped by women. Women's perceived responsibility for water has led them to organise around resistance to tourism development. For example, this resistance is illustrated by the Nimboyores project, which constituted one of the most high-profile conflicts over water in tourism development in Costa Rica. This case, recounted through a YouTube video (Nimboyores, 2002), highlights how women's activism prevented a multinational hotel chain from over-using a local water supply. It indicates how women used their labour on a local level to upset the international political economy and disrupt the power structures of international finance over local resources. When the Meliá hotel group tried to pipe water from the village of Lorena to a coastal tourism development, María Rosa organised the protest and stopped the project, calling it illegal and fearing the village would be left with insufficient water. For eight years she fought institutions at the local, provincial and national level and then 'by protests, blocking the streets, and we sat on the Meliá's pipes for a month, so they could not begin construction'. The parties eventually accepted the allocation of a smaller quantity of water for the pipeline and agreeing that the water would be distributed to the broader coastal community rather than solely tourism development (Kusdaz, 2012). This example reflects the power struggles between corporations and communities over water distribution, as it was the focus of women's activism over water allocation in the region. Women are the driving force for change coalescing around the unequal power relations between multinational enterprises and communities over water allocation (Cole and Ferguson, in press).

Unfortunately, the same level of proactive local participation and collective community action needed to ensure fair governance is lacking in Bali. Despite the widespread recognition of government disorganisation and inefficiencies, relative immunity is granted by the cultural norm of collectivism, whereby authority is revered and uncontested (Erb, 2000; Kling, 1997; Raka, 2000). There is a small but growing civil society movement in Bali that has coalesced against mega developments, most recently against a massive reclamation project (*Tolak Reclaimasi*), but the political authoritarianism of the New Order Government stifled democratic mobilisation (Suasta and Connor, 1999). There is no equivalent women's collective activism as seen in Costa Rica and elsewhere, for example Mexico (Hanson, 2015) and Bolivia (Laurie, 2010). My Balinese informants offer two possible reasons: 'Balinese women will is more attracted to economic activity, so the movement issues are considered "too much talk"' or 'They are busy with the daily ritual habits which are very time consuming. Most women are also not very aware of the "rights" as they usually do not have "a say" in the community, they are considered "domestic"'.

From intimate to international

Many of the stories women wanted to share were about their domestic daily lives: the struggles they had 'collecting drips', 'filling jars', of caring for sick children and tending to the needs of elderly parents. They did not talk about the international economic system that, in just one generation, had changed their village to an urban sprawling resort. One of the remaining elderly original inhabitants of Tamarindo told us 'almost all the people have sold their land, spent their money and are finished, without ground and without money they turn into a slave'. 'Foreigners make up 90 percent of the population and none of them care', said another. Yet another local explained: 'Investors who arrived in Tamarindo, are all foreigners: Europeans and Americans . . . they were investors who came, bought and left'. This transience of investors – the easy come, easy go – means that they do not put roots down in the community, they have no allegiance to the land and its ecology and little concern for the future of the destination. In Bali as well, most of the water is consumed by tourists and tourism investors and developers, but they are not affected by water shortages. This transience of water users was considered critical to the unsustainability of Bali's tourism-water socio-ecological system (Cole and Browne, 2015).

It is increasingly clear that the world's water resources will come under increasing stress, and water security is now recognised as a systemic global risk factor. Water is a critical cross-sectorial resource (Gössling et al., 2015). With global warming and increasing urbanisation and population pressure, governments will have to make critical choices about their water resources. In neither of these case studies did the Indonesian or Costa Rican governments sufficiently fulfil their duty to protect the residents' right to water, nor does the tourism industry respect destination communities' rights to water.

Conclusions

In both Bali and Tamarindo, environmental and political factors articulate with different social actors in different ways. Yet, in both cases they result in the distribution of water being diverted away from the quotidian needs of women towards the tourism industry and tourists. The mobilisation of water in Bali, Tamarindo and beyond is indeed a conflict-ridden process. Women – who are often primarily responsible for providing their families with water for domestic purposes – usually bear the brunt. Yet, in some places, women are also those who are mobilising against international powers. As this chapter illustrates, women can indeed be a force for change in the political ecology of tourism, taking on the might of multinationals and shifting the power status quo. International and local businesses are able to bore to ever greater depths to access underground water supplies and pay for private supplies while the most marginalised populations find their hand-dug wells have run dry and are unable to afford (often unreliable or unclean) piped/bottled supplies. As previous studies have shown (Robbins, 2004), those worst affected are the marginal communities at the fringes of social power, with little bargaining strength at the market, and little force in the political process.

This gendered political ecology analysis has highlighted how global economic interests are impacting local-level lived, emotional and material realities. We are reminded that the hegemony of the dominant capitalist discourse of economic growth stands in sharp contrast to women's experiences of hardship and struggles for environmental justice. Minorities by gender, race, class and ethnicity are already unfairly disadvantaged in the face of the global political economy, as well as increasingly those hardest hit by climate change. If justice for the women who so readily shared their stories with me – as well as the millions of others in coastal destination communities around the world who suffer from competition over their water resources – is to be realized, then an alternative discourse to the neoliberal growth of tourism will be required.

References

Ahlers, R. and Zweeten, M. (2009) The water question in feminism: water control and gender inequities in a neo-liberal era. *Gender, Place and Culture*, 16 (4), pp. 409–426.

Antlov, H. (2001) *Village governance and local politics in Indonesia*. Conference paper EUROSEAS, SOAS, London University.

Auskar Surbakti (2012) Bali's poor suffer despite booming tourism *ABC News*, 6 July. Available from: http://www.abc.net.au/news/2012-07-06/an-balis-poor-suffer-despite-booming-tourism/4114064 [Accessed 30 August 2015].

Bakker, K. J. (2003) A political ecology of water privatization. *Studies in Political Economy*, 70, pp. 35–58.

Baum, T. and Cheung, C. (2015) Women in tourism and hospitality: unlocking the potential in the talent pool. Diago White Paper.

Benda-Beckmann, F. and Benda-Beckmann, K. (2001) Recreating the Nagari: decentralisation in West Sumatra. Conference paper EUROSEAS, SOAS, London University.

Blackstock, M. D. (2005) Blue ecology: a cross-cultural approach to reconciling forest-related conflicts. *BC Journal of Ecosystems and Management*, 6 (2), pp. 38–54.

BPS (2010) *Statistics Indonesia*. Available from: http://www.bps.go.id/ [Accessed June 15, 2011].

Buechler, S. and Hanson, A-M. S. (2015) *A Political Ecology of Women, Water and Global Environmental Change*. Abingdon: Routledge.

Chant, S. (2000) Men in crisis? Reflections on masculinities, work and family in north-west Costa Rica. *The European Journal of Development Research*, 12, pp. 199–218.

Chant, S. (2008) The curious question of feminising poverty in Costa Rica: The importance of gendered subjectivities. *Gender Institute: New Working Papers Series*, 22, pp. 1–96.

Cohen, M. (2008) Voting in paradise. *The Guardian*, 30 October. Available from: http://www.theguardian.com/commentisfree/cifamerica/2008/oct/30/barack-obama-absentee-voting-bali [Accessed 30 August 2015].

Cole, S. (2012) A political ecology of water equity and tourism: A case study from Bali. *Annals of Tourism Research*, 39 (2), pp. 1221–1241.

Cole, S. (2014) Tourism and water: from stakeholders to rights holders, and what tourism businesses need to do. *Journal of Sustainable Tourism*, 22 (1), pp. 89–106.

Cole, S. and Browne, M. (2015) Tourism and water inequality in Bali: a social-ecological systems analysis. *Human Ecology – An Interdisciplinary Journal*, 43 (3), pp. 439–450.

Cole, S. and Ferguson, L. (2015) Towards a gendered political economy of tourism and water. *Tourism Geographies* (ahead of print), pp. 1–18.

Crase, L., O'Keefe, S. and Horwitz, P. (2010) *Australian tourism in a water constrained economy*. Gold Coast: CRC for Sustainable Tourism.

Creese, H. (2004) Reading the *Bali Post*: women and representation in post-Suharto Bali. *Intersections: Gender, History and Culture in the Asian Context*, 10 (25), pp. 2–18.

Crenshaw, K. (1989) Demarginalizing the Intersection of Race and Sex: A Black Feminist Critique of Antidiscrimination Doctrine, Feminist Theory and Antiracist Politics. In: Phillips, A. (ed.), *Feminism and Politics*. Oxford: Oxford University Press, pp. 314–343.

Crow, B. and Sultana, F. (2010) Gender, class, and access to water: three cases in a poor and crowded delta. *Society and Natural Resources: An International Journal*, 15 (8), pp. 709–724.

Dahles, H. and Bras, K. (1999) Entrepreneurs in romance: tourism in Indonesia. *Annals of Tourism Research*, 26 (2), pp. 267–293.

Dávila-Poblete, S. and Nieves Rico, M. E. (2005) Global water and gender policies: Latin American challenges. In: Bennett, V., Dávila-Poblete, S. and Nieves Rico, M. E. (eds), *Opposing Currents: The politics of water and gender in Latin America*. Pittsburgh, PA: University of Pittsburgh Press, pp. 30–49.

Deng, S. and Burnett, J. (2002) Energy use and management in hotels in Hong Kong. *International Journal of Hospitality Management*, 21 (4), pp. 371–380.

Derman, B. and Ferguson, A. (2000) *The Value of Water: Political ecology and water reform in Southern Africa*. San Francisco, CA: Annual American Anthropological Association.

De Stefano, L. (2004) *Freshwater and Tourism in the Mediterranean*. Rome: WWF Mediterranean Programme.

Duffy, R. (2002) *A Trip Too Far: Ecotourism, politics and exploitation*. London: Earthscan.

Enloe, C. (1989) *Bananas, Beaches and Bases*. London: Pandora.

Erb, M. (2000) Understanding tourists, interpretations from Indonesia. *Annals of Tourism Research*, 27 (3), pp. 709–736.

Essex, S., Kent, M. and Newnham, R. (2004) Tourism development in Mallorca: is water supply a constraint? *Journal of Sustainable Tourism*, 12 (1), pp. 4–28.

Ferguson, L. (2010) Tourism development and the restructuring of social reproduction in Central America. *Review of International Political Economy*, 17 (5), pp. 860–888.

Ferguson, L. (2011) Promoting gender equality and empowering women? Tourism and the third Millennium Development Goal. *Current Issues in Tourism*, 14 (3), pp. 235–249.

Frew, E. and Shaw, R. (1999) The relationship between personality, gender, and tourism behavior. Tourism Management, 20 (2), pp. 193–201.

Garcia, C. and Servera, J. (2003) Impacts of tourism development on water demand and beach degradation on the island of Mallorca (Spain). *Geografiska Annaler*, 85A (3–4), pp. 287–300.

Gentry, K. M. (2007) Belizean women and tourism work: opportunity or impediment? *Annals of Tourism Research*, 34 (2), pp. 477–496.

Gössling, S. (2001) The consequences of tourism for sustainable water use on a tropical island: Zanzibar, Tanzania. *Journal of Environmental Management*, 61 (2), pp. 179–191.

Gössling, S. (2003) Tourism and development in Tropical Islands: a political ecology perspective. In: Gössling, S. (ed.), *Tourism and Development in Tropical Islands: A political ecology perspective*. Cheltenham: Edward Elgar.

Gössling, S., Peeters, P., Hall, M. C., Ceron, J., Dubois, G., Lehmann, L., *et al.* (2012) Tourism and water use: supply, demand, and security. An international review. *Tourism Management*, 1–15.

Gössling, S., Hall, M. and Scott, D. (2015) *Tourism and Water*. Bristol: Channel View Publications.

Hanson, A. (2015) Shoes in the seaweed and bottles on the beach: global garbage and women's oral histories of socio-environmental change in Coastal Yucatan. In: Buechler, S. and Hanson A-M. (eds), *A Political Ecology of Women, Water and Global Environmental Change*. London: Routledge.

Hanson, A-M. and Buechler, S. (2015) Introduction towards a feminist political ecology of women, global change and vulnerable waterscapes In: Buechler, S. and Hanson A-M. (eds), *A Political Ecology of Women, Water and Global Environmental Change*. London: Routledge.

Harrill, R. and Potts, T. D. (2003) Tourism planning in historic districts: Attitudes toward tourism development in Charleston. *Journal of the American Planning Association*, 69 (3), pp. 233–244.

Harris, L. M. (2009) Gender and emergent water governance: comparative overview of neoliberalized natures and gender dimensions of privatization, devolution and marketization. *Gender, Place and Culture*, 16 (4), pp. 387–408.

Harrison, D. (2008) Pro-poor tourism: a critique. *Third World Quarterly*, 29 (5), pp. 851–868.

Hauser-Schäublin, B. (2011) Land donations and the gift of water. On temple landlordism and irrigation agriculture in pre-colonial Bali. *Human Ecology*, 39 (1), pp. 43–53.

Honey, M. (2008) *Ecotourism and Sustainable Development: Who Owns Paradise?* (2nd edn). Washington, DC: Island Press.

Howe, L. (2005) *The Changing World of Bali: Religion, society and tourism*. London: Routledge.

Hussey, A. (1989) Tourism in a Balinese village. *Geographical Review*, 79 (3), pp. 311–325.

INEC (National Institute of Statistics and Censuses of Costa Rica) (2012) Available from: http://www.inec.go.cr [Accessed 30 August 2015].

Kent, M., Newnham, R. and Essex, S. (2002) Tourism and sustainable water supply in Mallorca: a geographical analysis. *Applied Geography*, 22 (4), pp. 351–374.

Kim, D. Y., Lehto, X. Y. and Morrison, A. M. (2007) Gender differences in online travel information search: implications for marketing communications on the internet. *Tourism Management*, 28 (2), pp. 423–433.

King, M. (2005) *Water and violent conflict*, OECD. Available from: http://www.globalpolicy.org/images/pdfs/052605waterconflict.pdf [Accessed 30 August 2015].

Kinnaird, V. and Hall, V. (1996) Understanding tourism processes: a gender aware framework. *Tourism Management*, 17 (2), pp. 95–102.

Kling, Z. (1997) Adat: Collective self-image. In: Hitchcock, M. and King, V. (eds), *Images of Malay-Indonesian Identity*. Oxford: Oxford University Press, pp. 45–51.

Kuntjoro, J. (2009) *Bali tourism expiring, claims economic guru. Jakarta Post* 5/2/2009.

Kuzdas, C. (2012) *Unpacking water conflict in Guanacaste, Costa Rica* (GWF Discussion Paper 1242) Canberra: Global Water Forum. Available from: http://www.globalwaterforum.org/wp-content/uploads/2012/10/Unpacking-water-conflict-in-Costa-Rica-GWF-1242_.pdf [Accessed 30 August 2015].

Lansing, J. S. (2007) *Priests and Programmers: Technologies of power in the engineered landscape of Bali*. Princeton, NJ: Princeton University Press.

Laurie, N. (2010) Gender water networks: femininity and masculinity in water politics in Bolivia. *International Journal of Urban and Regional Research*, 35 (1), pp. 172–188.

Lehmann, L. V. (2009) The relationship between tourism and water in dry land regions. *Proceedings of the Environmental Research Event*, Noosa, Queensland, Australia.

Lewis, J. and Lewis, B. (2009) *Bali's Silent Crisis: Desire, tragedy, and transition*. Lanham, MD: Lexington Books.

Lorenzen, R. P. and Lorenzen, S. (2011) Changing realities – perspectives on Balinese rice cultivation. *Human Ecology*, 39, pp. 29–42.

MacRae, G. (2010) If Indonesia is too hard to understand, let's start with Bali. *Journal of Indonesian Social Sciences and Humanities*, 3, pp. 11–36.

MacRae, G. S. and Arthawiguna, W. A. (2011) Sustainable agricultural development in Bali: is the Subak an obstacle, an agent or subject? *Human Ecology*, 39, pp. 11–20.

Marshment, M. (1997) Gender takes a holiday: representation in holiday brochures. In: Sinclair, M. T. (ed.), *Gender, Work and Tourism*. London: Routledge, pp. 16–34.

Mason, P. and Cheyne, J. (2000) Residents' attitudes to proposed tourism development. *Annals of Tourism Research*, 27 (2), pp. 391–411.

McTaggart, D. (1988) Hydrological management in Bali. *Singapore Journal of Tropical Geography*, 9 (2), 96–111.

Merit (2010) Professor *Dr. Ir. I Nyoman Merit, M.Agr. Udayana University*. Interview July 2010, Pers com.

Mollett, S. and Faria, C. (2013) Messing with gender in feminist political ecology. *Geoforum*, 45, pp. 116–125.

Mosedale, J. (2011) *Political Economy of Tourism: A critical perspective*. London and New York: Routledge.

Nakatani, A. (2004) *Perempuan Bali dalam Tiga Peran*. In: *Majalah Mingguan Tokoh. No. 298*/Tahun *VI*, 29 Agustus–4 September, pp. 23–29.

Nimboyores (2002) Available from: http://www.youtube.com/watch?v=rCj6YQC1m5I and http://www.youtube.com/watch?v=iFzrQVew7aI [Accessed 30 August 2015].

Nunkoo, R. and Ramkissoon, H. (2010) Gendered theory of planned behaviour and residents' support for tourism. *Current Issues in Tourism*, 13 (6), pp. 525–540.

OECD (2009) *Latin American Economic Outlook 2010*. Available from: http://www.oecd.org/dev/americas/44535774.pdf [Accessed 30 August 2015].

OECD/UNEP (2011) *Climate Change and Tourism Policy in OECD Countries*. Available from: http://www.unep.fr/scp/publications and www.oecd.org/cfe/tourism [Accessed 30 August 2015].

Ostrom, E. (2009) A general framework for analyzing sustainability of social-ecological systems. *Science*, 325, pp. 419–422.

Pdam (2015) Diskusi Air dan Peradaban. Denpasar Film Festival, Denpasar 14 April 2015.

Peterson, V. (2005) How (the meaning of) gender matters in political economy. *New Political Economy*, 10 (4), pp. 499–521.

PHRI (2015) Diskusi Air dan Peradaban. Denpasar Film Festival, Denpasar 14 April 2015.

Picard, M. (1997) Cultural tourism, nation-building, and regional culture: the making of a Balinese identity. In: Picard, M. and Wood, R. (eds), *Tourism ethnicity and the state in Asian and Pacific societies*. Honolulu, HI: University of Hawai`i Press, pp. 181–214.

Pigram, J. (2001) Water resources management in island environments: the challenge of tourism development. Tourism (Zagreb), 49 (3), pp. 267–274.

Pritchard, A. and Morgan, N. (2000) Constructing tourism landscapes – gender, sexuality and space. *Tourism Geographies: An International Journal of Tourism Space, Place and Environment*, 2 (2), pp. 115–139.

Raka, G. (2000) Entrepreneurship for tourism: Issues for Indonesia. In: Bras, K., Dahles, H., Gunawan, M. and Richards, G. (eds), *Entrepreneurship and Education in Tourism*. Proceedings ATLAS Asia Inauguration Conference. ITB, Bandung.

Rico-Amoros, A. M., Olcina-Cantos, J. and Sauri, D. (2009) Tourist land use patterns and water demand: evidence from the Western Mediterranean. *Land Use Policy*, 26 (2), pp. 493–501.

Ritzdorf, M. (1995) Feminist contributions to ethics and planning theory. In: Hendler, S. (ed.), *Planning Ethics: A reader in planning theory, practices and education*. New Brunswick, NJ: Centre for Urban Policy Research.

Robbins, P. (2004) *Political Ecology: A critical introduction*. Chichester: Wiley-Blackwell.

Rocheleau D., Thomas-Slayter, B. and Wangari, E. (eds) (1996) *Feminist Political Ecology: Global issues and local experiences*. London and New York: Routledge.

Sanchez Taylor, J. (2001) Dollars are a girl's best friend? Female tourists' sexual behaviour in the Caribbean. *Sociology*, 35, pp. 749–764.

Sanchez Taylor, J. (2006) Female sex tourism: a contradiction in terms? *Feminist Review*, 83, pp. 42–59.

Sanchez Taylor, J. (2010) Sex tourism and inequalities. In: Cole, S. and Morgan, N. (eds), *Tourism and Inequality: Problems and prospects*. Oxford: CABI, pp. 49–66.

Schellhorn, M. (2010) Development for whom? Social justice and the business of ecotourism. *Journal of Sustainable Tourism*, 18 (1), pp. 115–135.

Sinclair, M. T. (1997) Issues and theories in gender and work in tourism. In: Sinclair, M. T. (ed.), *Gender, Work and Tourism*. London: Routledge, pp. 1–15.

Stonich, S. (1998) The political ecology of tourism. *Annals of Tourism Research*, 25 (1), pp. 25–54.

Straus, S. (2011) Water conflicts among different user groups in South Bali, Indonesia. *Human Ecology*, 39, pp. 69–79.

Suasta, P. and Connor, L. (1999) Democratic mobilization and political authoritarianism: tourism developments in Bali. In: Rubinstein, R. and Connor, L. (eds), *Staying local in the global village: Bali in the twentieth century*. Honolulu, HI: University of Hawai`i Press.

Sundra, K. (2007)Water quality degraded in Bali's tourism areas. *Xinhua News Agency*, 7 December.

Swyngedouw, E. (2009) The political economy and political ecology of the hydro-social cycle. *Journal of Contemporary Water Research &Education Issue*, 142, pp. 56–60.

Tortajada, C. (1998) Water supply and wastewater management in Mexico: an analysis of the environmental policies. *International Journal of Water Resources Development*, 14 (3), pp. 327–337.

Tortella, B and Tirado, D. (2011) Hotel water consumption at a seasonal mass tourist destination. The case of the island of Mallorca. *Journal of Environmental Management*, 92 (10), pp. 2568–2579.

Tourism Concern. (2012) *Water Equity in Tourism: A human right – a global responsibility*. London: Tourism Concern. Available from: http://www.tourismconcern.org.uk/uploads/Water-Equity-Tourism-Report-TC.pdf [Accessed 30 August 2015].

Tribe, J., Dann, G. and Jamal, T. (2015) Paradigms in tourism research: a trialogue. *Tourism Recreation Research*, 40 (1), pp. 28–47.

Truelove, Y. (2011) (Re-)conceptualizing water inequality in Delhi, India through a feminist political ecology framework. *Geoforum*, 42 (2), pp. 143–152.

Truong, T. (1990) *Sex, Money and Morality: Tourism and prostitution in South-East Asia*. London: Zed Books.

Tucker, H. and Boonabaana, B. (2012) A critical analysis of tourism, gender and poverty reduction. *Journal of Sustainable Tourism*, 20 (3), pp. 437–455.

Usman, S. (2001) Indonesia's decentralisation policy. Initial experiences and emerging problems. Paper Presented at EUROSEAS Conference London SOAS.

Vandegrift, D. (2008) 'This isn't paradise – I work here': global restructuring, the tourism industry and women workers in Caribbean Costa Rica. *Gender and Society*, 22, pp. 778–798.

van Noorloos, F. (2011) Residential tourism causing land privatization and alienation: new pressures on Costa Rica's coasts. *Development*, 54 (1), pp. 85–90.

Warren, C. (1998) Tanah lot: the cultural and environmental politics of resort development in Bali. In: Hirsch, P. and Warren, C. (eds), *The politics of environment in Southeast Asia: Resources and resistance*. London: Routledge, pp. 229–259.

Watts, M. J. (2000) Political Ecology. In: Sheppard, E. and Barnes, T. (eds), *A Companion to Economic Geography*. Malden, MA: Blackwell Publishers.

Wheeler, B. (2003) Alternative tourism: A deceptive ploy. In Cooper, C. (ed.), *Classic Reviews in Tourism*. Bristol: Channel View.

2 Ngarrindjeri Authority
A sovereignty approach to tourism

Ron Nicholls, Freya Higgins-Desbiolles and Grant Rigney

Introduction

This chapter examines the experience of the Ngarrindjeri Aboriginal community of South Australia in offering pioneering educational tourism and ecotourism ventures while trying to transition away from the devastating impacts of colonization to create sovereign and self-determining futures. We focus on the use and significance of the contemporary debates on the allocations of waters from the vital Murray–Darling River Basin (MDRB) system as a path to affirming Ngarrindjeri aspirations through the advocacy of cultural flows. The community argues that cultural flows[1] or Ngarrindjeri water moves beyond the articulation of environmental flows and are defined as water entitlements that are legally and beneficially owned by Indigenous Nations and are of an adequate quantity and quality to maintain and improve the spiritual, cultural, natural, environmental, social and economic conditions as inherent rights of indigenous Australians. Hence, the Ngarrindjeri community engages with Commonwealth and State governments of Australia to secure appropriate water flows that not only support a range of practical initiatives such as farming, education, commercial activities, natural resource management, and internationally recognized tourism ventures such as Camp Coorong Race Relations and Cultural Education Centre and the Coorong Wilderness Lodge, but also are a powerful means to assert Ngarrindjeri ways of being and sovereignty that challenge the dominance of settler-colonial Australia.

Central to the realization of these projects are Ngarrindjeri perspectives of interconnectedness and relationships to land (Ruwe) and waters (Yarluwar-Ruwe). These standpoints are underpinned by cosmological and epistemological foundations that hold significant challenges to the worldview of settler-colonial Australia and offer alternative lifeways, not only for Ngarrindjeri peoples, but also to others in search of alternative futures. In this collaborative work the authors offer insights into these developments regarding a number of issues that arise from their political activism approach. Following a brief overview of relevant literature, we present a case study of the Ngarrindjeri experience before exploring how the advocacy of both cultural flows and water sovereignty illuminates a political approach to tourism with important implications for understanding the value and possibilities of tourism.

Tourism and political ecology

Arguably, tourism scholars' engagement with politics has been limited at best. This may be the case for a number of reasons, not least of which is the takeover of the field of tourism studies by those keen to restrict it to its business or industrial attributes (Higgins-Desbiolles, 2006; Tribe, 1997). This has had a number of spin-off effects, including: an over-shadowing of the social capacities of tourism; an abandonment of social tourism with its concern for equity in access and benefits; a narrowing of focus to the interests of the tourism industry, tourists and compliant governments; and a stubborn refusal to engage with the politics of tourism with its questions of who benefits and how power is perpetuated. What has largely resulted from these framings is a concern to grow tourism as part of the neoliberal growth paradigm, the spreading of the ability of multinational and corporate tourism entities to seek global profits with minimal regulation and taxation as globalized trading arrangements are enacted in the service sector and, perhaps most importantly, a disempowerment of local communities. From India to Thailand to Hawai`i, the dynamics of tourism for elite profit-making at the expense of the subsistence, well-being and ecological sustainability of local peoples has been clear since the introduction of the neoliberal project in the 1980s (Harvey, 2006; McLaren, 2003). At a time when political analyses are most needed, the discipline of tourism studies has largely abandoned politics, leaving critical analyses of tourism to NGOs such as Equitable Tourism Options (EQUATIONS), the Tourism Investigations and Monitoring Team (TIM-Team) and Tourism Concern.

For these reasons, we argue that an examination of the political ecology of tourism is critical to tourism studies. According to Robbins (2012) political ecology links political economy, cultural ecology and activism, and addresses questions of who has access to resources, how they are able to wield such access for accumulating power and wealth and who bears the consequences of these actions, particularly the negative impacts. As Bryant states (1998, p. 86): '[r]unning through most political-ecology research is the notion of social and environmental conditions constituted through unequal power relations. At one level, power is reflected in the ability of one actor to control the environment of another'. While Bryant's work focuses on power articulations in the developing world, it holds wider relevance beyond this specific context:

> Political ecology examines the political dynamics surrounding material and discursive struggles over the environment in the third world and the role of unequal power relations in constituting a politicized environment is a central theme. Particular attention is given to the ways in which conflict over access to environmental resources is linked to systems of political and economic control first elaborated during the colonial era. Studies also emphasize the increased marginality and vulnerability of the poor as an outcome of such conflict. The impact of perceptions and discourses on the specification of environmental problems and interventions is also explored leading on to debates about the relative merits of indigenous and western scientific knowledge.
>
> (1998, p. 79)

The issues over struggles for access, use of and care of natural resources and natural environments is a key issue in tourism despite the tourism industry's best efforts to present itself as a 'smokeless' or benign 'industry'. The industry is known for its appropriation of the commons and the externalizing of the negative impacts of its activities whenever possible, due to poor regulatory regimes or the reduction of protections due to the neoliberalizing of economies (Hall, 2008). Briassoulos (2002) offers an effective analysis of how this 'tragedy of the commons' undermines the local community and its sovereignty over its local environment. As she explains:

> Social justice ... is not well served as the problems discussed affect the distribution of the costs and benefits at the local, regional, and higher spatial levels. The more socioeconomically fragile an area becomes, the more its self-sufficiency is placed at risk. In the broader spatial context, these areas are losers because their development hinges on decisions of powerful interests that have no interest in their long-term vitality and viability. Under these conditions, it is questionable if local participation in decision making contributes essentially to promoting local interests.
>
> (2002, p. 1074)

Briassoulos (2002) also offers a conceptual solution of governance for sustainable tourism:

> Governance systems for the tourism commons as a whole and their individual components are needed to establish rules of collective behavior and principles of resource allocation in host areas and offer the requisite authority that legitimizes and protects the rights and duties of local and nonlocal resource appropriators and users ... Nonlocals include foreign tourism producers (investors, developers, tourism agents) and consumers (the tourists). They should advance local autonomy and provide for the two basic requirements for solving CPR [common pool resource] problems: restricting or controlling access to sensitive resources and creating incentives for users to invest in resource maintenance and enhancement ...
>
> (2002, p. 1081)

It must be stated that this is just abstract theorizing in the context of a corporatized tourism system where governments have ceded power to tourism enterprises in the pursuit of economic growth and unfettering of markets. Despite recent efforts to tame tourism and mitigate its negative impacts through the implementation of concepts of sustainability, corporate social responsibility and pro-poor tourism, the reality is that the logic of the tourism industry under neoliberalism is to strive to accumulate greater profits, externalize negative impacts, and resist regulations that limit its capacity to achieve these things (Higgins-Desbiolles, 2008). Given such a context, the tourism industry today has inordinate power with little interest in 'local autonomy'. In fact the industrial discourse of tourism has worked largely to eradicate local people from any consideration. Tourism is most often defined by a model which designates two central nodes: (1) The tourists and their demand; and (2) The industry that supplies this demand (e.g. Ryan, 1991; Tribe, 1997).

Via such modelling, the only way in which the local community is even considered is either as the attractive backdrop to the destination (when they are known as the 'host community') should they behave benignly, or as a problem to be managed should they get unruly and disruptive (through either 'consultative processes' or repression). A political ecology approach provides theoretical guidance that allows us to disrupt this erasure of the local people, their rights, their needs and their deep embeddedness in the local place, which is at risk of being disrupted beyond repair by a tourism industry they often cannot control.

Thus we offer a case study of the Ngarrindjeri community of South Australia, an Aboriginal nation that has suffered invasion, dispossession, colonization and attempted assimilation, and yet in recent decades has asserted a sovereignty approach that has rejected the dominance of capitalistic and neoliberal ideologies. This case study historically situates resource claims made by the Ngarrindjeri and the ways in which they assert authority in overturning settler-colonial exploitation. We particularly emphasize their efforts to argue for the recognition of cultural flows of water in political debates over water allocations from the vital Murray–Darling River Basin of Australia. Bryant has noted a tendency for 'land centrism' in political ecology in the 1990s, despite the fact that water is an 'essential material for maintaining social life' (1998, p. 89), while relevant tourism literature that speaks to political ecology approaches reveals a healthy concern with water issues (e.g. Cole, 2012; Stonich, 1998). With the Ngarrindjeri people being a water people (the Murray River, the Lower Lakes and Coorong are key features of their ecological context), the cultural and spiritual articulation they have given to the concept of 'cultural flows' provides invaluable insights into how indigenous sovereignty is a key to just political and ecological outcomes for Ngarrindjeri and other Indigenous Nations and how tourism might support such efforts.

Reinhabitation: learning to (re)live in place

While much earlier work on political ecology and tourism focuses on tourism as the catalyst to disempowerment over water and other resources (e.g. Cole, 2012; Stonich, 1998), the case of the Ngarrindjeri is very different and as such requires a different framework. Moreover, as we show in this chapter, water sovereignty can be a crucial aspect of a people's thriving, which, with access to the cultural flows mentioned above, can underpin a form of tourism that empowers communities to create futures of their own choosing. Accordingly, a key element of our analysis is the pluralistic use of environments as articulated by Guattari (2000, p. 28) and his concept of ecosophy, and includes a triad of ecological registers of the environment, social relations and human subjectivity, emphasizing the complexity associated with the multifarious issues that impact Ngarrindjeri peoples. For example, colonial incursion has largely been a contested space dominated by colonization, dispossession, marginalization, and racism alongside the imposition of knowledge systems and discursive practices that have actively subordinated and displaced indigenous ways of knowing and being.

Significantly, these colonizing knowledge systems and practices are the very same structures that are largely responsible for the processes of cultural imperialism and the present decimation of the Earth's natural and social systems. Accordingly, responses to these colonizing discourses and practices are clearly related to decolonizing and/or anti-colonial practices. However, alongside the concepts of decolonization or anti-colonization we suggest that it is possible to envisage a parallel concept in concert with decolonization that expresses alternative ways of thinking – one that opens up possibilities for collaboration between diverse ontologies and epistemologies, including the ecological implications of our mutual connectivity to the environment. As Rose (2007, p. 16) explains: 'In one sense, connectivity is a statement of the ecological fact that organism and environment permeate each other, are mutually constitutive, and thus mutually necessary and sustaining'. Although these assertions of interconnectivity may be seen as problematic in terms of the risk of appropriation, Māori academic Stewart-Harawira (2005, pp. 250–251) argues that 'a deep understanding of the interconnectedness of all existence is not only fundamental to indigenous ontologies but has been empirically demonstrated in the studies of leading quantum physicists such as David Bohm'.

Given these perceptions, a concept that might help us to envision new ways of thinking, and stimulate the new narratives needed to transcend the present expansionist and unsustainable global capitalist order, is the term 'reinhabitation'. Coined during the emergence of the bioregionalism movement in the United States, reinhabitation was first mentioned in a landmark paper by Peter Berg and Raymond Dasmann and defined as:

> ... an ability to understand the activities and evolving social behaviour that will enrich the life of that place, restore its life-supporting systems, and establish an ecologically and socially sustainable pattern of existence within it.
> (Berg and Dasmann, 1977, p. 399)

The notion of the multiple layers of environmental registers that Guattari suggests we all inhabit also provides space for a complementary conceptual category of decolonizing/reinhabitation that is characterized by practices that 'seek to establish a balance between the social, cultural and ecological features of a region or a space that has usually been associated with a watershed [or catchment area])' (Berg and Dasmann, 1977, p. 399). Accordingly, this 'refers both to a geographical terrain and a terrain of consciousness – to a place and the ideas that have developed about how to live [and relive] in that place' (Berg and Dasmann, 1977, p. 399).

With these ideas in mind, the term reinhabitation can be seen as a particular application of the concepts of reterritorialization articulated by Gilles Deleuze and Felix Guattari. Although used in a specific way in *A Thousand Plateaus* (2004), Deleuze and Guattari encourage the use of their concepts in other contexts and the term deterritorialization has been used in anthropology to refer to the weakening of ties between culture and place. Moreover, deterritorialization and reterritorialization are necessarily interconnected, and have also been used to describe something of the processes involved as local communities become part of the global culture. In the present context of economic neoliberalism and the contested nature of relationships between notions of a globalized and borderless societies and the significance of place, this becomes a particularly pertinent issue.

In exploring the idea of connectivity to the natural world in the context of the reinhabitation of the three ecological registers outlined by Guattari (2000), theologian and cultural historian Thomas Berry has proposed that developing suitable alternative practices will first require a new story. Arguing that 'there can be no peace among humans without peace with the planet' he offers a belief that sees the Earth as a living being fundamental to human existence (Berry cited in Nicholls, 2006). Central to this vision, and contrary to the western mechanistic worldview, is Berry's call for a participatory consciousness that recognizes and embraces the need for a new relationship with the Earth 'as a communion of subjects rather than a collection of objects' (Berry, n.d.), that will bring forth new (or old) knowledges and practices modelled on natural systems.

The Ngarrindjeri and their care of Ruwe and Yarluwar-Ruwe

Ngarrindjeri country covers a large section of South Australia, and with water as a defining aspect encompasses the Lower Murray River, Lakes and Coorong and is bounded on the south-west by the Southern Ocean. Before European occupation, the Ngarrindjeri nation comprised 18 Lakalinyerer (or clans) and featured a sophisticated form of governance. Each lakalinyeri had its own defined territory and its own governance structures called the Tendi. Ngarrindjeri oral traditions, as well as accounts by missionaries and anthropologists confirm that the Ngarrindjeri possessed a democratic form of governance which predated the evolution of European government by many thousands of years (Jenkin, 1985, p. 13). The Tendi of each lakalinyeri also acted as a high court; men were elected to represent each lakalinyeri, and the Rupulle (or president) was elected by that body. As Jenkin states, '[T]o discuss and settle matters which affected the nation as a whole (such as disagreements between two Lakalinyerer) there was a grand or combined Tendi' (1985, p. 13). Therefore the description of the nation as a confederacy would appear to be an accurate one (Jenkin, 1985, p. 13) and it certainly represented a 'unified system of governance' (Bell, 1998, p. 137). It is this historical experience of sophisticated governance that laid the foundations for a political revival beginning in the 1980s as the Tendi system was revived and a Rupulle (George Trevorrow) elected.

Although similar to many indigenous groups in South Australia, the rich resources of the Lower Murray region and its lakes afforded the Ngarrindjeri peoples a wealth of food and other necessities, and supported a large population with a prosperous and comfortable standard of living. They built permanent or semi-permanent dwellings, were outstanding craftspeople, used woven fibres for basketry, netting and matting, and possessed technologies for living sustainably in their environment. Their adherence to a worldview that fostered a firm belief in the relationship between the human, natural and supernatural spheres and a practical and integrated system of governance and values provided a sound basis for a largely peaceful and orderly social system (Jenkin, 1985, p. 16). However, beginning with the invasion of the settler-colonists in South Australia (from 1836) this severely impacted (and continues to impact) on Ngarrindjeri peoples. This ultimately resulted in the fragmentation of many of these coherent structures, large-scale dispossession from their lands and waters and marginalization to missions and fringe camps with an aim to break connections to country and alienate the wealth of the land from its indigenous peoples.

In the contemporary era, Ngarrindjeri and other indigenous Australian nations have maintained the hope that a dismantling of oppressive laws and practices might result in some recognition of ongoing indigenous claims to sovereignty, self-determination and self-governance. High points such as the 1967 referendum, land rights legislation from the 1980s, recognition of ongoing Native Title following the Mabo and Wik High Court decisions and the 2008 apology given by then Prime Minster Rudd to the Stolen Generations raised some hopes among some Australians that an era of post-colonial (re)conciliation is possible. But, as recent events such as the Northern Territory Intervention and the decision to close some Aboriginal homelands through the de-funding of essential services indicate, such conciliation is far from within reach.

The Ngarrindjeri know this perhaps more than any other nation, considering the direct attack against their culture and spirituality in an episode that came to be known as the Hindmarsh Island Bridge conflict. Emerging in the 1990s, this conflict resulted from a plan to expand a marina and residential development on Hindmarsh Island (Kumarangk), with the South Australian government insisting that a bridge replace the small car ferry in order to cater to the expanding population on the island. When Ngarrindjeri spoke out against the siting of the proposed bridge:

> ... with their opposition based on sacred sites and the spiritual significance of the island (which became labelled as 'sacred women's business') ... the conflict drew national and international attention. A Royal Commission called by the State government in 1995 found that the Ngarrindjeri proponents of 'sacred women's business' were fabricators and questioned their attachment to the island.
>
> (Higgins-Desbiolles, 2004, p. 10)

This damaging event took longer than a decade and resulted in five government inquiries and more than 30 legal cases. Although the bridge was eventually built and opened in 2001, it resulted in the Ngarrindjeri making important decisions to become politically proactive and assertive of their rights. One initial focus was on asserting authority on country with the Alexandrina Council (the local government authority where Kumarangk is located) through a Kungan Ngarrindjeri Yunnan ('listen to Ngarrindjeri People speaking') Agreement in 2002, which committed the Council to extensive engagement with Ngarrindjeri rights and authority (Higgins-Desbiolles, 2007). This coincided with the establishment of the

Ngarrindjeri Regional Authority (NRA) through which the political structures of the nation are organized for full consultation, development of economic opportunities and management of external relations. The NRA was established in 1997 as an incorporated association in order to maintain a range of benefits which collectively benefit the Ngarrindjeri people. Key tenets of the Association include the welfare of Ngarrindjeri peoples, health and well-being, economic activities with a view to alleviating poverty, Native Title and intellectual property rights regarding traditions, languages, spirituality, traditional economies, resource management and ecological knowledges (NRA, 1985).

Through the efforts of the NRA, and a concerted approach to overturn the power relationships between Ngarrindjeri peoples and governments of all types, universities and other organizations, a new era is apparent. As a result, in 2009:

> The Ngarrindjeri nation in South Australia negotiated a new agreement with the State of South Australia that recognised traditional ownership of Ngarrindjeri lands and waters and established a process for negotiating and supporting Ngarrindjeri rights and responsibilities for country (Ruwe) . . . In line with Ngarrindjeri political and legal strategies, it takes the form of a whole-of-government, contract agreement between the Ngarrindjeri nation and the State of South Australia.
>
> (Hemming and Rigney, 2011, p. 351)

This has profound importance in terms of asserting Ngarrindjeri knowledges and authority over projects such as 'caring for country' and, as concerned in this analysis, in dealing with attempts to address management issues over the Murray–Darling River Basin and its water allocations. Before addressing this, however, it is necessary to address in more detail Ngarrindjeri worldviews.

As mentioned above, it is particularly important in this context to have an understanding of the Ngarrindjeri worldview and the present political structures that have survived and continue to strongly influence and underpin their relationships to land and waters at present. At the height of the Hindmarsh Island Bridge conflict, the Ngarrindjeri communicated this in their submission to the Ramsar consultation process (the Coorong is listed as a Ramsar Wetland of international importance). The Ngarrindjeri community's Ngarrindjeri Ramsar Working Group (NRWG) presented a submission that attempted to explain the nature of their attachments to their country:

> The Ngarrindjeri lands – in particular the River, Lakes and the Coorong are crucial for the survival of the Ngarrindjeri people. They have a spiritual and religious connection with the land and the living things associated with it. The sense of feeling, sense of belonging, sense of responsibility for the River, Lakes and Coorong experienced by Ngarrindjeri people has survived occupation, dispersal and attempted assimilation. It continues to exist irrespective of where Ngarrindjeri people currently live. The link with the land lies at the heart and soul of Ngarrindjeri culture.
>
> (NRWG, 1999, p. 61)

The landmark Ngarrindjeri Nation Yarluwar-Ruwe Plan (2006) also outlined their specific relationship to land and waters (Ruwe) as a living and interconnected entity and indicates the critical importance of the social, spiritual, cultural and environmental significance of the land and waters to Ngarrindjeri people. As the plan states:

> The waters of the seas, the Kurangk (Coorong), the rivers and inland waters are life-giving waters. Our Ancestors taught us how to respect and understand the connections between the lands, waters and beyond. The places where fresh and salt waters mix, are places of creation, where life dwells in abundance, and where Ngaitjies [Ngartjies] breed. Ngarrindjeri people have rejoiced to see the return to Ngarrindjeri Seas of Kondoli our Whale Ancestors. Some of our Ngaitjies have not returned to our land and waters. We mourn the loss of our friends. We fear for animals, fish, birds and all living things in our seas and water-ways.
>
> The land and Waters is a living body. We the Ngarrindjeri people are part of its existence. The land and waters must be healthy for the Ngarrindjeri people to be healthy. Ngarrindjeri say if Ruwi (land) dies, the waters die, our Ngaitjies (friends) die, and then the Ngarrindjeri surely will die.
> (Ngarrindjeri Nation, 2006, pp. 224–225)

Thus, the rules set out by the ancestral spirits and ancestors, and embodied in such sources as the Dreaming gave them and still now give guidance on how to live properly in relation with all entities (human and non-human). One example is a simple lesson recounted by Elder and leader of Camp Coorong Tom Trevorrow when he stated 'don't be greedy – only take what you need' (Tom Trevorrow, pers. comm. 20 January 2013), which was a key lesson to come out of the Thukeri (Bony Bream) Ngarrindjeri Dreaming Story (Ngarrindjeri Nation, 2006, p. 8).

While this worldview has similarities with other indigenous peoples' worldviews (see Grieves, 2008; Cajete, 2000) we want to be careful neither to essentialize nor romanticize Ngarrindjeri people and their approaches to total engagement with their ecologies. However, as articulated by Rigney (pers. comm. 6 March 2015) Ngarrindjeri worldviews 'encapsulates my nation, my people, my land and my waters – that is my world'. This confluence provides an intriguing insight to a social and cultural identity that Parajuli (2001, pp. 89–90) characterizes as an 'ecological ethnicity'. This concept refers to a people who share a common and ecologically distinct region, the development of mutually nurturing relationships to natural phenomena and a commitment to interacting with local ecosystems in a sustainable manner. In other words: a familial positionality or critical ecology of place that allows for beneficial and inclusive developmental processes that are mutually advantageous and sustainable, and underpinned by Ngarrindjeri worldviews and values.

In particular, we note that Ngarrindjeri assert the right to manage the 'natural resources' and the economic opportunities for Ngarrindjeri community members to create a sustainable future for themselves and their children; this occurs in a market context where employment, income earning opportunities and enterprise development are essential because the pre-invasion way of life is no longer possible. In terms of understanding Ngarrindjeri futures from a political ecology lens, Ngarrindjeri community leaders are forced to deal with a neoliberal capitalist system that can be hostile to their worldviews and so are in a constant struggle to negotiate these divides. For while Ngarrindjeri have been trying to assert their rights as an indigenous people in the settler-colonial context of Australia, the ideology of neoliberalism has increased its grip on the Australian political and economic system, and in turn, the Ngarrindjeri people.

Neoliberalism as political rationality

Neoclassical economics and neoliberalism as a political rationality (Beeson and Firth, 1998) can be said to presume a series of fundamental assumptions regarding human nature and the ways in which individuals behave under a given set of social and economic conditions. Among these assumptions is the idea that individuals, regardless of race, ethnicity or an underlying system of beliefs, are driven by a desire to maximize their level of individual happiness through the acquisition of more and more goods and services. This particular school of thought argues that market mechanisms, competition and efficiency are the most viable means by which to determine the allocation of resources; and the welfare of both the individual and society will be maximized when markets are free or unregulated. Thus, as Ngarrindjeri peoples work with government agencies to realize their aspirations relating to water allocations and their management in the Murray–Darling River Basin we suggest that neoliberal ideas guided by the 'belief that the market should be the organising principle for all political, social, and economic decisions . . . ' (Giroux and Giroux, 2006, p. 22) provide the presuppositional and conceptual framework that underpins the terms of the debate.

This is particularly pertinent from within the present-day focus on economic growth as a panacea for all societal (or community-based) ills. As noted by many historians, early ideas of economic expansion soon became intertwined with efficiency and a particular kind of religious zeal, and as Friedman (2005, p. 47) suggests, 'the material progress that it brought became central to the vision of moral progress' and has become the definitive apparatus for exploiting the Earth's resources in order to advance material wealth and human advancement (Rifkin, 2004, p. 115). A key aspect of this vision is the conflation of material wealth and human progress as a measure of human well-being through a reliance on neoliberal market-based economic and political narratives; a nexus stemming from a largely outdated modernist worldview, one that, we argue, can no longer be sustained in these critical times.

Since the 1980s, many governments – including Australia – have adopted neoliberal policies, espousing the need for economic growth, competition, free markets and privatization. Significantly, the perceived efficiencies that underpin neoliberal globalization are central to this doctrine and have been normalized and accepted as unquestionable, as underscored by the phrase 'there is no alternative'.

Analyst Stephen Gill (1995, p. 402) has described the power implications implicit here thus: 'the dominant forces of contemporary globalization are constituted by a neoliberal historical bloc that practices a politics of supremacy within and across nations'. This 'bloc' seeks to impose a prescribed economic structure, as well as a culture, ideology and a particular mythology of capitalist progress that has come about as a result of the globalization of liberalism. This ideology is put forward as the 'sole model of future development' and it is reinforced through the muscle of market discipline and political power (Gill, 1995, pp. 399 and 412). As Gill explains it, ' . . . a disturbing feature of market civilization is that it tends to generate a perspective on the world that is ahistorical, economistic, materialistic, "me-oriented", short-termist, and ecologically myopic' (1995, p. 399).

What we witness through initiatives such as Kungan, Ngarrindjeri Yunnan Agreements and the Ngarrindjeri Nation Yarluwar-Ruwe Plan is the NRA being able to successfully re-negotiate with governments on the basis of their holistic and interconnected worldview. This forms a foundation for assertions of sovereignty and provides possibilities for a sustainable, fair and nourishing economy. This was simply expressed in terms of Tom Trevorrow's philosophy above, 'Don't be greedy – only take what you need'; or as the Ngarrindjeri Nation Yarluwar-Ruwe Plan articulated it:

> Our Lands, Our Waters, Our People, All Living Things are connected. We implore people to respect our Ruwe (Country) as it was created in the Kaldowinyeri (the Creation). We long for sparkling, clean waters, healthy land and people and all living things. We long for the Yarluwar-Ruwe (Sea Country) of our ancestors. Our vision is all people Caring, Sharing, Knowing and Respecting the lands, the waters and all living things.
>
> (Ngarrindjeri Nation, 2006, p. 5)

However, development of such an economy can be addressed only within the present political and social justice arena, and the existing neoliberal exploitation of resource use characterizing the current globalizing economic order.

Context of water allocations and the advocacy of cultural flows

The Murray–Darling River Basin system is one of the most vital sources of water for the nation of Australia and thereby holds vital economic, social, cultural and environmental importance. It covers over 1,061,469 square kilometres, is home to 2 million people, provides drinking water to over 3 million people, provides for almost 45 percent of the value of Australia's agricultural production and supports tourism and recreation industries worth some AUS $800 million per year (Calma, 2009, p. 267). This vital environmental system has been brought to its knees following decades of poor management, flow restrictions, over allocation, drought and now the impacts of global climate change. Thus it became apparent to both State and Federal governments that a sound, integrated management plan, based on setting water allocations and reducing extractions from the system, was vital. This has not been easy considering the division of power between five of Australia's eight states and territories and the Commonwealth government, each with its own laws, policies and interests. However, as the Native Title Report of 2008 made clear, the Indigenous Nations of the region were also assertive of their political rights and cultural needs in this difficult policy arena (Calma, 2009). Weir (2007) noted:

> Massive extractions of water from the Murray River for irrigation have degraded the ecological health of the river country, transforming relationships previous sustained by the flow of the river water . . . The consequences of the over-extraction of water from the inland rivers are so serious that it is being experienced by the traditional Aboriginal land owners as a contemporary dispossession of their country.
>
> (2007, p. 44)

The Ngarrindjeri community bore the brunt of many of the problems and as a consequence of the declining state of the river and the factors outlined above, many of the Indigenous Nations of the Murray–Darling Basin

> resolved to develop a stronger voice for traditional owners in policy and management responses to the severely degraded Murray River, including strengthening the relationships between traditional owner groups through the development of 'Nation to Nation' protocols. This resolution resulted in the establishment of the Murray Lower Darling Rivers Indigenous Nations (Aboriginal Corporation) (MLDRIN), with an objective to represent traditional owners and be a platform to engage with government.
>
> (Calma, 2009, pp. 291–2)

Perhaps one of the most powerful signs of ecological devastation was the closing of the mouth of the Murray River where it feeds into the Southern Ocean,[2] a site of cultural significance for Ngarrindjeri people. In 2007, it was reported that 'the Coorong is dying' (Jenkin, 2007). National news featured numerous stories in those years (2004–2010) on issues that included the acid-sulphate soils being exposed in the drying lake beds, the rising salinity levels of the Coorong lagoons, the drying up of the water and the reductions in migratory birds visiting the area. A tragedy was unfolding on a national and international level but it was particularly an existential threat for the Ngarrindjeri with their interdependence and integration with this fragile environment. This is the context for their advocacy of cultural flows for the MRDB system. As then Aboriginal and Torres Strait Islander Social Justice Commissioner and Race Discrimination Commissioner Tom Calma noted, the degradation of the MRDB system was in fact a human rights issue and represented 'a new wave of dispossession for Aboriginal nations such as the Ngarrindjeri' (2008).

Ngarrindjeri Elder Matt Rigney, the inaugural Chair of MLDRIN, expressed this pain and damage in evocative words:

> We are of these waters, and the River Murray and the Darling and all of its estuaries are the veins within our body. You want to plug one up, we become sick. And we are getting sick as human beings because our waterways are not clean. So it is not sustaining us as it was meant to by the creators of our world.
>
> (cited in Calma, 2009, p. 289)

Interestingly, working with Ngarrindjeri leader Grant Rigney in exploring water, Ngarrindjeri rights to cultural flows, and a political ecology of tourism in this project also highlights a continuation of a family tradition. In this respect, Grant has followed in his father's example of leadership in political fora such as the NRA, MLDRIN and government negotiations and can enunciate the Ngarrindjeri view of sovereignty in all of its manifestations.

> I think what's happening in today's society is we have an environment and educational system that doesn't teach us about responsibilities in country. It doesn't teach us about responsibilities in country, it doesn't teach us about responsibilities or sustainability of the human race; because really that's what it was about, is about sustainability of the Ngarrindjeri. We had certain laws within these processes of sharing and caring, never taking more than what you need, and making sure everyone has [enough] and no one goes without. It was community orientated – it was about nationhood, it was about unification, it was about embodiment, and today's environment. In the western world we don't teach that; we teach about individuality, we teach about economics and basically about capitalism, they are the real things that drive the processes and affect our young ones
>
> So it's a different mindset and I think that we need to start changing that mindset by teaching these things through our educational systems. Education maybe has to start with the parents or the grandparents . . . but we know we have to live with each other. Ngarrindjeri in particular – we're not about greed and we're not about wanting anything extra – we want equity, we're wanting equilibrium across the board, where we have a right to be Ngarrindjeri because that's who we are and that is our sovereign right of this land, to be Ngarrindjeri and we want to practice our culture, and we still do strongly. But we want to make sure that it's out there and people know who we are and respect our country in the way that we respect our country and the way we be country.
>
> (Rigney, pers. comm. 6 March 2015)

This notion of 'we be country' (Rigney, 2015) is particularly powerful and clearly aligns with Parajuli's notion of ecological ethnicity as a politically charged concept that 'calls for a degree of autonomous governance for devising appropriate ownership over the biotic wealth, the commons, and the communities' (Parajuli, 2001, p. 90). As Rigney suggests regarding the water issues facing the Ngarrindjeri peoples:

water is the underlying denominator for all of this; we all know that you can't live without water. What we are trying to achieve through the water reform in Australia, particularly from the sovereign First Nations perspective of the Indigenous peoples of Australia, is to have an actual allocation, to have the right to water, and the right to have that allocation for the purposes that are required for the cultural strength and cultural connections through these particular nation groups right across Australia; it doesn't matter who they are. And water is the real dominator for all that, it gives life to everything and that's what supports the sustainability of all these nations. So we're really looking at trying to push that recognition, we are human beings and have a right to water. Prior to invasion, this was our water and has always been Aboriginal water. This has been taken away from us and they have regimented this [Murray–Darling] system now where it is so over regulated, and so over used, there might not be the sustainability of the surface water within this country for the next hundred years.

(Rigney, 2015, pers. comm 6 March)

Consequently, in the light of Ngarrindjeri peoples' fundamental relationships to country, prior occupation and sovereignty, non-indigenous recognition of Ngarrindjeri beliefs and traditions are fundamental to social justice and crucial to the success of any programmes aimed at positive and sustainable community development (such as 'Closing the Gap' between indigenous and non-indigenous Australians).[3] The severe drought devastating the Murray–Darling Basin between 2002 and 2010 reinforced these ongoing challenges with a reinvigorated desire for a range of long-term strategies centred on a range of Caring for Country programmes through the Kungun Ngarrindjeri Yunnan Agreements (KNY). Key elements of this approach are the development of a healthy future for Ngarrindjeri lands, waters and all living things predicated on respect for traditions, cultural responsibility and a sustainable regional economy. Crucial to these ongoing efforts have been partnerships between various Australian and international Indigenous Nations, the efforts of certain universities, and ongoing negotiations regarding legal contractual agreements with Australian governments. Central to these negotiations is the significance of Ngarrindjeri sovereignty and the inherent right of water entitlements that are 'legally and beneficially owned by Indigenous Nations – entitlements that are of a sufficient and adequate quantity and quality to improve the spiritual, cultural, environmental and economic condition of Indigenous Nations' (Calma, 2009, p. 284).

Indigenous Australian Nations from across the Murray–Darling Basin have articulated these entitlements as 'cultural flows' and 'indigenous water', a concept that incorporates and moves beyond environmental flows and sustains ongoing cultural practices and relationships to country. As a MLDRIN document on 'Cultural Flows' stated:

The difference between environmental and cultural water is that it is the Indigenous peoples themselves deciding where and when water should be delivered based on traditional knowledge and their aspirations. This ensures Indigenous peoples are empowered to fulfil their responsibilities to care for country.

(cited in Calma, 2009, p. 284)

Accordingly, cultural flows have potential benefits for indigenous Australians on multiple levels, including improved health and well-being, empowerment from being able to care for their country, ability to undertake cultural activities and ultimately also gift benefits for the wider community in the form of further comprehensive and holistic environment outcomes.

Returning to MLDRIN, in 2007 there were 11 Nation groups and the Ngarrindjeri were one of the main founders of this body. Nevertheless, it has grown since then and there are now 24 Nations involved. This early confederation initiated the Echuca Declaration[4] and what came out of this agreement was a definition of cultural flows, and the declaration that these are legally beneficial waters owned by those Indigenous Nation groups for the purpose of their ecological, spiritual purposes with the right to utilize the water as they deem fit for their sustainability (Rigney, pers. comm 6 March 2015).

> But knowing very well that having the responsibility for water flowing into country around country and through country . . . there is a sustainability of holistic connection right through country from the top of the system to the end of the system, and every nation group knew very well that they had a responsibility for the water to flow.
> (Rigney, pers. comm 6 March 2015)

Rigney's overview provides a significant understanding of the nature of Ngarrindjeri thought regarding the Murray–Darling River Basin as a holistic system rather that the fragmented approach associated with the present segmented and compartmentalized management approach enacted by Commonwealth and State/Territory governments. Thus, as he describes, it is a worldview or vision, because we know that for their worlds (the Indigenous Nations) and their country to be happy and sustainable, they know that the country down further needs to be happy and sustainable for that to happen because of the holistic connections through groundwater connectivity, through service water. But also through creation stories that connect all of our countries all the way through the whole beautiful landscape – so we have that responsibility to keep those processes and those cultural knowledge transmissions coming down through our different nation groups . . . (Rigney, pers. comm 6 March 2015).

Implications for Ngarrindjeri tourism ventures

The Ngarrindjeri community is known for an active engagement with the capacities for tourism to promote cross-cultural understanding and to foster greater ecological awareness, particularly through Camp Coorong and the Coorong Wilderness Lodge (CWL) (Higgins-Desbiolles, 2003). This engagement notwithstanding, environmental attributes that make Ngarrindjeri country attractive as an ecotourism and cultural tourism destination have nevertheless been severely damaged with the degradation of Ngarrindjeri ecology that has occurred with exploitation of the Murray–Darling. This is evocatively communicated by

Ngarrindjeri Elder George Trevorrow (2009) as he describes these impacts on the operation of the CWL:

> We built here on the basis that we're on a beautiful pristine area of the Coorong; where there's not much traffic, there's no boating, there's hardly anything here, it's really a wilderness area, the southern lagoon, and if we're going to lose that, well then we've lost everything that we built this for... You know, a little while back we could hardly do a kayaking tour, there was that little water here.
>
> (pers. comm. 18 July 2009)

However, while these impacts seem intractable, as Rigney indicates, country can be recovered, and country will always try to recover itself in some shape or another, whether it be through flood, fire or drought, country always tries to heal itself in one way or another, but we as the human race are making impacts so severe that nature's finding it very hard to actually recover, from what it can do to what we impact it on (Rigney, pers. comm. 6 March 2015).

However, in this analysis of the Ngarrindjeri assertion to proactively self-determine their future, tourism is used as a tool to share Ngarrindjeri worldviews and foster a site of engagement to make commensurability possible with non-indigenous others. According to Rigney:

> ... tourism is about nationhood and it's a different type of nationhood than what's, I suppose the norm as you would see it in the western society today, because our nationhood's about our country, it's about our people and our stories and our ceremonies and our songlines and our connection to those places. Tourism, like any other business enterprise that we as a nation would be looking to get into, is a means to develop socio-economic opportunities within our own communities and not just limited to Ngarrindjeri, we're talking general communities as well. It helps with economics but also health indicators and educational indicators ... and it creates so many other spinoffs in the process: skills training, building capacity, building self, feeling good, feeling wanted and connected to something and being a part of it, it builds those attributes into the people under training, but it also really helps develop that Ngarrindjeri nation and it's about having that capacity to have input into the sustainability of our Ruwe [country].
>
> (Rigney, pers. comm. 6 March 2015)

As a cultural educator at Camp Coorong for years before taking over his current roles in political leadership, Grant is also reflective on how tourism interpretation works in Ngarrindjeri enterprises. He noted that cultural interpretation was used as 'space' for conversations across diverse views and that tourism could be 'used as an educational tool, ... as a viewing platform for sustainability, of the ecologies and biodiversity [of the tourism site], as a platform for people viewing landscape completely differently ... [by] looking at connectivity' (Rigney, pers. comm. 6

March 2015). Returning to the politics of tourism, tourism interpretation can be used to assert sovereignty over country, enact care for country and share one's worldview with a view to bridging the divides or incommensurability to arrive at shared understandings for building sustainable futures.

Conclusion

Ngarrindjeri leaders argue that non-indigenous recognition of Ngarrindjeri beliefs and traditions is fundamental, not only for their efforts to care for Ngarrindjeri country and Ngarrindjeri people, but also for building a sustainable and equitable future. In the twenty-first century, Ngarrindjeri have identified the crucial challenge to be the creation of a future centred on Caring for Country, a future which incorporates respect for traditions, cultural responsibility, self-determination and economic development. The severe drought devastating the Murray–Darling Basin in the 2000s reinforced these ongoing community challenges but also offered potential for rethinking possibilities of building a more positive future through the rebuilding of relations between indigenous and non-indigenous Australians. Ngarrindjeri leaders sought a path through this environmental disaster that brought with it a greater opportunity for Ngarrindjeri to develop a long-term Ngarrindjeri Caring for Country Programme aimed at education, training, employment and a sustainable Ngarrindjeri regional economy. Complementing this strategy has been a vision to use tourism as a tool for education, the sharing of Ngarrindjeri worldviews and an assertion of sovereignty. In a time of grave crises and challenges for the spiritual, cultural, environmental, social and economic conditions for Ngarrindjeri peoples, we ask what we might demand from tourism in helping us meet and resolve such challenges. In adopting a political ecology approach we offer this case study of the Ngarrindjeri as a striking example of the need to disrupt the erasure of the local people that occurs in many tourism sites. This recognition of the significance of a respect for Ngarrindjeri rights and needs and an engagement with their deep embeddedness in and connections to local places is crucial as we learn to reinhabit our world and build sustainable futures.

Notes

1 Cultural flows are water entitlements that are legally and beneficially owned by the Indigenous Nations and are of a sufficient and adequate quality to improve the spiritual, cultural, environmental, social and economic conditions of these Indigenous Nations. This is our inherent right (MLDRIN 2007).
2 This continues to be problem and State and Federal governments are continuing to fund the dredging of the mouth in order to keep it open to the Southern Ocean (ABC News, 8 June 2015).
3 Closing the Gap is a strategy that aims to reduce indigenous disadvantage with respect to life expectancy, child mortality, access to childhood education, educational achievement, [and] employment outcomes. Endorsed by the Australian Government in March 2005, Closing the Gap is a formal commitment developed in response to the call of the

Social Justice Report (Calma, 2005) (1) to achieve indigenous health inequality within 25 years (HealthInfoNet).
4 The Echuca Declaration maintains 'that each of the Indigenous Nations represented within [the] Murray and Lower Darling Rivers Indigenous Nations is and has been since time immemorial sovereign over its own lands and waters and that the people of each indigenous Nation obtain and maintain their spiritual and cultural identity, live and livelihood form their lands and waters' (MILDRN 2007).

References

Beeson, M and Firth, A (1998) Neoliberalism as a political rationality. *Journal of Sociology*, 34 (3), pp. 215–231.
Bell, D. (1998) *Ngarrindjeri Wurruwarrin: A world that is, was, and will be*. Melbourne: Spinifex Press.
Berg, P. and Dasmann, R. (1977) Reinhabiting California. In: Berg, P. (ed.), *Reinhabiting a Separate Country: A bioregional anthology of Northern California*. San Francisco, CA: Planet Drum, pp. 217–220.
Berry, T. (n.d.) Earth Voices. Retrieved 3 July 2014 from http://earthheart.org/url/berry.htm
Briassoulos, H. (2002) Sustainable tourism and the question of the commons. *Annals of Tourism Research*, 29 (4), pp. 1065–1085.
Bryant, R. (1998) Power, knowledge and political ecology in the third world: a review. *Progress in Physical Geography*, 22 (1), pp. 79–94.
Cajete, G. (2000) *Native Science: Natural Laws of Interdependence*. Santa Fe, CA: Clear Light Publishers.
Calma, T. (2005) Social Justice Report 2005: The indigenous health challenge. Retrieved 1 October 2015 from https://www.humanrights.gov.au/publications/social-justice-report-2005-indigenous-health-challenge. Accessed 1st October 2015.
Calma, T. (2008) The future. A lecture in the Essentials for Social Justice series, 12 November, Adelaide, University of South Australia.
Calma, T. (2009) *Native Title Report 2008, Australian Human Rights Commission*. Retrieved 9 April 2015 from http://apo.org.au/node/14604.
Cole, S. (2012) A political ecology of water equity and tourism: a case study from Bali. *Annals of Tourism Research*, 39 (2), pp. 1221–1241.
Deleuze, G. and Guattari, F. (2004) *A Thousand Plateaus: Capitalism and schizophrenia*, translated by Brian Massumi. London: Continuum.
Friedman, B. (2005) *The Moral Consequences of Economic Growth*. New York: Knopf.
Gill, S. (1995) Globalisation, market civilisation, and disciplinary neoliberalism. *Millennium: Journal of International Studies*, 24 (3), pp. 399–423.
Giroux, H and Giroux, S. (2006) Challenging neoliberalism's new world order: the promise of critical pedagogy. *Cultural Studies-Critical Methodologies*, 6, pp. 21–32.
Grieves, V. (2008) A baseline for Indigenous knowledges development in Australia. *The Canadian Journal of Native Studies*, 28 (2), pp. 363–398.
Guattari, F. (2000) *The Three Ecologies*, translated by Ian Pinder and Paul Sutton. London: The Athlone Press.
Hall, C. M. (2008) *Tourism planning: Policies, processes and relationships*, 2nd edn. London: Pearson.
Harvey, D. (2006) Neo-Liberalism as creative destruction. *Geografiska Annaler*, 88 (2), pp. 145–158.

Hemming, S. and Rigney, D. M. (2011) Ngarrindjeri Ruwe/Ruwar: Wellbeing through Caring for Country. In Shute, R. H., Slee, P. T., Murray-Harvey, R. and Dix, K. L. (eds), *Mental Health and Wellbeing: Educational perspectives*. Adelaide, SA: Shannon Research Press, pp. 351–354.

Higgins-Desbiolles, F. (2003) Reconciliation tourism: tourism healing divided societies? *Tourism Recreation Research*, 28 (3), pp. 35–44.

Higgins-Desbiolles, F. (2004) Unsettling intersections: a case study in tourism, globalisation and indigenous peoples. *Tourism, Culture & Communication*, 5 (1), pp. 3–21.

Higgins-Desbiolles, F. (2006) More than an industry: tourism as a social force. *Tourism Management*, 27 (6), pp. 1192–1208.

Higgins-Desbiolles, F. (2007) Kungan Ngarrindjeri Yunnan: A case study of Indigenous rights and tourism in Australia. In: Buultjens, J. and Fuller, D. (eds), *Issues in Sustainability: Indigenous tourism case studies*. Lismore, NSW: Southern University Press, pp. 139–186.

Higgins-Desbiolles, F. (2008) Justice tourism: a pathway to alternative globalisation. *Journal of Sustainable Tourism*, 16 (3), pp. 345–364.

Jenkin, C. (2007) The Coorong is dying. *The Advertiser* (online), 9 January. Retrieved 6 June 2010 from http://www.adelaidenow.com.au/news/coorong-is-dying/story-e6freo8c-1111112805839.

Jenkin, G. (1985) *Conquest of the Ngarrindjeri: The story of the Lower Murray Lakes tribes*. Point McLeay, SA: Raukkan Publishers.

McLaren, D. (2003) *Rethinking Tourism and Ecotravel*, 2nd edn. West Hartford, CT: Kumarian.

MLDRIN (n.d.) Murray Lower Darling Rivers Indigenous Nations. Retrieved 3 March 2015 from http://www.mldrin.org.au.

Ngarrindjeri Nation (2006) Yarluwar-Ruwe Plan: Ngarrindjeri Nation Yarluwar-Ruwe Plan, Caring for Ngarrindjeri Sea Country and Culture, Kungun Ngarrindjeri Yunnan (Listen to Ngarrindjeri People Talking). Ngarrindjeri Tendi, Ngarrindjeri Heritage Committee, Native Title Management Committee.

Ngarrindjeri Ramsar Working Group (1999) Ngarrindjeri perspectives in Ramsar Issues. In: Draft Coorong and Lakes Alexandrina and Albert Management Plan, Appendix 8. Adelaide: South Australian Department for Environment, Heritage and Aboriginal Affairs.

Ngarrindjeri Regional Authority (NRA) (1985) Retrieved 25 March 2015 from http://www.ngarrindjeri.org.au.

Nicholls, R. (2006) Toward an ecology of peace. *World & I: Innovative Approaches to Peace*. New York, pp. 70–79.

Parajuli, P. (2001) No nature apart: Adivasi cosmovision. In: Arnold, P. P. and Grodzins Gold, A. (eds), *Sacred Landscapes and Cultural Politics*. Aldershot: Ashgate, pp. 83–113.

Rifkin, J. (2004) *The European Dream*. New York: Tarcher-Penguin.

Rigney, G. (2015) Personal communication, 6 March.

Robbins, P. (2012) *Political Ecology: A critical introduction*, 2nd edn. London: Wiley-Blackwell.

Rose, D. B. (2007) Justice and longing. In: Potter, E., Mackenzie, S., Mackinnon, A. and McKay, J. (eds), *Fresh Water: New perspectives on water in Australia*. Melbourne: Melbourne University Press, pp. 8–20.

Ryan, C. (1991) *Recreational Tourism: A social science perspective*. London: Routledge.

Stewart-Harawira, M. (2005) *The New Imperial Order: Indigenous responses to globalisation*. London: Zed Books.
Stonich, S. C. (1998) A political ecology of tourism. *Annals of Tourism Research*, 25 (1), pp. 25–54.
Tribe, J. (1997) The indiscipline of tourism. *Annals of Tourism Research*, 24 (3), pp. 638–657.
Weir, J. (2007) The traditional owner experience along the Murray River. In: Potter, E., Mackenzie, S., Mackinnon, A. and McKay, J. (eds), *Fresh Water: New perspectives on water in Australia*. Carlton: Melbourne University Press, pp. 44–58.

3 Co-management of natural resources in protected areas in 'postcolonial' Africa

Chengeto Chaderopa

Introduction

The management of national parks in Africa has evolved through various models, including twentieth-century approaches inspired by the philosophy that people and protected areas are incompatible. The 1980s, however, witnessed an ideological shift towards co-management of natural resources between conservation authorities, business and rural communities as the best way to address nature conservation and community development goals. The ability of co-management to adequately address the conservation-development nexus must, however, be appraised from the perspective of an operating environment that is characterized by neoliberalization trends in nature management and a proliferation of 'market based instruments' (Levine and Wandesforde-Smith, 2004). The debate rages on as to whether or not these neoliberal trends are consistent with the conditions that are necessary to promote local community inclusive involvement and participation. It is in this context that this chapter analyzes the co-management arrangements between the Makuleke community – who own a Contractual Park in Kruger National Park (KNP), South Africa – and their conservation and development partners. The chapter explores whether or not social justice, mutuality of interests, equitable sharing of benefits, fairness and joint decision-making characterize this co-management 'model'. The chapter raises key questions about the role of stakeholders in setting community and conservation goals and the trade-offs that are necessary and acceptable in order to establish viable co-management relationships.

The seemingly local decisions made by the local communities and local elites in the co-management of the Makuleke Contractual Park (MCP) are, in some cases, only local as far as the physical location of the people who made them is concerned. In reality, the genesis, texture and soul of the decisions are extra-local. The pressure to be politically, economically and environmentally correct in the eyes of the global sustainable development community as well as funding agencies influences the ways in which local communities relate with the natural benefits of their land. Although this chapter's unit of analysis is a rural community in South Africa, the forces shaping the operations of this community in the context of how they socio-economically and environmentally relate to their land are generated by larger global political, economic and environmental ideological fundamentals. In line with what Robbins (2012, pp. 21–22) describes as the '[f]ive dominant narratives in political ecology', this chapter therefore offers a historically situated, place-based and multi-scalar analysis of how conservation space is produced and controlled, and how power mediates the relationship between multiple stakeholders, sometimes resulting in 'degradation and marginalization' and 'environmental conflict and exclusion' (Robbins, 2012, p. 21). New nature conservation approaches tend to be sensitive to the global political, economic and political community's perspective of what needs to be 'preserved' as well as what is sustainable. In the process this sometimes results in the dislocation of 'local systems of livelihoods, production and socio-political organization . . .' (Robbins, 2012, p. 21).

Methodologically, most of the research presented in this chapter is based on a larger project, which consisted in part of in-depth interviews conducted in 2009 with 25 members of the Makuleke community including Chief Maluleke, past and present Community Property Association (CPA) executive members and several village representatives. The chapter also makes use of information contained in the latest *Makuleke Conservation and Development Framework* (April 2012) prepared by Environmental Resources Management Ltd (ERM) and Partners, a company that was commissioned by the Makuleke CPA. The choice of data analysis method was based largely on the author's perceptions of suitability and ease of use, as well as the ability to generate the necessary themes to answer the research questions. This is because qualitative research is 'largely intuitive, soft and relativistic' (Creswell, 1998, p. 142), and analysis of the data usually relies on 'insight, intuition, and impression' (Dey, 1995, p. 78) of the researcher. The analysis of the data was organized against the backdrop of the major divisions of the research questions such as access to MCP resources, devolution of decision-making, opportunities for entrepreneurial development and access to job opportunities.

The Makuleke community

Like most rural communities in colonial and apartheid Africa, the Makuleke suffered huge losses when, in 1969, they were forced to burn down their own homes at gunpoint in Old Makuleke (earlier known as Pafuri) and displaced to an area called Ntlaveni in order to make way for the expansion of the KNP (Cock and Fig, 2002, p. 137).

After their forced removal, the Makuleke quickly noticed that their resettlement reserve, Ntlaveni, 'lacked the rich game, wild fruits, lala palm, and fish' (Turner, 2004, p. 168) of the Old Makuleke. At Ntlaveni, they were 'concentrated on a . . . small parcel of land comprising 5000 hectares and divided into three villages' (Turner, 2004, p. 168), called Makahlule, Mabiligwe and Makuleke (Collins, 2010). Their allotted agricultural plots were also not big enough compared to the Old Makuleke. Predictably, this scenario resulted in a sharp decline in their livelihoods, causing significant discontent. The Makuleke also endured the humiliation of seeing their chief losing his chieftaincy and having to report to another chief (Friedman, 2005).

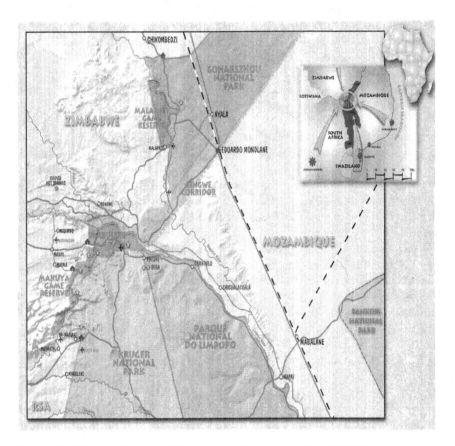

Figure 3.1 Old Makuleke/Pafuri Triangle, now the location of the 26,500-hectare Makuleke Contractual Park (MCP).

Source: JMB (2005). Reproduced with the permission of Gezani Lamson Maluleke, Maluleke CPA.

In 1998, after a protracted land claim case, the Makuleke won back title to their lost land. The 1998 Settlement Agreement between the CPA and the State transferred the ownership of the Makuleke Region of the KNP to the CPA, giving birth to the MCP. In addition to the MCP, SANParks has also established a number of other community 'contract parks' such as Kgalagadi San Mier and Richtersveld Contract National Parks, where rural communities are directly involved in conservation and ecotourism projects (SANParks, 2012) (see Figure 3.1).

One of the key provisions of this 1998 Settlement Agreement was that the Makuleke had 'full rights' to commercialize their land by entering into partnerships with private interests to build and operate game lodges, but any of these had to adhere to a sustainability framework of development (JMB, 2000). This was intended to address the nature conservation-community development nexus by both socio-politically and economically empowering communities through ecotourism development, and simultaneously protecting the 'integrity of ecosystem structure . . . ' (JMB, 2000, p. 28). This balancing act is consistent with the essence of co-management. It was envisaged that co-management would avail opportunities to the Makuleke community to make themselves whole again, by addressing the historical imbalances that were ushered into their lives when they were removed from Pafuri and resettled in Ntlaveni.

Co-management in a 'postcolonial' Africa context

Collaborative management is 'the sharing of power and responsibility between the government and local resource users' (Berkes *et al.*, 1991, p. 12). It is mostly used with reference to 'a situation in which two or more social actors negotiate, define and guarantee amongst themselves a fair sharing of the management functions, entitlements and responsibilities for a given territory, area or set of natural resources' (Borrini-Feyerabend *et al.*, 2000, p. 1). It occurs when 'a group of autonomous stakeholders of a problem domain engage in an interactive process, using shared rules, norms, and structures, to act or decide on issues related to that domain' (Wood and Gray, 1991, p. 146). Co-management is 'equated with joint management, shared management, participatory management, multi-stakeholder management' (Carlsson and Berkes, 2005, p. 2) and seeks social justice and 'democracy' in the management of natural resources (Borrini-Feyerabend *et al.*, 2000). The World Bank's (1999) definition emphasizes aspects such as sharing of responsibilities, rights and duties, joint decision-making and equality, while Gray (1985, p. 912) views the desire to pool resources so that the partners can 'solve a set of problems which neither can solve individually' (Gray, 1985, p. 912) as the essence of co-management. Co-management facilitates the accentuation of each party's 'strengths as a way of mitigating the weaknesses of each' (Singleton, 1998, p. 7). Complementary resources provide the potential for collaborative advantage (Huxham, 1993). For example, it has been observed that 'commercial success in mainstream hunting and tourism businesses requires large up-front investment, commercial experience and substantial risk-taking capacity' (Ashley and Elliot, 2003, p. 5). These requirements act as barriers to entry into the industry for local poor inexperienced communities. Being involved in co-management with experienced, skilled and knowledgeable partners can become a source of collaborative advantage for such groups.

Western values of nature and 'postcolonial' Africa

However, to better appreciate the performance of the co-management model in 'postcolonial' Africa in general and the MCP in particular, it is important to explore the philosophy that originally guided the establishment of parks in apartheid South Africa. The rationale is that the philosophy may be so engrained into the socio-economic and political fabric of 'postcolonial Africa' that it continues to influence the role played by previously marginalized black rural local communities in co-management of natural resources. White (2010) posits that the use of the word 'post' in postcolonialism, implies that 'it comes after colonialism', which effectively implies that it is something that happened in the past and is no longer in existence. Ashcroft *et al.* (2003, p. 1) argue that 'postcolonial' refers to 'all the culture affected by the imperial process from the moment of colonization to the present day'. This implies that there is continuity with practices of the past. Most Indigenous communities are still affected by colonialism and a colonial mindset (Moreton-Robinson, 2004). In turn this implies that it is more likely than not that some elements of colonial approaches to nature conservation continue to strongly influence how 'postcolonial' African communities interact with their natural resources. In the case of South Africa, research shows that its nature governance structures, systems and beliefs were largely inspired by Western values and ethics imported by colonists (Reid, 2001). Western nature values conceptualized wilderness preservation and human development as incompatible. This belief that human activities are inherently harmful to nature led to Africa being treated as the biblical pristine and fragile Eden that needed to be protected from desertification as well as from human abuse (Anderson and Grove, 1989). This, in turn, justified the implementation of a 'fortress' conservation natural resource management model (Adams and Hulme, 2001; Turner, 2004) with huge implications for the livelihoods of many black communities.

The negative impacts of this romantic Western conceptualization of nature on the livelihoods of black communities were exacerbated by the then South African government's apartheid governance ideology of segregation that overtly promoted racially skewed ownership of land and access to natural resource benefits. The disenfranchisement of local communities living near protected areas (PAs) was further accomplished through a two- pronged approach: (1) the promotion of conservation and wildlife management practices such as trophy hunting, in which the local community lacked skills (Makombe, 1993); and (2) simultaneously criminalizing the economic activities local people used to practice, such as hunting and honey collection in the forests (Leader-Williams, 2000).

In 'postcolonial' Africa, these historical injustices warranted the adoption of people-centered conservation management approaches. However, Büscher advises caution for those optimistically viewing co-management as a panacea for all these injustices:

> A paradox thus emerges, in that colonial suffering and inequalities are superficially recognised in policy and rhetoric of major conservation actors through a focus on 'community-conservation' and 'sustainable development', while the latter simultaneously retain many of the 'aesthetics, symbols, and fables' of conservation as 'white privilege'.
>
> (2011, p. 88)

Büscher's (2011) views here reinforce the argument that postcolonial natural resources co-management approaches that are intended to be people-inclusive have not exactly exorcised the ghost of a romanticized Western conceptualization of nature as fragile, and requiring, for its protection, to be separated from people.

Ecotourism and neoliberalism: co-management contradictions

One would also be remiss to discuss co-management without situating it on a landscape of changing environmental governance associated with neoliberalization of nature and the use of market-based instruments. In the context of SANParks, neoliberalism was a policy of the Mandela government that withdrew all operational funding for the parks around 1996. The democratization of South Africa meant that scarce government funds went to areas that the government perceived to be better placed to servicing the needs of the poor, of which Parks Management was not. The only alternative for SANParks was to make the tourists pay the bill for park management and using ecotourism as a community development strategy. It is important therefore to analyze how a neoliberalization nature-orientation, and the use of market-based instruments in the management of Pas, fit within the broad aim of co-management to empower the rural poor as well as to enhance and diversify their means of livelihood security.

Neoliberalism is a 'political economic approach that posits markets as the ultimate tool for achieving optimal use and allocation of scarce resources' (Mansfield, 2004, p. 43). It involves 'restructuring the world to facilitate the spread of free markets' (Igoe and Brockington, 2007; Harvey, 2005; McCarthy and Prudham, 2004) and is defined by deregulation, and reregulation in which the state's role in natural resource management is reduced (Igoe and Brockington, 2007). 'The mainstream argument in favor of neoliberalism is framed in terms of the efficiency of the market in contrast to the inefficiencies and high costs of government interventions' (Liverman and Vilas, 2006, p. 328). Government intervention in market processes is considered intrinsically dysfunctional (Mansfield, 2004). In this neoliberal environment, the development of ecotourism as a viable community empowerment and conservation strategy was spearheaded by the private sector precisely because the private sector was endowed with the requisite ecotourism business management knowledge and skills.

However, although ecotourism is presented as environmentally sustainable, it is very problematic in that, as a 'source of neoliberalising nature . . . it exists in a context of global neoliberalism; it is part of it . . . it is entirely compatible with it' (Duffy, 2005, p. 330). Ecotourism has increasingly been neoliberalized by being subjected to an expanding variety of market-based systems, a process that is also referred to as commodification of nature (Castree, 2003; Duffy, 2005).

Against the backdrop of the argument that the market tends not to place a high enough value on environmental quality or ecosystems, neoliberalism promises to 'infuse new types of resources into biodiversity conservation . . . increase democracy and participation . . . and protect rural communities by guaranteeing their property rights and helping them enter into conservation-oriented business ventures' (Igoe and Brockington, 2007, p. 434). By protecting nature through the mechanisms of investment and consumption, neoliberalism claims that it can establish a world in which one can 'eat one's conservation cake and have development desert too' (Igoe and Brockington, 2007, p. 434). In this regard, for communities to be competent in ecotourism management, they must be 'brought out of nature and into the market so that they can return to nature as competent conservationists' (Igoe and Brockington, 2007, p. 442). Members of potential ecotourism destination communities must become 'eco-rational subjects' aware of the dynamics of ecology and economy (Goldman, 2001). This eco-rationalization of communities seems to be consistent with local participatory forms of resource management (Mansfield, 2004) since co-management involves resource users, state agency officials and private-sector agents sharing responsibilities in the areas of decision-making, information gathering and enforcement (Jentoft, 2000; Singleton, 1998).

Amid these neoliberal claims of the potency of ecotourism and market-based instruments to achieve local community economic development, there are many contradictions it brings to the MCP co-management arrangement. The major aim of co-management is to facilitate the closing down of what Biermann *et al.* (2007) describe as 'the participation and implementation gaps' in natural resource management. The lack of adequate resources in developing economies has led to an implementation gap in PA management that necessitates the invitation of the more resourced private-sector players. On the other hand, the participation gap in PA requires the involvement and engagement of local black communities who, in the case of colonial communities, were excluded from national parks, either as consumers or decision makers (Cock and Fig, 2002). Today, these communities do not always have adequate modern tourism management skills or knowledge and are in a 'state of non-decision making' (Joppe, 1996) and constrained therefore to contribute to effective conservation management as a way of addressing both the implementation and participation deficits in natural resource management.

To address the participation deficit seems to demand the creation of suitable conditions that enable the adequate transference to communities of skills, knowledge and confidence needed for commercial success. However, it has been observed that community participation has serious time and administrative costs (Davis, 1996). Addressing these deficiencies requires that the private-sector partners dedicate large amounts of their budgets to the community's training needs. However, in an era of 'market triumphalism', where ideologies and the bottom line overshadow social issues, potential investments in natural resources often do not give due attention to rural peoples' poverty concerns (Dressler and Büscher, 2008). Due to financial implications, private-sector players or concessionaires involved in co-management of parks may be deterred from giving adequate attention to the development of human or community capital. Dressler and Büscher (2008) give the example of so-called 'people-oriented' conservation projects near the buffer zone of KNP where they discovered that the private-sector owned enterprises rarely reinvested in the natural assets such as goats, land and fuel wood, yet these are necessary for reducing rural peoples' livelihood vulnerability.

It is important to acknowledge that the outcomes of neoliberalism are not always problematic for conservation and local livelihoods (Castree, 2007).

Neoliberalism should be conceptualized as a set of processes, rather than an end point contingent on history and place, producing diverse rather than generic results (Peck and Tickell, 2002). However, on balance, neoliberal market-based instrument ideology seems parallel to the agenda of social justice, mutuality of interests, transparency and joint decision-making. It seems more inclined towards the disempowerment and 'de-responsibilization' (Banuri and Amalric, 1992) of local communities. As Duffy (2005) observes, the powerful networks that direct the pace and development of ecotourism have seen ecotourism being developed in a manner that conforms to externally produced and driven models that suit more the global market place rather than the local people's unique needs.

The Makuleke experience of the politics of co-management

The following two sections of the chapter present the views of the Makuleke regarding their perception of the manner of their involvement and participation in the Makuleke model of natural resource management. The framework of evaluation used here draws on a synthesis of diverse definitions of co-management, largely captured in The United Nations Economic and Social Council resolution 1929 (LV111):

> ... participation requires the voluntary and democratic involvement of people in (a) contributing to the development effort (b) sharing equitably in the benefits derived there from and (c) decision-making in respect of setting goals, formulating policies and planning and implementing economic and social development programs.
>
> (cited in Tosun, 2005, p. 335)

It has to be appreciated, however, that co-management is an expansive and complex concept whose essence goes beyond the confines of this definition.

The socio-economic utility of the MCP landscape

One of the issues that has generated controversy in the MCP relates to the divergent perceptions held by different members of the Makuleke community about the best land use options for the Old Makuleke. While some of the interviewees supported the use of the MCP to support non-consumptive forms of wildlife and nature-based tourism, others were against the use of 'our land' to support such activities. This latter group preferred to refer to the MCP as Pafuri or Old Makuleke in place of the 'Makuleke Contractual Park' in a seemingly defiant tone that seemed to emphasize their disconnect to the current socio-economic identity of the MCP. In apparent reference to the paucity that characterizes Ntlaveni, they also described the Old Makuleke as a 'fertile' place that provided 'us with a lot of meat, honey, fish, fruits, firewood ... ' They associated their success in regaining ownership of the Old Makuleke with reverting to the 'old way of doing things', through which almost everyone had access to natural resource benefits of the land – the Old Makuleke. They therefore preferred using the MCP for hunting and other agricultural activities to 'putting productive land to waste' by dedicating large tracts of 'fertile land' to 'protect animals and forests'. Wildlife-based tourism was described in derisory manner as an activity for 'our white colleagues . . . not for us Africans'. Agriculture and hunting were also in harmony with their survival skills, unlike tourism, which forced them to enter into partnerships with other people 'who seem to have no interest in our development'. Using the land to support tourism rendered them passive participants in the project because they did not have the necessary knowledge. Others fondly reminisced about a time soon after their land claim success when they sold hunting rights and managed to raise large sums of money, which was used to fund diverse community development projects.

Interviewees indicated that they were prepared to support the idea of using the MCP for nature-based tourism but they were irked by what they perceived as their conservation and development partners' lackadaisical attitude towards equipping them with the necessary skills to enable them to sustainably manage wildlife-based tourism independently. They suspected that their partners were afraid to genuinely upskill the communities out of fear that if the communities became well trained, the CPA would not renew their operational permits at the expiry of their 15-year tenure of operation. They also believed since their partners were generating huge profits in the MCP, they would do 'everything in the power' to delay their eventual departure from the park. Additionally, they complained that 'very few of our people are employed . . . doing small jobs . . . ' in the MCP. They believed that 'well-paying, good jobs' were not always finding their way to the community members but were taken up by people from elsewhere.

Other interviewees complained that the stipulation in the Settlement Agreement that stated that the reclaimed land was to be used for conservation did not originate with them but was rather imposed on them by the government and private-sector players. They accepted the stipulation simply because they were tired of arguing and had realized that without conceding to the requirement to use the reclaimed land for conservation purposes, their land claim would have dragged on forever:

> It's not like we had a choice . . . in a war you reach a stage where you can't go any further . . . Half a loaf is better than nothing . . . When we look at it now, the settlement agreement did not give us a lot of power . . .

On the other hand, there was also a distinct group, mostly CPA executives past and present, who did not feel at all disadvantaged as there was a lot of evidence to show that the community was benefitting immensely from the MCP ecotourism proceeds. They supported ecotourism development in the MCP as the best way to achieve community development goals such as poverty reduction. They saw no reason to doubt the intentions of their conservation and development partners. Rather, they asserted that there was no evidence to support the view that their partners were not committed to helping them to develop the necessary wildlife

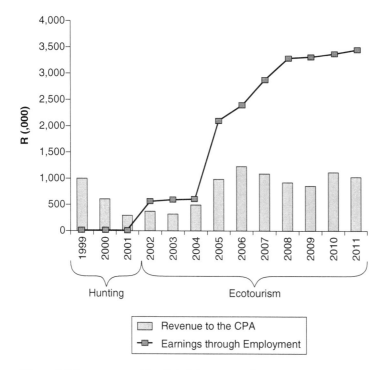

Figure 3.2 Income earned by Makuleke community.
Source: ERM (2012, pp. 2–12).

and nature-based tourism business management knowledge and skills. One interviewee used marriage as an analogy to explain the MCP co-management operational environment:

> Like any normal marriage, there are ups and downs ... but I have no regrets about the agreements we made ... where else in Africa do you find a rural community doing what we are doing here ... people from all over the world have come here to learn about our model ... of course there will always be challenges ...

Therefore, they attributed the slow rate at which the Makuleke were being trained in wildlife-based tourism business management to the diminishing support that the NGO community had previously extended to the Makuleke. Throughout the process of reclaiming their land, and post 1998, the Makuleke received support from NGOs such as GTZ, the Ford Foundation and Endangered Wildlife Trust, as well as from other technical advisers known as Friends of Makuleke (Grossman and Holden, 2002).

This decision to use the MCP to support ecotourism seems sound in light of the evidence presented in Figure 3.2 showing that ecotourism has presented the communities with more benefits over an extended period of time through employment creation as well as through direct revenue to the CPA, where concessionaires pay a proportion of turnover to the CPA on an annual basis.

Significantly, although the possession of different resources by partners provides an incentive to collaborate in an attempt to benefit from them, these different resources are also:

> the result of differences in organizational purpose. Although partners ostensibly may agree on a broad label for the collaboration's purpose – at least to the extent that they are willing to be involved – they each have different reasons for being there.
>
> (Vangen and Huxham, 2003, p. 18)

SANParks' position, to use the MCP for conservation and ecotourism, epitomizes their philosophical understanding of how best to achieve nature conservation and community development goals. This reflects that in co-management, '[e]ach party [may] want to pursue its own interests to the full, and in so doing ends up contradicting, compromising, or even defeating the interest of the other' (Ochieng, 2000, p. 8). Ultimately, these conflicting perspectives regarding the MCP's socio-economic utility raise fundamental questions about collaboration, and the conditions that are necessary for it to succeed in achieving conservation and community development goals. Although it is undeniable that co-management has improved the socio-economic well-being of the Makuleke, it can be argued that the establishment of vibrant co-management has been compromised by these conflicting constructions of the landscape upon which the co-management process is based. These conflicting perspectives, although not unexpected, suggest the absence of mutuality of interest, which also reflects a deep lacuna between the perceptions of the diverse MCP 'stakeholders' regarding the collaborative advantages that each partner wishes to draw from one another in this co-management.

Co-management and the emergence of a 'supra'-community group: implications for MCP benefits sharing

The operation of the MCP co-management model has predictably facilitated the emergence of a supra-community group with huge implications for the manner in which other community members evaluate the flow of the MCP benefits. Indeed, it is ironic that while some community members felt that there was inadequate community capacity building to facilitate meaningful local community involvement in the MCP's governance, the 'little' that has taken place has had the effect of creating a new elite socio-economic and political group that is environmentally conscious and, therefore, overtly proud of the vast knowledge it has accumulated about tourism and natural resource management through attendance at national and international conferences that deal with community capacity building for effective natural resource management. They took every opportunity during conversations with the author to eloquently show their understanding of various community participation models such as Pretty's (1995) typology of participation and they also generously coated their explanations with Brundtland Commission (WCED, 1987) sustainable development diction.

The emergence of this new elite group was also evidenced in the controversy surrounding the call for more operators in the MCP, instead of the few in operation during 2009, as illustrated in Figure 3.3. This group, consisting mostly of current and past CPA executive members, has developed fairly robust knowledge about tourism markets, demand and supply issues, and the elusiveness of the concept of sustainable development, among other important issues to do with the conservation-development discourse. They understood how overcrowding in the MCP can lead to the deterioration of the quality of the attractions, resulting in fewer tourists visiting the MCP. They also cogently argued that the selling of hunting rights in the MCP would inevitably attract the ire of the global environmentally conscious community who consider selling hunting rights an immoral and unsustainable economic activity. One senior member explained how since South Africa was a signatory to CITES (The Convention on International Trade in Endangered Species), it was important for the MCP to abide by the CITES governance principles. Pursuing consumptive forms of tourism was, in his view, against the tenets of CITES. In the same context, other interviewees also rationalized that the backlash from the international community would ultimately impact negatively on the viability of the MCP with huge repercussions for the whole community: an unacceptable fate especially in light of 'how we toiled to get our land back . . . '

Existing tourism developments in the Makuleke Contractual Park

This relatively cogent business reasoning regarding the long-term impacts of unplanned developments in the sustainability of the MCP project clearly shows how community capacity building workshops had unwittingly created an elite group among the Makuleke who, for all intents and purposes, can no longer be placed into the category of the general Makuleke villagers who want the MCP to support consumptive forms of tourism. The emergence of this new group also highlights the complex nature of 'community' and the dangers of treating the Makuleke or any community, as simply communities of place or of identity that are 'extraordinarily cohesive' (Steenkamp and Uhr, 2000, p. 5). To some extent, co-management is ill-prepared to deal with the complexities that define community.

The refusal to support hunting and other common agricultural activities in the MCP as per the wishes of some of the interviewees reflects Robbins's 'conservation and control thesis' (2012, p. 21). The reference to CITES and the use of Brundtland Commission (1987) sustainability discourse by local elites in their

Name of facility	Operator	Developer	Number of beds	Number of vehicles
Pafuri Camp	Wilderness Safaris Group (Wilderness Adventures)	Wilderness Safaris Group	52 plus 8 trail beds	6
The Outpost	Rare Earth Retreats (KLPG)	Matswani Safaris	24	3
Eco Training	Eco Training	Eco Training	20	2
Elsmore	N/A	To be developed	28	3

Figure 3.3 Lodge operations in the MCP.
Source: ERM (2012, p. 2).

justification for refusing to support the practice of wildlife consumptive tourism in the MCP shows how local decisions can be 'influenced by regional policies, which are in turn directed by global politics and economics' (Robbins, 2012, p. 20). At another level, they reflect the 'relationship between power and knowledge' (Robbins, 2012, p. 24) as well as how conservation is produced, resulting in some sections of the community being marginalized. In the context of local rural communities and national parks, 'knowledge' and 'differential power' determine winners and losers as well as the production of 'social and environmental outcomes' (Robbins, 2012, p. 20).

Joint decision-making and exercise of power in the MCP

Co-management implies 'a joint decision-making approach to a problem resolution where power is shared and stakeholders take collective responsibility for their actions and subsequent outcomes from those actions' (Selin and Chavez, 1995, p. 191). The assumption is that sharing power and decision-making will augment the process of resource management and make it more receptive to diverse needs (McCay and Jentoft, 1998).

In this regard, the Settlement Agreement seems to have safeguarded the rights of the Makuleke by clearly stating that the Joint Management Board (JMB) has the right to determine levels of resource utilization and that the CPA and the community shall always have right to retain access to the Makuleke region, from time to time, as determined by the JMB (JMB, 2000). The MCP's JMB comprises three SANParks and three community representatives (ERM, 2012; Reid, 2001). 'Decision-making requires a 2/3 majority from each group' (Turner, 2004, p. 170).

Natural resources in protected areas 83

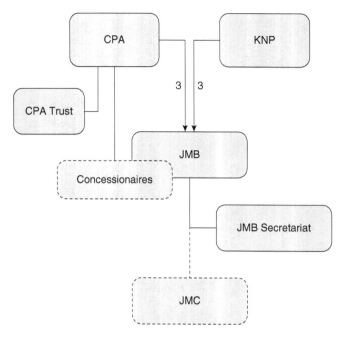

Figure 3.4 Existing institutional arrangements.
Source: ERM (2012, pp. 6–62). Reproduced with the permission of Steve Collins, ERM.

It is however, ironic that this created structure of democratic decision-making, the JMB, has proved counterproductive from the perspective of some of the Makuleke's wishes to exercise control over MCP land use options. They were not impressed by the fact that SANParks has veto power over any decisions made, and complained that the provisions of the Settlement Agreement that obligate them to seek approval for any proposed MCP land use plans from partners was compromising their stranglehold over the Old Makuleke. They felt that by placing the burden of proof in the hands of the Makuleke community regarding the sustainable nature of their ideal development projects in the MCP, their conservation partners were subverting the originally envisaged collaborative advantages into tools that facilitated the Makuleke's socio-economic and political marginalization. The Makuleke's CPA executive committee members who comprise local primary and secondary school teachers, ordinary village men and women, with 'little' technical sustainability discourse in comparison to SANParks, have not been able to match SANParks in this debate, leaving some of the community members believing that their co-management partners are seeking advantages from the partnership that are not in harmony with local community empowerment. This issue also highlights the role of power in co-management: 'While power relations are included within collaborative theory, it is frequently assumed that collaboration can overcome power imbalances by involving all stakeholders in a

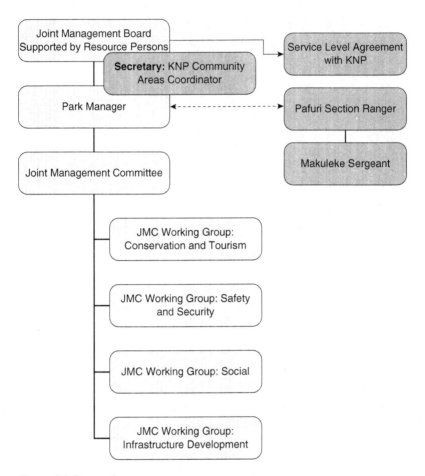

Figure 3.5 Proposed new governance arrangement.

process that meets their needs' (Jamal and Getz, 1995, p. 567). However, power is generally not equally distributed among the stakeholders in any partnership.

In the same vein, it is important to note that the inadequacies of these existing institutional arrangements (Figure 3.4) have been acknowledged and a new governance structure has been proposed as shown in Figure 3.5.

It is however debatable whether or not this proposed new governance structure will improve the bargaining position of the communities through their representation in the JMB and JMC.

Co-management, community inclusive representation, involvement and participation

Successful collaboration depends on the nature of representation of the diverse stakeholders and therefore 'as much as possible, coalitions should attempt to include all individuals and organizations that have a direct stake in the issue or issues being addressed by the coalition' (Legler and Reischl, 2003, p. 55). To this end, the Makuleke community has an administrative structure, the CPA, which is made up of nine executive committee members, four village representatives and four general representatives with the chief as the ex-officio chair. The CPA (Figure 3.6) is designed to be a tool of community empowerment as it allows the representation of various segments of the community.

It may be argued that this is the most comprehensive community representative structure possible and that it ideally should comprehensively represent the diverse interests and perspectives of the villagers. Compared to other communities in 'postcolonial' Africa who have suffered the bitterness of land dispossession, this structure places the Makuleke in a special class whose membership includes the once vibrant Zimbabwean 'CAMPFIRE' (Communal Areas Management Program for Indigenous Resources) (Murombedzi, 1999). However, some community members hold little faith in the ability of CPA to represent their socio-economic interests and felt that the Tribal Authority had more power over the CPA. They accused the CPA of rubber-stamping decisions made by the Tribal Authority. These complaints about (mis)representation in the CPA extended to the way the CPA presided over the sharing of MCP natural resource benefits. They complained that the chief's family members and close associates treated the CPA and MCP as a 'family affair' and benefitted more from the proceeds of the MCP. They suggested that there was a need to separate tribal matters from the CPA business.

It is however important to acknowledge that the Makuleke co-management model has directly and indirectly benefitted the majority of the Makuleke community. Various community members have benefitted from capacity building and training programs. Most CPA Executive Committee members have been empowered through various training programs to improve their decision-making in community-based tourism and natural resource management. About 100 Makuleke are employed at the lodges as tour guides, room attendants, housekeepers and general hands.

Despite using a lottery system to allocate job opportunities, there were, however, some serious reservations from some interviewees regarding fairness and transparency in the allocation of jobs opportunities. The CPA executive responsible for employment confirmed that although they tried to be transparent by adopting the lottery system in their hiring process, 'we are still accused of nepotism and favoritism'. Other interviewees argued, however, that there was nothing sinister behind the involvement of the chief's family and that the benefits the chief and his family were enjoying were in keeping with their cultural position and role.

It can be argued that the controversy regarding the nature of access to the MCP benefits reflects frustrations by some community members at the failure of the MCP to generate as abundant benefits as some communities had envisaged when they won back the land in 1998. Support for this line of thinking is evident in the results of a SWOT analysis conducted by ERM, a consulting firm commissioned by the Makuleke which identified the following constraints, among others, that are hindering the MCP from supporting profitable tourism businesses and

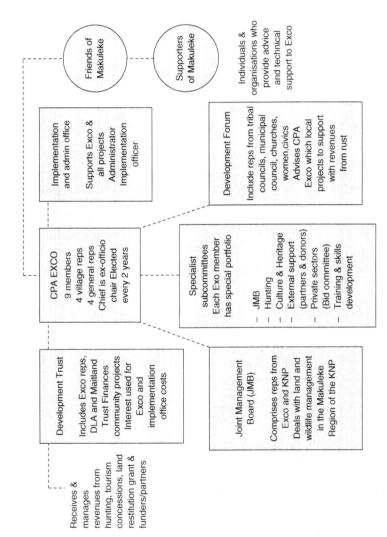

Figure 3.6 Makuleke residents as members of the Community Property Association (CPA).
Source: Koch and Collins (n.d., p. 4). Reproduced with the permission of Steve Collins of ERM.

therefore sustainable benefits to the community: 'Limited socio-economic benefits for the Makuleke community, both in income and other, less tangible terms [...] Problems in achieving sustainable tourism because of access, marketing, demand and infrastructure related issues . . . ' (ERM, 2012, p. 11).

Conclusion

This chapter aimed to examine the extent to which co-management has facilitated the participation and engagement of local communities, as the case study indicates that it may actually be difficult, especially for an outsider, to assess the socio-economic and political significance of the Makuleke's involvement in the co-management of the MCP. Analysis of the Makuleke co-management model can therefore be polarizing. However, when this analysis is placed in the context of Scheyvens' (1999) empowerment framework, it could be argued that the Makuleke community's participation in the MCP co-management has brought them diverse tangible and intangible psychological, social, political and economic benefits. Their involvement in the MCP co-management has given them confidence in their ability to manage their resources. This is an important achievement, especially when considered from the viewpoint of the many years of apartheid's racially skewed socio-economic and political policies under which the black communities had operated. These policies were intended to undermine the black communities' confidence and self-belief to socio-economically and politically chart their development trajectory without the help of the settler community. The mere success of the Makuleke in reclaiming the land they lost in 1969 is an achievement of immense proportion. It could therefore be argued that their involvement has raised their status globally as a 'model' for joint management of national parks to achieve both development and conservation goals. The Makuleke village is also a fairly developed rural village compared to other communities in rural South Africa. The Makuleke village was among the first to have access to electricity in South Africa because of, among other considerations, their capacity to fund the project with benefits accruing from the MCP. The significance of such achievements cannot easily be quantitatively or qualitatively comprehended, especially by an outsider such as myself.

This, however, does not obviate observations that the MCP co-management model has facilitated more access of the MCP natural resource benefits to the Makuleke community's conservation and development partners, in addition to the newly created elites who are mostly CPA executive members. This reflects a significant chasm between the rhetoric and practice of co-management. One of the major challenges facing the co-management model seems to be that the co-management participants appear to share diverse conservation-development ideological orientations.

At another level, the Makuleke case highlights the realities that communities have to contend with when they get involved in the tourism industry. 'For most communities . . . success at nature tourism requires new sets of relationships with extra-local actors, but on uneven terrain' (Turner, 2004, p. 178). From the perspective of poor rural communities, it is unfortunate that the tourism industry is structured in a manner that facilitates the benefits of nature-based tourism to flow away from, rather than towards, the local communities (Turner, 2004).

Another argument could be that co-management, operating in a neoliberal ideological framework, does not seem to be capable of transforming the traditionally demeaning means through which local communities engage with tourism. Consequently, most Makuleke continue to relate to their land, largely as job seekers or as sellers of cultural artifacts and providers of entertainment to tourists, thereby reinforcing the stereotype of rural communities as part of the tourism product, rather than its rightful owner. Despite the myriad advantages associated with MCP co-management, it is clear that more effort needs to be put towards adequately responding to the Makuleke community's interests of poverty reduction and the engagement of, especially, traditionally marginalized members of the community such as youth, women and the poor.

References

Adams, W. and Hulme, D. (2001) Community and conservation: changing narratives, policies and practices in African conservation. *African Wildlife and Livelihoods: The Promise and Performance*, pp. 24–38.

Anderson, D. and Grove, R. (eds) (1989) *Conservation in Africa: Peoples, policies and practice*. Cambridge: Cambridge University Press.

Ashcroft, B., Griffiths, G. and Tiffin, H. (2003) *The Empire Writes Back: Theory and practice in post-colonial literatures*. Abingdon: Routledge.

Ashley, C. and Elliot, J. (2003) Just Wildlife? or a Source of Local Development? *Natural Resource Perspectives*, 85, pp. 1–6.

Banuri, T. and Amalric, F. (eds) (1992) *Population, Environment and De-Responsibilisation: Case studies from the rural areas of Pakistan*, Sustainable Development Policy Institute, Working Paper, Islamabad: POP 1.

Berkes, F., George, P. and Preston, R. (1991) Co-Management: The Evolution in Theory and Practice of the Joint Administration of Living Resources. Paper presented at the Second Biennial Conference of the International Association for the Study of Common Property, Winnipeg, Manitoba, September 26–29.

Biermann, F. M., Chan, A. M. and Pattberg. P. (2007) Multi-Stakeholder Partnerships for Sustainable Development: Does the Promise Hold? In: P. Glasbergen, F. Biermann and A. Mol (eds), *Partnerships, Governance and Sustainable Development: Reflections on theory and practice*. Cheltenham: Edward Elgar Publishing, pp. 239–260.

Borrini-Feyerabend, G., Farvar, M. T., Nguinguiri, J. C. and Ndangang, V. (2000) *Co-Management of Natural Resources, Organizing Negotiation and Learning by Doing*. Heidelberg: Kasparek.

Brundtland Commission (1987) *Report for the World Commission on Environment and Development: Our common future*. United Nations.

Büscher, B. (2011) The Neoliberalisation of Nature in Africa. In: T. Dietz (ed.), *New Topographies of Power? Africa negotiating an emerging multi-polar world*. Leiden: Brill, pp. 84–109.

Carlsson, L. and Berkes, F. (2005) Co-management, concepts and methodological implications. *Journal of Environmental Management*, 75 (1), pp. 65–76.

Castree, N. (2003) Commodifying what nature? *Progress in Human Geography*, 27 (2), pp. 273–92.
Castree, N. (2007) Neo-liberalizing Nature 1: The logics of de- and re-regulation. *Environment and Planning A*, 40 (1), pp. 131–152.
Cock, J. and Fig, D. (2002) From colonial to community-based conservation. Environmental justice and the transformation of national parks (1994–1998). In: D. McDonald (ed.), *Environmental justice in South Africa*. Cape Town: University of Cape Town Press, pp. 131–155.
Collins, S. (2010) Makuleke Concession Tourism Development Workshop presentation, July, 2010. Pafuri Camp, Makuleke Contractual Park, ASL Foundation, pp.1–9.
Creswell, J. W. (1998) *Qualitative Inquiry and Research Design: Choosing among five traditions*. Thousand Oaks, CA: Sage Publications.
Davis, G. (1996) *Consultation, Public Participation and the Integration of Multiple Interests into Policy Making*. Paris: Organization for Economic Co-operation and Development (OECD).
Dey, I. (1995) Reducing Fragmentation in Qualitative Research. In: U. Kelle (ed.), *Computer-Aided Qualitative Data Analysis: Theory, methods and practice*. London: Sage Publications.
Dressler, W. and Büscher, B. (2008) Market triumphalism and the CBNRM 'crises' at the South African section of the Great Limpopo Transfrontier Park. *Geoforum*, 39 (1), pp. 452–465.
Duffy, R. (2005) The politics of global environmental governance, the powers and limitations of transfrontier conservation areas in Central America. *Review of International Studies*, (31), pp. 307–327.
ERM (Environmental Resources Management Ltd (ERM) and Partners) (2012) *Makuleke Contractual Park Conservation and Development Framework*. Unpublished report, Makuleke, South Africa.
Friedman, J. (2005) *Winning isn't Everything: What the Makuleke lost in the process of land restitution*. Unpublished BA Thesis, Environmental Studies. South Africa: University of KwaZulu-Natal.
Goldman, M. (2001) The birth of a discipline: producing authoritative green knowledge. World Bank-style. *Ethnography*, 2 (2), pp. 191–217.
Gray, B. (1985) Conditions facilitating inter-organizational collaboration. *Human Relations*, 38, pp. 911–936.
Grossman, D. and Holden, P. (2002) *Contract parks in South Africa*. Available from: http://www.Conservationdevelopment.net/Projekte/Nachhaltigkeit/CD1/Suedafrika/Literatur/PDF/Grossmann.pdf [Accessed August 2009].
Harvey, D. (2005) *A Brief History of Neoliberalism*. Oxford: Oxford University Press.
Huxham, C. (1993) Pursuing collaborative advantage. *The Journal of the Operational Research Society*, 44 (6), pp. 599–611.
Igoe, J. and Brockington, D. (2007) Neoliberal conservation: A brief introduction. *Conservation and Society*, 5 (4), pp. 432–449.
Jamal, T. B. and Getz, D. (1995) Collaboration theory and community tourism planning. *Annals of Tourism Research*, 22 (1), pp. 186–204.
Jentoft, S. (2000) The community: a missing link of fisheries management. *Marine Policy*, 24 (1), pp. 53–60.
JMB (2000) *Master Plan*. Makuleke: Makuleke Community Property Association.
JMB (2005) *Pafuri Area TFCA: Approach to an integrated land use and tourism plan draft, May 2005*. South Africa: Makuleke CPA.

Joppe, M. (1996) Sustainable community tourism development revisited. *Tourism Management*, 17 (7), pp. 475–479.
Koch, E. and Collins, S. (n.d) *Removals, Restitution and Development in the Makuleke Region of the Kruger National Park: Our hearts are healing*. South Africa: Alex Hetherington Media.
Leader-Williams, N. (2000) The Effects of a Century of Policy and Legal Change on Wildlife Conservation and Utilization in Tanzania. In: H. Prins, J. Grootenhuis and T. Dolan (eds), *Wildlife Conservation by Sustainable Use*. Dordrecht: Kluwer, pp. 219–245.
Legler, R. and Reischl, T. (2003) The relationship of key factors in the process of collaboration a study of school-to-work coalitions. *The Journal of Applied Behavioral Science*, 39 (1), pp. 53–72.
Levine, A. and Wandesforde-Smith, G. (2004) Wildlife, markets, states, and communities in Africa: Looking beyond the invisible hand. *Journal of International Wildlife Law and Policy*, 7 (3–4), pp. 135–142.
Liverman, D. M. and Vilas, S. (2006) Neoliberalism and the environment in Latin America. *Annual Review of Environmental Resources*, 31, pp. 327–63.
Makombe, K. (ed.) (1993) *Sharing the Land: Wildlife, people and development in Africa*. IUCN/ ROSA Environmental Series No. 1. Harare: IUCN.
Mansfield, B. (2004) Rules of privatization: contradictions in neoliberal regulation of North Pacific fisheries. *Annals of the Association of American Geographers*, 94 (3), pp. 565–584.
McCarthy, J. and Prudham, S. (2004) Neoliberal nature and the nature of neoliberalism. *Geoforum*, 35 (3), pp. 275–283.
McCay, B. J. and Jentoft, S. (1998) Market or community failure? Critical perspectives on common proerty research. *Human Organization*, 57 (1), pp. 21–29.
Moreton-Robinson, A. (2004) *Whiteness epistemology and Indigenous representation*. Canberra: Aboriginal Studies Press.
Murombedzi, J. C. (1999) Devolution and stewardship in Zimbabwe's CAMPFIRE programme. *Journal of International Development*, 11 (2), pp. 287–293.
Ochieng, O. M. (2000) *Oxfam Karamoja Conflict Study: A Report*. Kampala: Oxfam.
Peck, J. and Tickell, A. (2002) Neoliberalising space. *Antipode*, 34 (3), pp. 380–404.
Pretty, J. N. (1995) *Regenerating Agriculture: Policies and practice for sustainability and self-reliance*. Washington, DC: Joseph Henry Press.
Reid, H. (2001) Contractual National Parks and the Makuleke Community. *Human Ecology*, 29 (2), pp. 135–155.
Robbins, P. (2012) *Political Ecology: A critical introduction*, 2nd edn. Malden, MA: John Wiley & Sons.
SANParks (2012) Report on corporate social investment programs: From Fortress Conservation to People and Conservation – SANParks and a Developmental Approach to Conservation in the 21st Century. Available from: http://www.sanparks.org/assets/docs/about/reports/social-investment-report-2012.pdf [Accessed 20 October 2014].
Scheyvens, R. (1999) Ecotourism and the empowerment of local communities. *Tourism Management*, 20, pp. 245–249.
Selin, S. and Chavez, D. (1995) Developing an evolutionary tourism partnership model. *Annals of Tourism Research*, 22 (4), pp. 844–856.
Singleton, S. (1998) *Constructing Cooperation: The evolution of institutions of comanagement*. Ann Arbor, MI: University of Michigan Press.

Steenkamp, C. and Uhr, J. (2000) The Makuleke land claim, power relations and community based natural resource management. *Evaluating Eden Series*, Discussion Chapter No 18. London: IIED.

Tosun, C. (2005) Stages in the emergence of a participatory tourism development approach in the developing world. *Geoforum*, 36 (3), pp. 333–352.

Turner, R. L. (2004) Communities, wildlife conservation, and tourism-based development: can community-based nature tourism live up to its promise? *Journal of International Wildlife Law and Policy*, 7 (3), pp.161–182.

Vangen, S. and Huxham, C. (2003) Enacting leadership for collaborative advantage: dilemmas of ideology and pragmatism in the activities of partnership managers. *British Journal of Management*, 14 (1), pp. 61–76.

WCED (1987) *Report of the World Commission on Environment and Development: Our common future*. Published as Annex to General Assembly document A/42/427. Available from: http://www.un-documents.net/wced-ocf.htm [Accessed 10 July 2009].

White, N. (2010) Indigenous Australian women's leadership: staying strong against the post-colonial tide. *International Journal of Leadership in Education*, 13 (1), pp. 7–25.

Wood, D. J. and Gray, B. (1991) Toward a comprehensive theory of collaboration. *The Journal of Applied Behavioral Science*, 27 (2), pp. 139–162.

World Bank (1999) Report from the International Workshop on Community-Based natural Resource Management (CBNRM), Washington, DC, 10–14 May 1998. Available from: http://www.worldbank.org/wbi/conatrem/ [Accessed 10 July 2009].

4 'Few people know that Krishna was the first environmentalist'

Religiously motivated conservation as a response to pilgrimage pressures in Vrindavan, India

Tamara Luthy

Introduction

While *Guru Purnima* is not the most important holiday of the Hindu calendar focused on the sacred hill of Govardhan, tens of thousands of people, myself included, turned out that day in 2014 to complete the circumambulation, or *parikram*, of this sacred natural site. A carnivalesque atmosphere pervaded the whole affair. Bus after bus drove away from the religious tourism path. Even though this act of religious tourism celebrates the sacred hill of Govardhan, the event should not be construed as a serene hike or leisurely stroll through nature. Dozens of *pujasthalas* only feet from each other blasted fast-paced Hindi music while volunteers thrust out plastic cups full of consecrated food (*prasad*) to pilgrims who race down the asphalt religious tourism path. These cups are then unceremoniously discarded on the ground and accidentally stomped underfoot by the crowds. Temples dot the main road every hundred yards or so, along with chai stalls and vendors selling religious paraphernalia. Pilgrims raced past us to complete the 21-kilometer religious tourism path. Near some of the more popular *pujasthalas*, male pilgrims are crowded in circles to dance ecstatically to the music.

Govardhan is a sacred hill and, according to local lore and religious texts, the location of numerous sacred groves. It is one of many such areas that are circumambulated by pilgrims every year in the Braj of Uttar Pradesh in northern India. Save for some cowherds and village children, few individuals stray from the paved road to wander into the fields immediately adjacent to the mountain. However, for those who do traverse the less-beaten path, small signs designating each sacred grove on the route provide a sense of the sacred geography described in religious texts. Some of these groves are groves in name only; other still show lush vegetation. Some bear little resemblance to their scriptural namesakes. Yet, just as in the scriptures, cowherds continue to tend to their animals grazing in the fields.

The *Govardhan parikram* is just one among many extremely popular religious tourism destinations in the Braj. The town of Vrindavan is considered to be one of the seven holiest cities for Hindus (cf. Haberman, 1994). Together with the nearby city of Mathura, and many outlying temples, groves, and temple water features (*ghats*), the Braj cultural region has become an important nexus for Hindu pilgrims. While each religious tourism site is unique, the ecological issues facing sacred natural land features like Govardhan Hill are played out daily across India, especially during festival times. Religious tourism is here defined as tourism that is motivated, either fully or partly, for religious reasons (Rinschede, 1992). The phenomenon of religious tourism involves at least 100 million people visiting holy sites scattered across the subcontinent every year (NCAER, 2003). The number of pilgrims has increased sharply in recent decades, especially with the advent of affordable public transportation and new roads to previously inaccessible places (Gladstone, 2005; Shinde, 2012). Places of pilgrimage in India often experience environmental problems as a result of pilgrimage pressures (Alley, 2002). Vrindavan is no different (cf. Shinde, 2012; Sullivan, 1998).

Religious tourism-related environmental pressures in the cultural area of the Braj have enabled local activists to forge new alliances with the Uttar Pradesh Forest Department, scientists, international NGOs, wealthy devotees, and temple authorities. This chapter focuses primarily upon the Environmental NGO *Friends of Vrindavan* and the *Braj-Vrindavan Heritage Alliance*. Though their missions do not always coincide, environmental activists and temple authorities have frequently met together to protest against the felling of trees, or to celebrate new restoration projects. In this chapter I argue that religiously motivated conservation in the Braj area, a prominent religious tourism destination in north India, focuses on an image of Krishna as an environmental deity to shore up concern for modern-day ecological issues facing communities and the ecosystem within this religious tourism area. This is part of a broader trend in Indian environmentalism to focus on the ecosystem services and biodiversity role of sacred groves (cf. Khan *et al.*, 2008). By merging scientific discourses with popular beliefs about the deity Krishna, these activists create a case for conservation and restoration of the sacred groves in the area of Braj, primarily in the Mathura district of Uttar Pradesh. This discourse is intended to induce behavioral change among stakeholders who are economically reliant upon the religious tourism industry, and the pilgrims themselves. The use of religious beliefs requires a careful selection of both religious and ecological tropes amid myriad conflicting viewpoints among interested actors. The activists must present the case of the god Krishna as an 'environmental deity' who maintains a vested interest in the material state of sacred natural land features, such as the Yamuna River, Govardhan Hill (*Giriraj*), sacred groves (*vanas*), ponds (*kundas*), and ghats. Local environmental activists appropriate the language of religion to compel government agencies, local politicians, and pilgrims to pay more attention to environmental issues in the religious tourism sites of the Braj.

This chapter stems from a larger research project investigating the impact of religious discourse, belief, and practice on environmental restoration movements in sacred forests in Vrindavan. Here, I draw primarily upon interviews and participant observation conducted over a total of nine months during the summers of 2010, 2013, and 2014. I also engage in textual analysis of written sources, such as sacred texts, the official websites of NGOs, pamphlets published by temples, newspaper articles, and blog posts of devotees of Krishna. Robbins (2012, p. viii) states that political ecology is not just a theoretical approach or academic discipline, but rather is 'an urgent kind of argument ... that surveys both the status of nature and the stories about the status of nature.' This work engages in the broader discussions of political ecology by taking seriously the use of religious discourse by environmental NGOs to effect tangible social and political changes in land management. Throughout this chapter, I use the term 'religiously motivated conservation' and/or restoration. Other scholars use the terms spiritual ecology (Sponsel, 2012), religious environmentalism (Nanda, 2004), nature spirituality (Taylor, 2009), moral ecology (Apffel-Marglin and Parajuli, 2000) and, in India, dharmic ecology (Prime, 1992) to describe similar phenomena. I prefer the term 'religiously motivated conservation' because it highlights the distinction between the philosophical viewpoints that motivate restoration efforts, and the efforts themselves.

The viewpoints of various environmental activists exist in dynamic tension with the competing belief that sacred natural land features have an unchangeable, eternal divine nature that cannot be sullied by changes in the material world (cf. Alley, 2002). Pilgrims themselves revere these groves for very different reasons than do religiously motivated conservationists (cf. Haberman, 1994; Shinde, 2012). My research suggests that the pilgrims themselves do not necessarily see themselves as environmental subjects (Agrawal, 2005), meaning that they do not see themselves as actors with an awareness of, or concern for, environmental issues. Nor do they view their religious tourism activity in relation to greater 'ecological' issues in the Braj area. While they often remark on the 'degraded' state of the sacred natural land features they do not relate these problems to larger political-economic issues. Instead, they often gloss ecological issues in religious terms, but in a markedly different way than religiously motivated conservationists. While few pilgrims to Braj seem to exhibit 'environmental subjectivity' in the religious tourism context, this may change as NGOs, environmentalists, and scientists increasingly utilize this type of discourse in relation to sacred groves.

The political ecology of religious tourism

In this chapter I focus on the sacred sites in and around the city of Vrindavan, located in the Mathura district of Uttar Pradesh. The area of Vrindavan is described in the *Bhagavata Purana* as the childhood home of the deity Krishna. This mythologized place was located geographically when the Bengali saint Chaintanya (1486–1533) travelled to the Mathura district to rediscover Krishna's childhood home. Chaintanya experienced visions of the deity in the uninhabited forest (*jangala*) near several small agricultural villages in the area of current-day Vrindavan (Ghosh, 2002; Haberman, 1994). Chaintanya and his followers set out to recover deities, construct temples, and create the conditions for pilgrimage. Major religious tourism routes within the forests of Vrindavan were first established in the sixteenth century, based upon the Sanskrit text *Vraj Bhakti Vilasa* written by the poet-seer Narayan Bhatt (Shah, 2006). Bhatt is largely responsible for the specifics of today's sacred geography. Bhatt mapped the place-names found in the Bhagavata Purana onto the physical terrain, essentially establishing the locations of the current sacred groves (Ghosh, 2002; Haberman, 1994). This tradition of nature reverence and praising the natural beauty of the forests was at least partly an outcome of the Sanskritic poetic tradition's romantic emphasis on nature (cf. Banerjee, 1987). Since that time, countless devotees have made the trek on foot, which takes between 20 and 40 days, to experience the sacred geography. It has become one of the seven most important Hindu places of pilgrimage or *tirthas* (Haberman, 1994).

Whereas Braj was once entirely forested, with tiny settlements, today only 1.8 percent of the Mathura district is covered by forest (Forest Survey of India, 2003). The name Vrindavan literally means 'forest of *tulsi*,' referring to holy basil (species *Ocimum sanctum*). Naming the town after this plant, which is almost universally revered throughout Hindu India, is a testament to the importance of sacred natural land features in the popular religion of this region. Yet the tourism-related construction and infrastructure development often impacts local, sacred natural landscape features. Vrindavan, although lacking sufficient infrastructure, has developed rapidly. To match the demands of pilgrims, *dharmashalas*, *ashramas*, yoga studios, hotels, vacation homes, and restaurants are increasingly common in holy cities.

From the perspective of political ecology, the unchecked and unregulated growth of pilgrimage has undoubtedly shaped the direction of local conservation and restoration movements. Peet and Watts (2004) note that entities at various levels must engage in the 'cultural production of conceptual devices' that allow them to discursively battle the (seemingly) opposing forces of growth and environmental conservation. I suggest that religiously motivated conservation draws upon just these sorts of conceptual devices to inspire conservation in the face of pilgrimage pressures. Environmental activists and concerned members of the temples have noted that sacred groves have been cut down to build hotels (Prime, 1992). Some of the 12 sacred forests, for which the town is famous, are now forests in name only, with very sparse vegetation. The groves are not the only sacred natural features to have been degraded. The Yamuna River is reportedly one of the most polluted water sources in India (Haberman, 2006). Many devotees lament that the goddess of the river is dying. Yet, in spite of these issues, pilgrims come to Vrindavan primarily to experience the sacred geography. While not all of the ecological problems facing the area are due to religious tourism, religious tourism undoubtedly contributes significantly to ecological problems. After each festival, the countless plastic cups discarded by pilgrims are swept up and burnt. The city of Vrindavan does not have a municipal garbage disposal system in place, something that the group *Friends of Vrindavan* is attempting to remedy.

While religiously motivated conservation discourse draws upon language and imagery that is highly local and contextual, these sacred groves are embedded in a broad array of ecological and social issues. Climate change, loss of dryland forest systems, the ongoing water crisis, the dumping of industrial effluent into rivers, and erosion due to illegal mining are just some of the pressing ecological problems facing much of rural north India. These ecological problems are exacerbated by widespread structural problems including changing land-tenure patterns, lack of enforcement of environmental laws, corruption of local politicians, the unequal impacts of climate change on subsistence farmers, the rash of farmer suicides related to avaricious money-lending practices, and issues of resource access stemming from class and caste inequality (Saha *et al*., 2010).

Tourism adds to the already palpable ecological problems facing rural north India. Admittedly these issues are not unique to sacred cities or religious tourism sites. Though common, they are especially problematic in areas with religious tourism, which face huge influxes of visitors every year and increasing demands on local ecosystems. In a sense, tourism may be the straw that is breaking the jewel-bedecked back of Krishna's prized camel. Wealthy foreign devotees, especially members of the International Society for Krishna Consciousness (ISKCON), have bought up a great deal of property in the surrounding area. Wealthy Indian and foreign pilgrims alike prefer air-conditioned accommodation with hot water. Yet only recently have adequate sewage, waste disposal, and electricity come to many parts of the town. Lacking a robust industrial or financial sector, the Braj district arguably would not be witnessing such rapid urbanization were it not for the demands of a growing religious tourism industry.

In spite of these issues, Vrindavan remains the most important religious tourism destination for pilgrims. The exact extent of religious tourism in the Vrindavan-Braj area is unknown, as no rigorously acquired figures exist of the number of pilgrims who visit each year. *The Alliance of Religions and Conservation* (2011) projects a figure of 500,000 pilgrims to Vrindavan each year, though most residents and scholars believe the figures to be far higher. When I asked local residents how many people visit each year, many of them simply shrugged and said '*Lakh aur lakh*,' which essentially means countless hundreds of thousands. In fact, several sources quote a figure of 5 million or more per year (Walters, 2010). In a 2010 article entitled 'Religious Tourism in Govardhan: Economic Survey,' the *Friends of Vrindavan* staff cite an unpublished study by Girami MBA College students on the economic impact of religious tourism in the Govardhan area. Over the course of one month, these students report having counted 21,544,000 people who completed *parikram* on foot, 35,000 visitors in tourist buses, and 832,867 who completed *parikram* in four-wheeled vehicles. Within the same month, they tracked 96 million rupees spent on accommodation in guesthouses and 176 million rupees on food (Vrindavan Today Admin, 2010). While these figures should be evaluated with skepticism until the results can be confirmed within a validly published account, they nevertheless highlight the impact of religious tourism travel in this area. Yet the pilgrims themselves, as I shall demonstrate, largely remain uninvolved in conservation and restoration.

Bowers, *bhava* and bliss: Vrindavan's sacred geography

In his account of the practice of walking the *parikram marg*, or religious tourism circuit around the 12 sacred groves of Vrindavan, Haberman (1994) notes that religious tourism is a highly embodied, emotive experience for Gaudiya Vaishnavas. The forests were the sites for the erotic encounters of the principal deity, Krishna, with his favorite *gopi*, or cow-herd girl, Radharani Radhakrishna, and where they are understood to have engaged in their *lilas*, or transcendental pastimes, eternally on earth in the forests of Vrindavan, regardless of modifications to the physical terrain itself. According to ISKCON and various other Gaudiya Vaishnava sects, these particular pastimes are of the highest sacred importance, because they are the primary source of pleasure for Krishna. While he operates as the Supreme Controller in all other respects, in his affairs with Radharani, his pleasure potency, he is able to actually enjoy himself.

For these devotees, working for Krishna's pleasure is the only source of true eternal, spiritual pleasure. Therefore, spending time in the sacred geography of Braj is said to rid one of all of the sins of past lives and to open up the spirit-soul to experience transcendental bliss, which is its highest function. On the whole, these individuals come to participate in the mythologized landscape and sacred natural features of the area, not to change it. For them, the main role of the plants and trees in these groves is the pleasant atmosphere that they create for pilgrims and devotees. Perhaps for this reason, restoration projects in the area thus far have focused on pragmatic interests such as providing shade for pilgrims and enhancing the beauty of religious tourism areas. However, when pilgrims reflect upon or discuss the ecological issues in Vrindavan, they typically use depoliticized, religious frames of reference to describe the problems.

In the context of religious tourism sites, the pilgrims themselves are essentially depoliticized actors in the greater drama of environmental degradation. Among many pilgrims, religious explanations are the dominant explanatory model for the ecological crisis in Braj. However, these ecological problems do not occur outside of the sphere of political economy. Indeed, Independence witnessed dramatic changes in land tenure and property ownership. Many temple trusts lost large tracts of land under new laws. Many of these parcels of land, some of which contained sacred groves, were sold off to local elites, rather than being passed down to poorer farmers, as the architects of local laws had originally intended. The privatization of former temple lands was also accompanied by the acquisition of many sacred groves by the Forest Department (cf. Kent, 2013). Furthermore, members of the burgeoning middle class represent the bulk of the pilgrims, though exact figures are unknown. Although they create the largest impact of all of the stakeholder groups, the fact that every other group of stakeholders depends on pilgrims for their livelihood constructs a situation in which they can neither be prevented from coming nor compelled to participate in a meaningful way. Nor would preventing religious tourism be seen as a desirable outcome by *Brajbasis*. Many *Brajbasis* are economically reliant upon religiously motivated tourism yet, nevertheless, lament the ecological consequences of economic 'development.' In spite of their depoliticized stance in relation to direct environmental action, it is important to explore their viewpoints, as these shed some light on the reasons why Vrindavan cannot legitimately be described as an 'ecotourism' site, even though many come to view the sacred natural land features.

Pilgrims' progress: explanatory models of environmental issues in Vrindavan

Pilgrims vary in their levels of awareness of the ecological consequences of the religious tourism industry itself. The reasons articulated by pilgrims for environmental degradation may exhibit class and caste bias. I interviewed one prominent Rajput politician who resided near Mathura and had participated in the Govardhan *parikram*. He suggested that the majority of the pilgrims were poorer people from rural areas. According to him, the reason for environmental degradation in this region, and others, was due to the ignorance and mismanagement of and by rural people. He stated that Hindu tradition has an intrinsic respect for sacred natural land features, but that rural people have 'forgotten' about the importance of conservation and instead only seek to gain a profit, even at the expense of felled trees.

The ecological impacts of religious tourism are compounded by a set of widely circulating religious tropes which facilitate a certain apathy toward the development of environmental subjectivity. The first is the belief in an eternal, perfect, transcendental nature of Braj, and the second is the belief in Kali Yuga. These two beliefs inform and interact with one another in a fashion that does not necessarily encourage the development of environmental subjectivities.

The belief that Vrindavan is part of a sacred geography that is both immanent and transcendental is common among both Brajbasis and pilgrims. I suggest that this belief has implications for pilgrims' experiences of the natural environment of the area. One research participant stated: 'You cannot see Vrindavan with your material eyes. In this Age of Ignorance called Kali Yuga you will only see the dirt and the sewage. Instead, you have to see Vrindavan with your spiritual eyes.'

It is understood that the physical terrain is indelibly imbued with the energy of the god Krishna. The teachings of A. C. Bhaktivedanta Swami demonstrate this belief. According to this religious teacher, one must be capable of seeing beyond the material degradation and see the eternal pastimes, or *nitya lila*, of the deity Krishna. According to this swami and others associated with the Gaudiya Vaishnavaa tradition, the dominant religious group in this area, only the truly spiritually elevated can see Vrindavan in its true, eternal state.

In a *Vrindavan Today* article entitled, 'The defects we see in the Dham are not real,' the author quotes a religious text to demonstrate the transcendental qualities of the landscape. It states, 'This Vrindavan is most perfect in all respects. Those who see flaws in it due to their imperfect vision and announce that they are real are fools' (Poddar, 2014). Another common explanatory model for the degradation of Vrindavan is that of Kali Yuga. According to many devotees, we currently live in Kali Yuga, the Age of Ignorance, which is characterized by a decline in morality and a rise in environmental degradation. Kali Yuga, the last of the four cycles of time, is believed to be the age of greed, ignorance, and loss of *dharma* or religious principles. In the age of Kali Yuga, people become selfish and greedy and live short, desperate lives (cf. Grodzins-Gold and Gujar, 2002; Narayanan, 1997).

Regardless of whether pilgrims exhibit environmental subjectivity, it is clear that many individuals are disturbed or even disappointed by the degraded state of the forests. My companion on the Govardhan Yatra expressed extreme dismay at the huge crowds in the temples and the degraded state of the forests. She even stated that she would rather visit other religious sites since, 'Krishna is everywhere.' This sentiment seemed especially prominent among foreign pilgrims. A pilgrim named Gokul, who spent his childhood in Vrindavan, stated: 'It used to be so forested, so full of trees. It is so different to see it today. When our guru was here, it was beautiful.' He went on to reiterate a common sentiment expressed by foreign pilgrims, namely that Vrindavan was a small town until about 30 years ago. While this sentiment is contradicted by accounts from previous centuries (cf. Growse, 1883; Robertson *et al.*, 1913), it is discursively related to the claim that Kali Yuga has begun to escalate more rapidly in the last few decades. Even the eternal sacred landscape of Vrindavan is thus subject to the degradation of this age of moral impurity. When asked about the state of the trees on the walking circuit, one member of my temple, named Syamasundari, replied, 'What trees? Vrindavan is so ugly.' She, like many others, decided to visit other sacred sites, such as Mayapur in West Bengal, to enjoy the scenery, spending most of her time in Vrindavan in the temples rather than the religious tourism circuits or groves. Many of these foreign pilgrims readily admit they dislike staying in Vrindavan.

Reworking Krishna as an environmental deity

For many *Brajbasis*, local environmental issues are seen to have cosmic significance (Haberman, 1994; Prime, 1992). *Brajbasi's* concerns about the environmental state of the area sit uncomfortably within the alternative, mainstream discourse, which suggests that Vrindavan is the physical embodiment of a transcendental, incorruptible reality. Nevertheless, this position is seen as philosophically justified. In a newspaper article in *The Harmonist*, one foreign devotee remarks: 'Devotees and others who feel that endeavors of this kind [i.e. environmental restoration] merely entangle us in undesirable worldly struggles should be allowed to have their point of view as well' (Walters, 2010).

On my first visit to Vrindavan, my riverboat guide urged me to dip my hand into the water and pour some of the water onto my head. The water from the Yamuna River is considered to bestow blessings and purification. However, he suggested that I shouldn't drink it, because the water was dirty and full of sewage. This example highlights the tension between pilgrims' views of this sacred area, juxtaposed against the frustrations of locals, or *Brajbasis*, who have witnessed the ecological degradation of the Braj area. A quote from a senior member of the Vrindavan Municipal Council's illustrates these sentiments: '[W]hen people from around the world visit Vrindavan they have a feeling that they will find a glimpse of a Vrindavan which is full of groves and trees, but when they come here they only find the concrete structures' (Poddar, 2013).

Some devotees and *Brajbasis* call for environmental action by suggesting that Krishna was himself an environmentalist. An article from the *Times of Agra* states:

> [E]ven though Lord Krishna is worshiped by hundreds of thousands of people the world over for being the Supreme Godhead, but few people know that he was the first environmentalist. Most of his pastimes were related with the environmental activism, right from containing the Kaliya from polluting the Yamuna to saving the Braj Mandal from excessive rain, or eating fire to save the forest. But his environmental legacy is being protected neither by the government nor by his followers.
>
> (Khandelwal, 2010)

In the aptly titled blogpost 'Jiva Garden – Krishna was the First Environmentalist,' one devotee of Babaji Satyanarayana Dasa writes: 'Krishna is always protecting nature because he loves nature'. Other ISKCON-affiliated groups, based more upon the notion of *seva*, or service, have also implemented proactive solutions to environmental problems. According to this paradigm, the truest form of devotion is to lovingly offer devotional service, or *seva*, to Krishna. The service of the *gurukulis*, or schoolchildren, is literally to Krishna's land itself. For these groups, such as the ISKCON *gurukul*, or school, the issue of Kali Yuga seems to have taken a back seat to the more pressing issue of creating a plantation. The students are taught about agroforestry via hands-on projects, while also being taught about the importance of rendering service to Krishna's sacred landscapes. Such efforts as these rely heavily upon volunteer labor on the part of visiting devotees. There is also a diversity of smaller temples and foundations apart from ISKCON involved in restoration in some form or another, adding other layers of participation to local environmental activism.

While these discourses have managed to attract the attention of state-level agencies and large international environmental NGOs they have not yet managed to filter into the consciousness of pilgrims themselves. Perhaps the trend toward a 'sustainable Braj' will eventually manifest in a greater involvement of pilgrims in the restoration of its sacred groves and forests. However, in the meantime, the demands of pilgrims and the interests of environmental NGOs remain at odds.

Capturing 'Krishna the environmentalist': secular environmental NGOs and religiously motivated conservation

To combat the inertia facilitated by a lack of environmental subjectivity among pilgrims and others, environmental activists in Braj have embraced religious practices and beliefs to justify their work. A strategy invoked by religiously motivated conservationists involves tree worship. Tree worship is a common phenomenon throughout India. In the context of religiously motivated conservation, tree worship is highly performative and decidedly context dependent. On several occasions, local activists, holy men, and even politicians have publicly worshipped *neem* trees that are slated to be cut down. The trees are garlanded during anti-tree felling demonstrations. These events have also occasioned *kirtan*, or devotional music performed in the streets. This type of tree worship demonstrates the cooperation between temples, *bhaktins* or lay-religious people, and environmental NGOs for a common purpose, namely, preventing trees from being felled due to urban expansion. These incidences of public tree worship have drawn the attention of the local press, which has perhaps raised the consciousness of the local populace regarding the importance of conservation. While these actions have met with mixed success, they are evidence of the strength of performing devotion as a political tactic.

One such incident occurred in 2013, on the day I returned to Vrindavan; it was sweltering, the monsoon season being only days away. My colleague Jagganath wiped sweat from his brow as he told me of the scene that had taken place a few hours prior to my arrival. There had been a great stir right in front of one of the major temples. With the massive influx of pilgrims in recent years and their attendant traffic, the municipality wanted to widen the road. The paved roads alongside some of the largest and most popular temples came at the cost of felling hundreds of old *neem* trees; the root systems of other trees were covered over by concrete, which is now cracking due to the growth of the trees over time. Widening the road meant removing even more of the neem trees lining the pavement. Not only do some people consider it a sin to cut *neem* trees (they are revered for their medicinal and spiritual properties) but it is also illegal to cut down trees of any sort in the city. Protestors had painted the trees with the name of the goddess Radharani. Since the name of the goddess is inseparable from the goddess herself, how could they dare put a chainsaw to these trees? The police were called in as protestors shouted the name of the goddess 'Radhe! Radhe!' over and over and threw themselves in front of the chainsaws.

After excitedly discussing the great clamor, Jagganath proudly handed me the newest pamphlet printed from his organization (Figure 4.1).

The pamphlet's cover proclaims its message with a clever play on words in Hindi and English. The word 'dham' means a sacred home or abode of a deity; 'seva' (pronounced 'save-ah') means to serve. Therefore, the pamphlet urges readers to save this sacred place through environmental service. This pamphlet describes an imperiled sacred landscape that requires action on the part of the devotees of Krishna, the principle local deity. Even within nominally secular NGOs, such as the Braj Foundation and Friends of Vrindavan, religious imagery drawn from the Krishna stories of the area influence institutional visions of the sacred geography. Although the Braj Foundation makes it clear on their website that they are a secular organization, they nevertheless draw upon religious texts like the *Bhagavata Purana* to support their case that Krishna the Environment Deity supports their cause.

While pilgrims are largely depoliticized actors, religiously motivated conservationists use religious tropes in highly politicized ways to attract attention to their interests. These and other groups have actively sought new alliances and partnerships with both government agencies and international NGOs. Most famously, Ranchor Prime, a Western ISKCON devotee, attracted the involvement of the World Wildlife Fund (WWF – now known outside North America as the World Wide Fund for Nature) in the early 1990s. The WWF, numerous local

Figure 4.1 The pamphlet cover: 'Dham Seva, Save a Dham'.

temples, ISKCON, and the nascent *Friends of Vrindavan* organization were all involved in planting thousands of trees on pilgrimage routes. Unfortunately, this restoration attempt was not as successful as the strength of its partnerships would have portended. When speaking of this early reforestation effort conducted by the WWF in affiliation with local temples, one environmental activist lamented that their group had gone out of their way to cater to the religious interests of the temple priests. Pujas were done; the temples were consulted heavily in the project. Nevertheless, the temples did not follow through by watering the fragile young saplings, leading to the death of most of the outplantings.

More recently, in 2011, when local environmental activists sought the help of the Uttar Pradesh Forestry Department to restore the Sunrakh Reserve Forest, the department claimed to have planted over 12,000 trees on the land, but less than 10 percent could be found two years later when the BVHA went to evaluate (Vrindavan Today Staff, 2015a, b). BVHA claims that the number of trees planted was far less than the 12,000 claimed by the Uttar Pradesh Forestry Department. BVHA then met with the District Magistrate to intervene by having more trees planted in the area. Without sufficient care and attention, many of these trees died just two years after planting. The conflict between environmental NGOs – such as *Friends of Vrindavan* and the Indian Forestry Department – reflects the sometimes fraught relationship between state-level interests and local interests with regard to forest commons. Environmentalists involved in the repeated efforts to restore sacred groves in Vrindavan note that the temple authorities and, later, the Forestry Service, simply do not sufficiently maintain the trees that have been outplanted. Instead of destabilizing the authority of state forms of knowledge and power, these organizations are trying to shame state agencies to act in a responsible manner in their own reforestation endeavors. In this case, I am less interested in the mismanagement of the Forestry Department than in how *Friends of Vrindavan* and others co-opt both religious and ecological discourses to show how the temples and the state have failed to address a crucial environmental issue in this religious tourism site. It is possible that these failures have contributed to the increasingly localized, NGO-based, restoration activities in the area, which have been much more successful than more organized attempts.

Contemporary debates about religion and ecology in India

There is a need for caution when describing religiously motivated conservation as an example of 'traditional' forest management. Scholars have noted that reverence of nature does not necessarily lead to conservation (Jacban and Chattopadhyay, 2000; Nanda, 2004). In fact, concern for the environment may have less to do with the maintenance of sacred groves than concerns such as avoiding divine retribution, achieving merit, avoiding illness, and ensuring good crops (Kent, 2013). To further problematize the view of Hinduism as a religion with an inherent ecological ethos Grodzins-Gold and Gujar (2002) note that people do not necessarily lament the decline of the forests that were maintained by exploitative local elites who prevented access and use of these forests.

I argue that religiously motivated conservation in Braj is a highly politicized, adaptive, and historically mediated phenomenon. In Vrindavan, the current actors are not members of a coherent 'traditional management system' as such, but are instead a group of actors who have become compelled to respond to contemporary challenges by selectively emphasizing facets of popular religion. While I do not dispute the existence or efficacy of traditional management systems in India, in this case, the current resurgence of religiously motivated conservation seems to be part of a diffuse network of individuals and entities that are increasingly coming together to form alliances at different scales to create radical change. At the present time, I am not aware of the presence of a local *van panchayat* (village forest council), a common feature of community-based management throughout India. The deforestation of groves on temple lands suggests something other than a coherent 'system' with checks and balances. To the contrary, an evaluation of the literature demonstrates that prior to the 1700s, local people had 'forgotten' that these forests had been the birthplace of the god Krishna. The forests were considered to be frightening jungles (Prabhupada, 2014). To add a further layer of complexity, Shinde (2012) argues that at no point in Vrindavan's history was Vrindavan actually the forest paradise that social conceptions of the sacred geography of the region suggest, but has rather been constructed as a sacred forested site due to its importance as a pilgrimage site.

Others note that religiously motivated conservation may contribute to highly radicalized social movements with strong political agendas and occasionally violent tactics (Taylor, 2009). This is increasingly the case in India (cf. Mawdsley, 2005; Nanda, 2003; Sharma, 2012; Tomalin, 2004). Some have argued that Hindu nationalists have a vested interest in demonstrating that Hinduism is an inherently ecological religion, leading to strange bedfellows between left-wing environmentalists and right-wing political entities like the Bharata Janiya Party and the Vishva Hindu Parishad (cf. Mawdsley, 2005). Whereas environmental activists, geographers, botanists, and ecologists have often emphasized the role of sacred groves in promoting biodiversity (cf. Khan et al., 2008), this discourse has only lately begun to affect the ways local and indigenous people articulate the reasons why they protect sacred groves and other sacred natural land features (Kent, 2013). Indeed, the 'NGO-ization' of local environmental initiatives may have had a profound impact on the frequency with which local activists employ scientific and environmental theories, methods, and discourses. Still, the increasing interest in religiously motivated conservation, and spiritual ecology more generally, demonstrates the charismatic appeal of searching for an ecological ethos within religious traditions.

The popular view that Hinduism promotes biological diversity, creates environmental subjectivity, and leads to conservation has become what Mathews (2009) calls a 'globally circulating environmental theory'. Rather than merely being a 'green development fantasy' (Tsing, 1999) of urban elites, I would posit that religiously motivated conservation is a vital amalgam of interests and motivations on many different levels that is a response to contemporary problems. Rooted in older practices, participating in local historical contingencies and borrowing liberally from both ecology and popular religious movements, religiously motivated conservation is becoming an increasingly fixed element of Indian conservation (cf. Naryanan, 2001). Whereas some scholars have presented the management of sacred groves as 'traditional' and thus in 'danger,' I focus instead on the dynamic use of religious discourse among conservationists and devotees alike as evidence of a greater cultural and environmental revitalization project. As noted by Mosse (2006, p. 696), community institutions can be 'objects of intervention,' meaning that they are capable of being 'upgraded, democratized, or modernized so as to meet new demands and fit within contemporary policy objections'.

Conclusion: religiously motivated conservation and political ecology

Mathews (2009) suggests that we should not stop at the 'empirical facts' when judging globally circulating environmental theories, but should instead interrogate the reasons why certain theories resonate with local people and the public. I see the global circulation of this concept as productive of new ways of relating to sacred natural land features, like sacred groves. Relationships among actors who sometimes have divergent interests can nevertheless generate new, mutually beneficial political possibilities (Agrawal, 2005; Agrawal and Sivaramakishnan,

2000; Mathews, 2009). However, the difference is that these entities are not exactly exemplars of 'traditional land management systems' that are struggling to adapt to the 'new' phenomena of pilgrimage. Instead, they represent contemporary actions to continue to re-assert the sacred geography and importance of these sacred natural land features, which are always already over-determined by the pilgrims' gazes. By focusing on the micropolitics of tree worship and other forms of environmental protests, I suggest that activists and devotees alike are beginning to rally around an image of 'Krishna as an Environmental Deity' in a move to create new management regimes. In the context of Vrindavan, sacred groves link a mythologized sacred geography to modern-day issues of desertification and environmental degradation facing this religious tourism site in a way that is politically productive. These discourses involve negotiations of new understandings of place and practice, which are endeavoring to attract the attention of extra-local agencies and engage them in new alliances to save the sacred landscape.

References

Agrawal, A. (2005) Environmentality: community, intimate government and environmental subjects in Kumaon, India, *Current Anthropology*, 46 (2), pp. 161–190.

Agrawal, A. and Sivaramakrishnan, K. (2000) Introduction: Agrarian Environments. In: Agrawal, A. and Sivaramakrishnan, K. (eds.), *Agrarian Environments: Resources, Representation, and Rule in India*. Durham, NC: Duke University Press.

Alley, K. (2002) *On the Banks of the Ganga: When Wastewater Meets a Sacred River*. Ann Arbor, MI: University of Michigan Press.

Alliance of Religions and Conservation (2011) Religious tourism statistics – Annual figures, available: http://www.arcworld.org/downloads/ARC%20pilgrimage%20statistics%2015m%2011-12-19.pdf [accessed 1 May 2014].

Apffel-Marglin, F. and Parajuli, P. (2000) Sacred Grove and Ecology: Ritual and Science. In: Chapple, C. K. and Tucker, M. E. (eds.), *Hinduism and Ecology: The Intersection of Earth, Sky, and Water*. Cambridge, MA: Harvard University Press, pp. 269–290.

Banerjee, S. R. (1987) *Indo-European Mood and Aspect in Greek and Sanskrit*. Sanskrit Pustak Bhandar.

FSI (2003) State of forest report 2003. Ministry of Environment and Forests, Government of India, Dehra Dun, India.

Ghosh, P. (2002) Tales, tanks, and temples: the creation of a sacred center in seventeenth-century Bengal, *Asian Folklore Studies*, 61 (2), pp. 193–222.

Gladstone, D. (2005) *From Pilgrimage to Package Tour: Travel and Tourism in the Third World*. New York: Routledge.

Grodzins-Gold, A., and Gujar, B. (2002) *In the Time of Trees and Sorrows: Nature, Power, and Memory in Rajasthan*. Oxford: Oxford University Press.

Growse, F. (1883) *Mathura: A District Memoir*, 3rd edn. Ahmedabad: The New Order Book Company.

Haberman, D. (1994) *Journey through the Twelve Forests: An Encounter with Krishna*. Oxford: Oxford University Press.

Haberman, D. (2006) *River of Love in the Age of Pollution: The Yamuna River of Northern India*. Berkeley, CA: University of California Press.

Jacban, C. and Chattopadhyay, M. (2000) Identities and Livelihoods: Gender, Ethnicity and Nature in a South Bihar Village. In: Agrawal, A. and Sivaramakrishnan, K. (eds.),

Agrarian Environments: Resources, Representation, and Rule in India. Durham, NC: Duke University Press, pp. 3089–3507.

Kent, E. (2013) *Sacred Groves and Local Gods: Religion and Environmentalism in South India.* Oxford: Oxford University Press.

Khan, M., Khumbongmayum, A., and Tripathi, R. (2008) The sacred groves and their significance in conserving biodiversity: An overview, *International Journal of Ecology and Environmental Studies*, 34 (3), pp. 277–291.

Khandelwal, B. (2010) Braj bhoomi cries for attention, *Times of Agra*, 8 October, available: http://news.vrindavantoday.org/?s=Braj+Bhoomi+Cries+for+Attention [accessed 1 May 2015].

Mathews, A. (2009) Unlikely alliances: encounters between state science, nature spirits, and indigenous industrial forestry in Mexico, 1926–2008. *Current Anthropology*, 50 (1), pp. 75–101.

Mawdsley, E. (2005) The abuse of religion and ecology: the Vishva Hindu Parishad and Tehri Dam, *Worldviews*, 9 (1), pp. 1–24.

Mosse, D. (2006) Collective action, common property, and social capital in South India: an anthropological commentary. *Economic Development and Social Change*, 54 (3), pp. 695–724.

Nanda, M. (2003) *Prophets Facing Backwards: Post-Modern Critiques of Science and Hinduism.* London: Rutgers University Press.

Nanda, M. (2004) Dharmic ecology and the neo-pagan international: the dangers of religious environmentalism in India, accepted for the *18th European Conference on Modern South Asian Studies,* July.

Narayanan, V. (1997) 'One tree is equal to ten sons': Hindu responses to the problems of ecology, population, and consumption, *Journal of the American Academy of Religion*, 65 (2), pp. 291–332.

Narayanan, V. (2001) Water, wood, and wisdom: ecological perspectives from the Hindu traditions, *Daedalus*, 130 (4), pp. 179–206.

NCAER (2003) *Domestic tourism survey: 2002–03*, National Council of Applied Economic Research and Ministry of Tourism and Culture, Government of India, New Delhi.

Peet, R. and Watts, M. (2004) *Liberation Ecologies: Environment, Development, Social Movements,* 2nd edn. London: Routledge.

Poddar, J. (2013) BVHA demonstrates for protecting the heritage trees of Vrindavan, *Vrindavan Today*, 8 August, available: http://news.vrindavantoday.org/2013/08/bvha-demonstrates-for-protecting-the-heritage-trees-of-vrindavan [accessed 1 May 2015].

Poddar, J. (2014) 1.11 The defects we see in the Dham are not real, *Vrindavan Today*, 4 August, available: http://news.vrindavantoday.org/vrindavan-mahimamrit/1-11/ [accessed 1 May 2015].

Prabhupada, S. (2014) *Śrī Caitanya-caritāmṛta. Bhaktivedanta VedaBase.* Bhaktivedanta Book Trust, available: http://vedabase.com/en/cc [Accessed on 28 March 2014].

Prime, R. (1992) *Hinduism and Ecology.* London: Cassell.

Rinschede, G. (1992) Forms of religious tourism, *Annals of Tourism Research*, 19, pp. 51–67.

Robbins, P. (2012) *Political Ecology: A Critical Introduction,* 2nd edn. Oxford: Wiley-Blackwell.

Robertson, J., Harriss, S., and Singh, T. (1913) *Report of the pilgrim committee United Provinces.* Simla: Government Central Branch Press.

Saha, S., Moorthi, S., Pan, H. L., Wu, X., Wang, J., Nadiga, S. . . . and Reynolds, R. W. (2010) The NCEP climate forecast system reanalysis. *Bulletin of the American Meteorological Society*, 91 (8), 1015–1057.

Shah, B. (2006) The pilgrimage of the groves: reconstructing the meaning of a sixteenth-century landscape, *Arnoldia*, 64 (4), pp. 39–41.

Sharma, M. (2012) *Green and Saffron: Hindu Nationalism and Indian Environmental Politics*. Ranikhet: Permanent Black.

Shinde, K. (2012) Place-making and environmental change in a Hindu pilgrimage site in India, *Geoforum*, 43 (1), pp. 116–127.

Sponsel, L. (2012) *Spiritual Ecology: A Quiet Revolution*. Santa Barbara, CA: Praeger.

Sullivan, B. (1998) Theology and Ecology at the Birthplace of Krishna. In: Nelson, L. (ed.), *Purifying the Earthly Body of God: Religion and Ecology in Hindu India*. Albany, NY: State University of New York Press, pp. 247–268.

Taylor, B. (2009) *Dark Green Religion: Nature Spirituality and the Planetary Future*. Berkeley, CA: University of California Press.

Tomalin, E. (2004) Bio-divinity and bio-diversity: perspectives on religion and environmental conservation in India, *Numen*, 51 (3), pp. 265–295.

Tsing, A. (1999) Becoming a Tribal Elder, and Other Green Development Fantasies. In: Li, T. (ed.), *Transforming the Indonesian Uplands: Marginality, Power, and Productions*. London: Harwood Academic Press, pp. 159–202.

Vrindavan Today Admin (2010) Religious tourism in Govardhan: Economic survey, *Vrindavan Today*, 10 June, available: http://news.vrindavantoday.org/2010/06/braja-news-digest-june-10-2010/ [accessed 1 May 2015].

Vrindavan Today Staff (2015a) Students concerned about the diminishing green cover in Vraja, *Vrindavan Today*, 28 April, available: http://news.vrindavantoday.org/2015/04/students-concerned-diminishing-green-cover-vraja/ [accessed 1 May 2015].

Vrindavan Today Staff (2015b) Natural heritage of Vraja should be preserved, *Vrindavan Today*, 8 February, available: http://news.vrindavantoday.org/2015/02/natural-heritage-vraja-preserved/ [accessed 1 May 2015].

Walters, K. (2010) Approaching Vrindavan's environmental conservation, *The Harmonist*, 8 December, available: http://harmonist.us/2010/12/approaching-vrindavans-environmental-conservation/ [accessed 6 May 2015].

5 Festive environmentalism
A carnivalesque reading of eco-voluntourism at the Roskilde Festival

Mette Fog Olwig and Lene Bull Christiansen

Introduction

In this chapter we provide a reading of popular forms of engagement in environmentalism that is alternative to familiar interpretations in the literature on ecotourism and voluntourism. In doing so, we endeavor to more fully understand the cultural meanings that are created in festive or celebratory versions of participation in popular environmentalism. We do this by applying the Russian literary philosopher Mikhail Bakhtin's idea of the *carnivalesque* (Bakhtin, 1984a) to a case study of eco-voluntourism at the Roskilde Festival, an international culture and music festival held each summer in Denmark. This chapter represents a view on what Robbins terms 'environmental subjects and identities' (2004, p. 215). This view focuses on a setting outside the geographical, political and historical contexts that are conventionally the subjects of such studies, that is, communities affected by colonial practices of exploitation and/or environmental degradation (e.g. Robbins, 2004, p. 220). This is primarily a place-based study, which applies the Bakhtinian concept *chronotope* (Holquist, 2002, pp. 109–113) to understand the historically situated cultural context of the Roskilde Festival and its relation to Danish societal norms and forms of public engagement (see: Olwig and Christiansen, 2015, pp. 187–188). Methodologically, the chapter is based on participant observation at the Roskilde Festival (henceforth referred to as 'the Festival'), as well as a textual analysis of official Festival communication and volunteer blogs on 'good cause' initiatives at the Festival, focusing especially on environmental causes. All interviews were conducted in Danish, and all translations appearing here are the authors' own.

> Roskilde Festival is the largest North European culture and music festival and has existed since 1971. We are a non-profit organisation consisting of about 50 full-time employees and thousands of volunteers.
> (Roskilde Festival, 2015a)

As the first lines in the 'About Roskilde Festival' section of the Festival's official website, this statement is important. Of course the Roskilde Festival website advertises which bands will appear next and thus join big names such as Prince, the Rolling Stones, Rihanna and Björk, who have played in previous years. The Festival website, however, also strongly emphasizes its donations and support to humanitarian, cultural and non-profit projects, and its non-profit and environmentally conscious status. A sense of community is actively promoted as an important aspect of the Festival and festivalgoers, especially the many volunteers among them, are often actively engaged in humanitarian and environmental initiatives. In fact, the more than 30,000 volunteers who participate are, according to the Festival, key to its success and to its goal of instilling a sense of socio-environmental solidarity in the festivalgoers that pervades the event. In the political ecology of tourism the concepts of 'the local community' and 'the visiting tourists' most often refer to separate categories of people, often representing different interests, divergent points of view on their encounter and different notions of the space they inhabit together. In our study, however, 'the community' is made up of the festival tourists.

Recent initiatives to engage festivalgoers in environmental issues include encouraging them to use bicycle-powered phone chargers, listen to solar-powered radios sewn into a jungle hat and recycled drinking cups and cans, the deposit fees from which are given to major Danish NGOs. This form of environmentalism takes its point of departure in the bodily performance of the festivalgoers. This argument has two interconnected levels. First, on a general level, the Festival (like many festivals of its kind) functions as a particular 'time-space relation', what Bakhtin calls a 'chronotope' (Holquist, 2002, pp. 109–110), with its own social codes and norms, such as dressing up in festive and/or silly costumes, drinking and dancing, changing daily rhythms and engaging in a form of sociality outside of ordinary social codes (e.g. Bakhtin, 1984b, p. 107). Second, the festivalgoers are encouraged to actively use their bodies when participating in the organized environmental and/or social initiatives (e.g. by wearing the jungle hats with sewn-in, solar-driven radios or pedaling the bikes to charge their phones). Bakhtinian thought and affect theory suggests that using the body in this way – experiencing environmentalism on and via the body – will enable one to better empathize with environmentalist initiatives in a non- or pre-intellectual manner (e.g. Hemmings, 2012; Jung, 1998). In this way, the Festival can be seen to be part of a general discourse where empathy is viewed as a solution to 'a wide range of social ills' (Pedwell, cited in Mostafanezhad, 2013, p. 494). Furthermore, we suggest that the alternative time-space relations created by the Festival enable new cultural meanings and a possible reimagining of dominant views on the sociality of environmentalism (Bakhtin, 2001, p. 30).

The bodily actions involved in the 'good cause' initiatives are often mediatized, celebritized,[1] and celebrated in reality-show style. Facebook pages, blogs and hashtags all accompany the environmental initiatives and thus, in addition to generating socio-environmental solidarity and encouraging bodily and affective experiences, these initiatives also feed into the growing trend of coupling the promotion of good causes to the individual, through social media. We call this form of environmentalism 'celebratory environmentalism'. While civil society has embraced mediatized, celebritized and celebratory calls for environmentalism, because they can engage the public and empower it to contribute to global environmental causes (Brockington, 2009; Mostafanezhad, 2013), critics have argued that such approaches oversimplify and depoliticize environmental problems.

By analyzing how the Festival unites good causes and volunteering with celebratory bodily performances, we illuminate the ways in which such celebratory environmentalism negotiates between two potentially conflicting realms. At one extreme, it is engaging and educating the public in environmental causes through empathy and affect; at the other, it could be argued to be providing merely a venue for the volunteers' self-promoting entertainment, thereby potentially obscuring (or simplifying, or essentializing) the political ecology of environmental degradation. In this chapter we do not seek to evaluate the Festival in the dichotomous terms of authentic altruism versus frivolous narcissism, a dichotomy that we suggest partially expresses a modern form of puritanism; instead, we see this apparent contradiction as being part of a long-standing historical synergy that has been characteristic of what has been termed the *carnivalesque*.[2] As Jung explains:

> The carnivalesque is the most radical aspect of the dialogics of difference because it serves as a *non-violent technique of social transformation* [. . .]. It is festive politics that is a communal celebration of festive bodies [. . .]
>
> (Jung, 1998, p. 104) [original emphasis]

Through this lens of the carnival, we are able to discover what we call *popular-festive eco-voluntourism*, a practice that we will discuss further below.

Celebratory environmentalism and voluntourism

In this section we will further describe the notion of celebratory – or popular-festive – eco-voluntourism and its performance at festivals. In her discussion of humanitarian mega event concerts such as Live 8, held simultaneously in London, Philadelphia, Berlin, Rome, Paris and Barrie (Toronto) on 2 July, 2005, Chouliaraki argues that such events are

> [. . .] crucial performances of the humanitarian imaginary, insofar as they use the global appeal of rock to disseminate and legitimize the moral imperative of solidarity – the imperative to act on vulnerable others without the anticipation of reciprocation.
>
> (Chouliaraki, 2012, p. 106)

Similarly, we view the Festival as an event in which trends from environmentalism, humanitarianism and volunteering intersect to become what we might term *popular-festive eco-voluntourism*. Our conceptualization of the 'popular-festive', which draws upon Bakhtin's *carnivalesque* analytics, helps us understand the Festival as a time-space relation in which new meanings are created out of popular-festive forms of participation and volunteering. We are in this regard particularly interested in Bakhtin's depiction of the ways in which the carnivals of the Renaissance, as described by the French author Rabelais, constituted a culture of popular-festive forms, in which 'multiple cultural forms combine' to form a specific site of carnivalesque energy (Dentith, 1995, p. 70).[3] We aim here to explore popular forms of participation in eco-voluntourism, which from a critical perspective might be dismissed as superficial, commercialized and even counterproductive to familiar political ecology analyses that see commercialization as part of the problem rather than part of the solution. However, when viewed from the Bakhtinian perspective, they may reveal new insights into the production of cultural meaning and practice in popular participation and socio-environmental solidarity.

For our purposes, volunteering has mainly been studied among youth volunteers who travel abroad. The ethics, consequences and justification of voluntourists have long been debated among scholars as well as in the press. According to Mostafanezhad, volunteer tourism is 'the fastest growing niche tourism market in the world' (2013, p. 485). While volunteer tourists often pay for their experience, in part or in whole, many forms of volunteering involve NGO and government funds and range from three-year Peace Corps stints to three-week English language teaching opportunities. As Tiessen and Heron point out, in the case of Canadian volunteers, this has led some to question whether it is ethical to spend government funds for poverty alleviation to enable youth to go abroad as volunteers (2012). Do volunteers help 'change the world' or is volunteering merely a form of 'personal growth and individual fulfillment' (ibid., 2012, p. 54), or 'a particular iteration of cosmopolitanism' that 'smoothes over the histories and specificities of the countries being visited' (Baillie et al., 2013, p. 127)?

The problems linked to volunteering as a purely individualized phenomenon are becoming even more pertinent as volunteers are increasingly inspired by individual celebrity humanitarianism. As Mostafanezhad argues, 'Angelina Jolie and Madonna have made international volunteering sexy' (2013, p. 485). Celebrities are powerful new actors in development (Richey and Ponte, 2014) and are promoted by aid organizations based on the belief that by coupling good causes with celebrities, the causes receive more attention in the media, and this increased attention leads to more popularity and greater levels of funding (Olwig and Christiansen, 2015). Still, critics worry that this process is carried out at the risk of simplifying and downplaying the underlying causes of the issues (Brockington, 2009; Kapoor, 2013). Not only do the celebrities generate media attention, they also inspire more people to volunteer – just as their favorite celebrities have done. Chouliaraki (2012) depicts how these celebrity performances of global caring, constitute an 'aspirational discourse', which places the celebrity narrator at the center of a universalizing claim of humanitarian dispositions 'by proposing an altruistic disposition for all to share' (Chouliaraki, 2012, p. 2). Thus, by acting like one's favorite celebrity, one can imagine sharing something with that celebrity: a cosmopolitan humanitarian disposition.

In some ways these volunteers mimic celebrity performances, often publicizing their experiences through individualized celebrity-style social media self-representations in the form of blogs, Instagrams and tweets. As Mostafanezhad argues, in the volunteers' quests to get in touch with their inner Angelina Jolie 'the political is displaced by the individual with celebrity sheen' (2013, p. 486). Volunteers mirror their favorite celebrities (Goodman, 2010) who are themselves mirrored or referential imaginaries. For example, photographs of Angelina Jolie holding her adopted daughter Zahara are often recreated in volunteer tourism imagery. Images of female volunteers holding a child have, in fact, 'become iconic Facebook profile pictures' (ibid., 2013, p. 491).

Celebratory individualized volunteerism, however, cannot be seen as *only* reproducing global inequalities and fostering narcissism. Christiansen and Frello have, for example, argued that alternative bodily performances can in fact pry open the mold of celebrity engagement with global issues of social justice (as depicted by Chouliaraki and others) and produce different meanings and affects related to issues of global justice (Christiansen and Frello, 2015, pp. 10–12). Further, Mostafanezhad argues that: '[b]y repositioning humanitarianism as part of mainstream popular culture, female celebrities have brought some of the most pressing agendas of our time to the masses, and volunteer tourists have engaged with development agendas that may not otherwise exist' (Mostafanezhad, 2013, p. 496). As noted, however, the question that is continuously reiterated in the literature is whether it is possible to promote these agendas without losing or ignoring the complexities and the structural relations of the underlying issues, for example the political, economic and social aspects of environmental degradation as highlighted when applying a political ecology approach (Brockington, 2009; Mostafanezhad, 2013). We argue that festivals like the Roskilde Festival may provide important insights into these tensions between volunteering as self-promoting entertainment, and volunteering as a venue for engaging with the political ecology of environmental issues and, not to forget, such festivals as events where people come to both feel part of a community and have a good time.

The celebratory as carnivalesque

On the face of it, the Festival in many respects embodies Bakhtin's description of the carnival, where festive forms of the carnival include parodies, dressing up in costumes, eating and drinking in abundance, satirical rendering of authority figures and an emphasis on carnal pleasures such as sexual experiences and eating. Historically, the carnival took place in an entire village or in particular locations in larger cities (usually the town square), and an array of activities, roles and figures were traditionally associated with the carnival – such as the jester, the fool, the magician, the grotesque figure of the hunchback – all of which had laughter as the organizing element. For Bakhtin, laughter ties together an ambivalent utopia, which is the carnival:

> Carnivalistic laughter likewise directed towards something higher – toward a shift of authorities and truths, a shift of world orders. Laughter embraces both poles of change, it deals with the very process of change, with *crisis* itself. Combined in the act of carnival laughter are death and rebirth, negation (a smirk) and affirmation (rejoicing laughter). This is a profoundly universal laughter, a laughter that contains a whole outlook on the world. Such is the specific quality of ambivalent carnival laughter.
>
> (Bakhtin, 1984b, p. 127) [original emphasis]

Thus, for Bakhtin, the carnival constituted a temporary suspension of the hierarchical structures, norms and privileges of medieval society; the social laughter of the carnival could relativize the hegemonic truths of the day and make joyfulness a property of the people (Bakhtin, 2001, p.30) – the carnival is a laboratory for a new society:

> Carnival is the place for working out, in concrete sensuous, half-real and half-play-acted form, a *new mode of interrelationship between individuals*, counterposed to the all-powerful socio-hierarchical relationships of noncarnival life.
> (Bakhtin, 1984b, p. 123) [original emphasis]

Our interest in Bakhtin thus emerges from a wish to understand how the seemingly nonsensical, joyful and silly displays during the Festival can be understood as something beyond less-than-serious distractions from the political ecology of environmental issues.

In Bakhtin's conceptualization, the carnival constitutes a particular time-space relation, a 'chronotope'. This chronotope is, for Bakhtin, the relation in literature between the 'cultural environment' in which a novel emerges with the time-space relation which forms a novel's narrative framework. In this sense, the chronotope is both a cultural analysis and an exploration of possible (new or other) time-space relations (Holquist, 2002, pp. 109–113; see also Turner, 1967). As such, we understand the utopian carnival as not only a narrative form, but also an exploration of the cultural context of Rabelais' authorship, and Bakhtin's exploration of possible (new or other) time-space relations.

A translation of Bakhtin's depictions of the Renaissance carnival into an analytical framework for a contemporary event naturally produces a number of problems, stemming not just from Bakhtin's interpretation of the social and political context of the Renaissance, but also from contextual differences between the medieval and present-day celebrations of carnivalesque events. Our use of the carnivalesque therefore depends on the formulation of a utopian vision, which Bakhtin derives from Rabelais, but in which carnivalesque popular-festive forms are seen not via historiography, but via what Bakhtin terms 'poetics' (Bakhtin, 1984b). That is, we use the popular-festive forms of the carnivalesque as 'figures that recur'; figures that are taken up over time in different genres and forms (Holquist, 2002, p. 108). In his introduction to Bakhtin's writings, Dentith describes the utopian vision of Bakhtin's carnival as an already realized utopian space:

> In at least some of his formulations, it is not that carnival looks forward to some distant prospect of social perfection, but that the space of carnival has already realized it. Carnival becomes a time outside time, 'a second life of the people, who for a time entered the utopian realm of community, freedom, equality, and abundance'.
> (Dentith, 1995, p. 76)

Bakhtin's carnivalesque utopia depicts a temporal and spatial relation that reconfigures itself, and occurs as figures over time; a utopia in which anti-authoritarian energies are released and are given the opportunity to create new cultural meanings. Furthermore, for Bakhtin, the carnival had the ability to foster critical consciousness:

> Parodic forms are, for Bakhtin, evidence of the growth of critical consciousness evident in the unfolding of symbolic forms, and the 'semantic clusters' that endure have the character of a priori elements necessary for thinking as such.
>
> (Brandist, 2002, p. 137)

We therefore turn to the carnivalesque in order to elucidate the elements of the Festival, which may, from a critical analytical point of view, appear nonsensical, unserious and distracting, but which might, from the Bakhtinian perspective, reveal different aspects of conscious-making practices. It is our hope that with this analysis we will push research toward taking 'just having fun' seriously by asking: 'What does this fun produce?' and 'Does "just having fun" raise ethical issues?'

Roskilde Festival

We examine the Roskilde Festival as our case study in part because of the Festival's long-standing social and environmental ethos. The Festival is one of the largest annual cultural events in Denmark, and in Europe more broadly and, as such, it has a significant impact on the European cultural imaginary – particularly for the many generations of youth (and others) who have participated in the Festival over the years.

In 2015, the Festival was held from June 27 to July 4 – the Festival lasts a total of eight days with the main musical events taking place over four of those days. In addition to approximately 170 bands that performed, approximately 31,000 people volunteered, and the Festival was covered by about 2,500 accredited journalists (Roskilde Festival, 2015c). The Festival – held across 2,500,000m² and attracting some 135,000 attendees – becomes Denmark's fourth largest city over the four days when concerts are held. The area is divided into the festivalgoers' campsites and the main festival site itself, which includes most of the music stages as well as scores of stalls selling food, drinks and a variety of merchandise. The festival site is cleaned every day and festivalgoers are checked on arrival to prevent glass bottles and other banned items from entering.

Apart from the tents where the participants live, the campsites include food and drink stalls, swimming and fishing lakes, saunas, communal barbeques, yoga areas, toilets, showers, supermarkets, pharmacies and ATMs. The campsites are also important sites for integration between different groups of festivalgoers; it is customary to meet and congregate with strangers (who often later become friends). This is something that sets the Festival apart as a particular social timespace, in which prototypical Nordic reserve is set aside to accommodate the party spirit of the Festival. Many dress up in costumes of superheroes or animals, or dress down *au naturel* when participating in the annual, much celebrated, Naked

Run, further contributing to the festive spirit. Likewise, the Festival supports participants' efforts to set up camps that have festive themes to attract visitors. This adds to the overall spirit of openness and playfulness of the event.

The Festival is often used to provide exposure to various humanitarian as well as environmental causes. This exposure takes place through both large-scale installations, such as a mock West Bank separation barrier (2012) and a bike-powered Ferris wheel (2009), and smaller events, such as the 2013 event hosted by the Danish NGO IBIS, where festivalgoers were encouraged to be photographed wearing a red mask in order to show their support for the indigenous populations of Ecuador (IBIS, 2013). Volunteers play an important role in organizing and realizing these initiatives, and in this way, the Festival becomes a destination for voluntourists, as we now describe them.

Figure 5.1 Bicycle-powered Ferris wheel at the 2009 Roskilde Festival.
Photo by Mette Fog Olwig.

The Roskilde Festival as voluntourism

Many participants come to the Festival as volunteers, often working for various NGOs that have a presence at the event. For example, in 2015 participants attended the Festival as volunteers for the humanitarian organization MS/ActionAid, the Danish Diabetic Union and a local Dog Friends' association, among other organizations. Many local non-profit organizations – such as sports clubs – engage their members in income-generating activities at the Festival. Volunteers also work directly for the festival association in jobs that are not directly linked to a good cause, such as checking parking permits; yet even these volunteer jobs indirectly enable the Festival to generate profits, which can be used for charity. It should be noted that all volunteers are given free entrance to the Festival (in 2015, full festival tickets cost 1,940 DKK, or around US$ 275). In return, they are required to work for a certain number of hours (32 hours in 2015) over the eight days of the Festival. Many volunteer, we have observed, because they cannot afford the price of a ticket, while others who have good jobs and who can easily afford the ticket still sign up to volunteer as part of social or political engagement. Several of the groups that organize volunteers often attempt to attract volunteers by presenting the volunteer positions as opportunities for socializing and partying with fellow volunteers, organizing parties, food and drinks, as well as setting up volunteer-designated campsites. Thus, volunteering is closely intertwined with the popular-festive practices of the Festival, and the Festival's non-profit character.

The Festival website repeatedly emphasizes the Festival's non-profit status, and the link between attending the Festival and supporting good causes:

> Roskilde Festival is non-profit. That means we're donating all profits to charity after each festival. We support humanitarian, cultural and non-profit projects all over the world. When you buy a ticket to Roskilde Festival or volunteer at the festival, you are a part of increasing the quality of life for many people around the world. Thanks for that!
> (Roskilde Festival, 2015a)

In one of its promotional videos aiming at attracting volunteers, shown on 'the official website of Denmark', Niels Bjerrum, identified as Head of Information, even asserts that '[o]ne of the reasons why a lot of people come together here as volunteers, is also because we are a humanitarian organization. All the money that is made by this festival is donated for humanitarian causes afterwards' (Denmark DK, 2011).

Since its inception in 1971 the Festival has been linked to humanitarianism by the two founding high school students Mogens Sandfær and Jesper Switzer Møller. They were 17 and 18 years of age and were inspired by the hippie movement and dreamed about creating a Danish version of Woodstock, emphasizing 'peace, love and understanding' (Pagh, 2013). They intended the profits to go to those in need, but many 1971 festivalgoers apparently forgot to pay for the ticket (Pagh, 2013). The two high school students almost failed their exams, and therefore had to give up organizing the Festival, which was taken over by the

Festive environmentalism 117

Figure 5.2 2015 Roskilde Festival poster listing the organizations that received donations in 2014.

Photo by Mette Fog Olwig.

Roskilde Festival Charity Society. According to the website, since 1971 the charity society has contributed more than €26 million to numerous organizations such as Doctors without Borders, Amnesty International and Save the Children, among others (Roskilde Festival, 2015e). However, it is not just through donations that the organizers of the Festival believe they can make a difference. Indeed, on the website they boldly state: 'We believe that a festival can trigger and reinforce a social movement of young people who want something more than just themselves' (Roskilde Festival, 2015e). One of the areas that the Festival focuses on as part of this social movement is the natural environment.

Roskilde Festival and the environment

Despite its links to liberal, progressive politics and various hippie movements, environmentalism has not traditionally been associated with the 'sex, drugs and rock and roll' of music festivals. Rather, it has been argued that environmentalism is inherently a modern form of puritanism akin to the Protestant ethics of the seventeenth and eighteenth centuries, as depicted by Max Weber (Steiner-Aeschliman,

1999). Likewise, it has been pointed out that environmentalism has its roots in religious puritanism and Protestant ethics of denying worldly pleasures in order to prove religious piety (Nelson, 2014). In Weber's construct, the value scheme of the puritanism of the ascetically directed religious movements in Holland and England, upon which he builds his theory of capitalism, the interconnection between religious idealism and ascetic practice was the determining factor (Ghosh, 2003; Weber, 2002). This is not unlike present-day environmentalists' attempts at encouraging eco-friendly, everyday and consumer practices.

While the popular-festive excesses of the music festival might appear contrary to environmentalist goals, the Festival was nevertheless in 2015 awarded the Green Operations Award (Green Operations Europe, 2015; Roskilde Lokalavis, 2015). According to the press release, the Festival in 2014: 'joined forces with the Stop Wasting Food movement, Denmark. We turned food waste into charity, and ultimately we produced 27.5 tons of food which was given to the homeless' (Roskilde Festival, 2015d). The press release emphasizes the Festival's debt to the volunteers: 'Thanks to all volunteers for making this possible!' (ibid.). Waste, however, remains one of the issues that creates most dilemmas for the Festival – and which may clash with other aspects of the popular-festive spirit of the event – namely the excess of food and drink and festive behavior that is at the core of the festival experience, and which does not necessarily lend itself well to environmentalism.

The Festival repeatedly emphasizes the strong desire of their volunteers and general festivalgoers to be part of making a change, and being environmentally conscious. According to an audience survey carried out by the Festival, 94 percent of the audience believes that it is (very) important that the Festival concerns itself with sustainability; 76 percent see themselves as someone who cares for the environment; 73 percent consider themselves as someone who respects the environment in their daily lives; and 42 percent think that they respect the environment when they are at the Festival (Roskilde Festival, 2015c).

Yet, the festivalgoers do not seem to always act on their concerns for the environment. A recent article in *Politiken*, one of Denmark's leading newspapers, carried the following headline: 'Festivalgoers do not bother living up to Roskilde's nice environmental profile: a trash mountain grows in the camping area, while the festival tries to get guests to clean up' (Hjortdal, 2014). According to the Festival website, 1,500 tons of trash were removed from the Festival in 2014 (Roskilde Festival, 2015c) and, as explained in the *Politiken* article, the Festival sorted trash into 13 categories, and all cutlery and plates from the food stalls were biodegradable. Nevertheless, the *Politiken* article depicted the festivalgoers as more concerned with having a good time than with sorting trash. One festivalgoer is quoted in the article as saying, '(w)e are having fun. That is the most important thing' (Hjortdal, 2014).

In the newspaper article, as in other critiques of the Festival's actual commitment to environmentalism, a dichotomy is emphasized between being environmentally responsible and having fun; they are perceived and presented as opposites that cannot be combined. It might be concluded that the Festival and its audience are merely paying lip service to environmental and humanitarian causes and that the event is largely about celebratory performance, with little consideration of the complex realities of the causes supported. However, the many different ways in which environmentalism is being performed as 'fun and festive' throughout the Festival is an attempt on the part of the organizers at redirecting the festivalgoers' bodily experience of environmentalism, thus creating not only a time-space relation of popular-festive environmentalism but also a chronotope, where environmentalism is not associated with puritanism.

In contrast to the apathetic festivalgoers described in the *Politiken* article cited above, there is a large group of volunteers who are engaged in environmentalism. One of the most recent initiatives at the Festival that addresses the overwhelming amounts of trash produced is called Clean Out Loud. According to the project's Facebook page, it focuses on:

> how to reduce the huge amounts of trash at Roskilde Festival and having a blast of a party at the same time. We are creating a community which welcomes everybody who's interested in supporting the project. Being a part of Clean Out Loud means that your camp will contribute to creating a better environment at the festival, and arranging and participating in events at the camping area.
>
> (Clean Out Loud, 2015)

This initiative is run by students from Vallekilde People's College, a special form of Danish school that provides a one-year independent liberal education at the college level. Such students want to be part of a community and contribute to creating a better environment as well as to party.

In the following section we provide an in-depth example of the kind of popular-festive eco-voluntourism that takes place at the Festival, in which volunteers play a key role and work toward something more than just themselves – yet also want to have fun.

Popular-festive ethical consumption at Roskilde Festival

Between 2008 and 2010 Mette Fog Olwig participated in the Festival as a volunteer for DanChurchAid, one of the largest humanitarian organizations in Denmark. Founded as part of the Danish state church *Folkekirken*, DanChurchAid has its roots in relief efforts immediately following the First World War, later becoming involved in development work in the 1970s. Since then the organization has become involved in disaster relief, development work and social and environmental activism, and is associated with the global ACT alliance (DanChurchAid, 2015a, 2015b). DanChurchAid was present at the Festival through a pop-up version of its Copenhagen-based shop and café called *fisk*, which sold[4] organic and fair trade items as well as vintage clothing – 'fisk' means fish in Danish. DanChurchAid's logo includes a fish, or rather the ichthus symbol, a referent for Jesus Christ or Christianity; see Figure 5.3. On their website DanChurchAid explains:

DanChurchAid is connected with the fish. It is an old Christian symbol that dates back to the first centuries of the church. The fish originates from the gospels' story about Jesus who gives the starving something to eat – five loaves of bread and two fish – and at the same time, the letters in the Greek word for fish – IKTYS – are synonymous with the words: Jesus Christ Son of God Saviour.

(DanChurchAid, 2015c)

Although DanChurchAid is rooted in the Danish National Evangelical Lutheran Church, the *fisk* booth displayed no overt religious references or activities beyond the fish symbol. In some ways this makes it an exemplar of Bakhtinian eco-voluntourism. As opposed to the more traditional piety associated with Protestantism, DanChurchAid's *fisk* booth takes a very different approach. There is no requirement that volunteers are religious and, by actively promoting the organization at the Festival, DanChurchAid buys into the celebration and festivity the Festival entails. The presence of *fisk* at the Festival is thus very much in line with Bakhtin's carnival, since it combines both serious Christian notions of charity, with the festive. Many religious festivals, for example, have historically combined celebrations of holy days (the origin of 'holidays') with festivities. Such carnivalesque holidays, according to Bakhtin (1984a), humorously play off the intellectual preoccupations of the mind against the physical demands of the body and, importantly, do so as part of a collective where all become equal, establishing a sense of community.

Figure 5.3 The *fisk* booth at Roskilde Festival 2009.
Photo by Mette Fog Olwig.

Furthermore, the fish symbol is in and of itself a symbol that has been used as part of the Christian response to the hippie New Age of Aquarius culture. There exists, for example, an Ichthus Music Festival (Creation Festivals, 2015), which 'started in 1970 when Asbury Theological Seminary professor Bob Lyon encouraged his students to develop a Christian answer to Woodstock' (Copley, 2009). As a symbol of charity, the fish symbol fits perfectly into the New Age of Aquarius focus on peace, love and harmony, and its affinity for occult-like, underground signs and symbols.

At the Roskilde Festival, the *fisk* booth was placed near one of the music stages in an area reserved for booths with an ethical agenda. *Fisk* had its own fair trade clothing collection, *SKIFt*[5] *trade fair*. When fair trade clothing items were first sold in Denmark they tended to be brown, itchy and shapeless – and thus not entirely fashionable. *SKIFt* was to be an antidote to this 'problem'. As Connie Yilmaz Jantzen, DanChurchAid national director, explained at the launch of the 2008 *SKIFt* collection:

> For years I have felt that fair trade clothing was an obvious and right choice considering what I usually stand for. But when I entered the stores that sold fair trade clothing, I just had to admit that I would not be the one to model these clothes. Maybe it's because fair trade clothes in the first years had to be authentic to be credible – creating quite a challenge within Danish fashion trends. SKIFt has effectively changed this – with SKIFt, fair trade and fashion are united in a very convincing way.
>
> (Jantzen, 2008)

As Jantzen explained, the fair trade clothes had to be authentic to be credible or, following the puritan logic, they had to be simple to be considered ethical. The *SKIFt* clothes were made of organic cotton and produced following fair trade principles. The idea behind *SKIFt trade fair* and *fisk*, however, was to make products that were fair trade, organic and vintage, cool and fun – or popular-festive – something that would be enjoyable to wear on the body and not just something purchased to support a good cause. Among the volunteers, it therefore seemed like a great idea to use popular celebrities for that purpose. The volunteers convinced Danish singer Karen Mukupa and actor Thure Lindhardt to be ambassadors for the collection. This involved, among other things, their modeling the clothes on their bodies. The two celebrities were also engaged at festivals – including the Festival – in 2009 where they were installed as sales staff at the booth in order to attract customers. This was covered in a news story on the Danish-language version of the DanChurchAid website entitled 'Fair Fashion, Festival and Celebrity-service: DanChurchAid's lifestyle and vintage store *fisk* has moved to the Roskilde Festival with star sales assistants and fashionable vintage clothing' (DanChurchAid, 2008). The story included a picture of Thure Lindhardt with then General Secretary of DanChurchAid, Henrik Stubkjær, and an image of a sign posted in the booth that read: 'Buy fair fashion from Thure Lindhardt and Karen Mukupa! Friday between 1 and 3pm, here at *fisk*'.

Figure 5.4 Karen Mukupa being interviewed by MTV at the *fisk* booth. Karen Mukupa is wearing clothing items from the *SKIFt* collection.

Photo by Mette Fog Olwig.

In the news story on the DanChurchAid website, the article pointed out that Thure Lindhardt recently had his international breakthrough in the US blockbuster movie *Angels and Demons*, quoting him as saying 'I use the celebrity effect for something sensible and very valuable to me'. In the same article, Karen Mukupa was quoted as saying 'it helps when Thure and I wear the clothes. It attracts more [people]'. At the Festival, sales soared as a result of the celebrity presence, with many people buying clothing items while the two celebrities were present. While the added funds were welcomed, Connie Yilmaz Jantzen, DanChurchAid national director, was aware of the potential critique the *SKIFt* collection might receive on account of its pandering to what she referred to as 'our creamy lifestyle'. As she put it,

> Of course one could become a bit critical of the fact that everything has to be connected to our creamy lifestyle in order for us to adopt it as modern people. Conversely, one can see the lifestyle wave as an expression of a mental and economic opportunity to put things on the agenda, which have been difficult to create awareness about for some years. It is most likely the reason why fair trade is booming in these years – because we have the surplus to buy goods – that are more than a product – but that also tell a story.
>
> (Jantzen, 2008)

There were several news items on DanChurchAid's website further explaining the details of the initiative and telling the story behind the clothes items. Clothing items also came with a small hangtag describing the initiative. One of the goals of *SKIFt* was to build a knowledge base of methods and approaches to ethical production that other designers and producers could use (Nielsen, 2008).

The combination of the popular-festive element of celebritizing and promoting DanChurchAid and the *SKIFt trade fair* clothing line at the Festival, coupled with the attempt at educating shoppers and volunteers about the plight of workers in the Global South, illustrates one of our arguments. The DanChurchAid national director echoed the academic debates about popularizing serious causes; this type of involvement might be superficial and consumerist – indeed, she concedes that the gains of fair trade might well be due to Danish economic prosperity. However, in addition to having fun at the Festival, the DanChurchAid volunteers (especially those working with *SKIFt*) were learning about the situation of the women sewing the clothes as well as the compromises and difficulties involved in creating a fair trade product. Marie Søegaard Tarpø, one of the designers of the *SKIFt* collection, for example, went several times to Burkina Faso, where one of the collections was made. In a news article on DanChurchAid's website, she described how the women sewing the collection worked 12 hours a day, sat on the floor and had to deal with many hours of power outages effectively preventing them from using the sewing machines (Lemvig, 2008). Maybe the question is, therefore, not how many additional people learned about fair trade and the plight of the workers that produce the clothes in the Global South, but whether those already engaged were further educated concerning the political ecology of global inequality.

Volunteering and community spirit at Roskilde Festival

When Olwig asked her first year students in social science at Roskilde University – the university closest to the Festival – how they would describe the Roskilde Festival to someone who had never participated in it, the first student responded 'a community'. As described above, this is manifested in aspects such as the campsites. Many festivalgoers spend a significant amount of time creating a camp theme and each year the Festival names one of the camps 'Camp of the Year', as explained on the 'Camp of the Year @ Roskilde Festival' website:

> Camp of the Year is for all the camps that throughout the festival create fantastic atmosphere, togetherness and love with creative, innovative experiences and festive engagement. And the winner will be the camp that is not just a collection of tents, but manages to create the best festival for themselves, neighboring camps and all those who drop by.
> (Roskilde Festival, 2015b)

In recent years the themes of winning camps have included Camp Find Waldo, Camp Crazy Legs, Camp Where is the Love, Camp Farmer Dating, Camp Burt Reynolds and Camp Ping Pong. These camp themes are often times played out by festivalgoers throughout their full participation in the Festival – for example by wearing costumes throughout the week, or putting together themed football teams for a charity football tournament (e.g. Hummel, 2013; Orange Karma, 2013). The importance of a sense of community at the Festival is thus greatly emphasized by the festival organizers, festivalgoers and the volunteers. Social norms, such as being open, and partying with friends and strangers alike (something that is widely assumed to be outside the comfort zone for Danes), is part of the ethos of the Festival. As Jesper Switzer Møller, one of the Festival founders, puts it: ' . . . I love, that we here at the Roskilde Festival learn that the community is bigger than the individual . . . ' (Pagh, 2013). The community spirit with which the Festival describes itself and in which festivalgoers see themselves as active participants, we argue, is closely linked to the carnivalesque – instilling different social norms, aesthetics and modes of sociality via the festival ethos and performed by festivalgoers. By performing these different forms of sociality and adhering to social norms that are different from everyday life, festivalgoers participate in a particular time-space relation in which new cultural meaning and ethics may emerge.

Conclusion – festivals as utopia

As a widespread tourist community, music festivals are in many ways microcosms of society itself, thereby providing a space- and time-confined experimental laboratory for the study of societal phenomenon. Yet, festivals are very particular places that facilitate out of the ordinary spaces where people often behave in out of the ordinary ways. This relationship between out of the ordinary spaces and behaviors is a critical point that Bakhtin repeatedly made with reference to the carnivals of the Renaissance (Dentith, 1995, p. 68). This is in part because, if the carnivalesque is suspended beyond the carnival itself, it would disrupt the social order. The popular-festive forms of social engagement of festivals such as the Roskilde Festival have a similar quality. People act in ways that in other contexts would be considered unusual or even socially unacceptable. If excessive drinking, wearing costumes or stripping and cavorting freely with strangers were practiced outside the Festival, people would most likely be sanctioned – not only socially, but possibly also legally. The various popular-festive performances practiced as an integral part of community-making during the Festival tend to stop once festivalgoers leave the Festival. Yet, the integration between these popular-festive forms and the practices related to global solidarity and environmentalism, which are also part of the festival chronotope, produce the possibility at least for imagining new forms of social engagement and new meanings ascribed to environmentalism, beyond the puritanical.

In this chapter we have described instances of festive eco-voluntourism at the Roskilde Festival and suggested that festivals may provide the venue required for successful popular-festive environmentalism. They do so by implicitly acknowledging a mind-body tension (as opposed to a fixed binary), as embedded in Western culture, and by offering a venue for both to be stimulated. In this way festivals move beyond the two extremes often emphasized in the literature of, on the one hand, emphasizing the value of engaging and educating eco-voluntourists about the political ecology of environmental issues through empathy and affect and, on the other hand, simply providing a venue for self-promoting entertainment.

Notes

1 That is to say, involving celebrities.
2 Translators of Bakhtin used the terms 'carnivalistic' (denoting carnival forms) and 'carnivalization' (denoting carnival consciousness in literature) (Bakhtin, 1984b, p.12; Platter, 2001, p.55). Critics have later referred to the time-space relation depicted by Bakhtin in his readings of carnival, as 'carnivalesque' in order to capture the time-space relation in which 'multiple cultural forms' carry 'traces of carnival energy' (Dentith, 1995, p.79; Jung, 1998, p.104).
3 Bakhtin introduces the carnivalesque in his analysis of Rabelais' novel *Gargantua and Pantagruel*.
4 *fisk* no longer exists.
5 'Skift' means change in Danish, but SKIF is also an anagram of *fisk*.

References

Baillie, G., Smith, M., Laurie, N. and Hopkins, P. (2013) International volunteering, faith and subjectivity: Negotiating cosmopolitanism, citizenship and development, *Geoforum*, 45, pp. 126–135.
Bakhtin, M. (1984a) *Rabelais and his World*. 1st Midland book edn. Bloomington, IN: Indiana University Press.
Bakhtin, M. (1984b) *Problems of Dostoevsky's Poetics*. Minneapolis, MN: University of Minnesota Press.
Bakhtin, M. (2001) *Karneval og latterkultur*. Copenhagen: Det Lille Forlag.
Brandist, C. (2002) *The Bakhtin Circle: Philosophy, Culture and Politics*. London: Pluto Press.
Brockington, D. (2009) *Celebrity and the Environment: Fame, Wealth and Power in Conservation*. London: Zed Books.
Chouliaraki, L. (2012) The theatricality of humanitarianism a critique of celebrity advocacy, *Communication and Critical/Cultural Studies*, 9 (1), pp. 1–21.
Christiansen, L. B. and Frello, B. (2015) Celebrity witnessing: Shifting the emotional address in narratives of development aid, *European Journal of Cultural Studies*, 2015 (1), pp. 1–16.
Clean Out Loud (2015) Clean Out Loud. *Facebook*, Available from: https://www.facebook.com/CleanOutLoud/info?tab=page_info (accessed 3 April 2015).
Copley, R. (2009) At 40, Ichthus isn't looking back. *kentucky.com*. Available from: http://www.kentucky.com/2009/06/11/826913_at-40-ichthus-isnt-looking-back.html?rh=1 (accessed 30 May 2015).
Creation Festivals (2015) Ichthus festival. *Asbury University*. Available from: http://ichthusfestival.com (accessed 30 May 2015).
DanChurchAid (2008) Fair Fashion, Festival og Kendisbetjening. *Dan Church Aid*. Available from: http://www.noedhjaelp.dk/sider_paa_hjemmesiden/i_danmark/kampagner/roskilde_festival/laes_mere/fair_fashion_festival_og_kendisbetjening (accessed 25 February 2015).

DanChurchAid (2015a) ACT alliance. *Dan Church Aid*. Available from: https://www. danchurchaid.org/donors/international-networks/act-alliance (accessed 30 May 2015).
DanChurchAid (2015b) Historien om Folkekirkens Nødhjælp. *Nødhjælpen DK*. Available from: https://www.noedhjaelp.dk/om-os/hvem-er-vi/vores-historie (accessed 30 May 2015).
DanChurchAid (2015c) Values and mission. *Dan Church Aid*. Available from: https:// www.danchurchaid.org/about-us/organisation-and-strategy/values-and-mission (accessed 30 May 2015).
Denmark DK (2011) Volunteer at Roskilde Festival. *Denmark DK Videos about Denmark*. Available from: http://video.denmark.dk/video/2208322/volunteer-at-roskilde-festival (accessed 30 May 2015).
Dentith, S. (1995) *Bakhtinian Thought: An Introductory Reader*. London: Routledge.
Ghosh, P. (2003) Max Weber's idea of 'Puritanism': A case study in the empirical construction of the Protestant ethic, *History of European Ideas*, 29 (2), pp. 183–221.
Goodman, M.K. (2010) The mirror of consumption: Celebritization, developmental consumption and the shifting cultural politics of fair trade, *Geoforum*, 41 (1), pp. 104–116.
Green Operations Europe (2015) Eurosonic 2015 – here we GO again! *Go Group*. Available from: http://go-group.org/2014/12/eurosonic-2015-here-we-go-again/ (accessed 21 June 2015).
Hemmings, C. (2012) Affective solidarity: Feminist reflexivity and political transformation, *Feminist Theory*, 13 (2), pp. 147–161.
Hjortdal, M. (2014) Festivalgæster gider ikke leve op til Roskildes flotte miljøprofil. *Politiken*, Copenhagen, 4th July. Available from: http://politiken.dk/kultur/musik/ roskildefestival/roskildenyheder/ECE2333423/festivalgaester-gider-ikke-leve-op-til-roskildes-flotte-miljoeprofil (accessed 30 May 2015).
Holquist, M. (2002) *Dialogism: Bakhtin and his World*, 2nd edn. London: Routledge.
Hummel (2013) Girl Power and Fisherman. Available from: http://www.hummel.net/ aa-DK/karma/page/373/GirlPowerandFisherman (accessed 16 March 2015).
IBIS (2013) Mød Yasuni på Roskilde Festival. *ibis.dk*. Available from: http://ibis.dk/event/ mod-yasuni-pa-roskilde-festival/ (accessed 20 June 2013).
Jantzen, C.Y. (2008) Skift – tale ved lancering d. 2. juni 2008. *Nødhjælpen DK*. Available from: https://www.noedhjaelp.dk/content/download/29945/319123/file/CYJs%20tale% 20SKIFt%20lancering%2008.pdf. (accessed 30 May 2015).
Jung, H. Y. (1998) Bakhtin's Dialogical Body Politics. In: Bell, M. M. and Gardiner, M. (eds), *Bakhtin and the Human Sciences*. London: Sage Publications, pp. 95–111.
Kapoor, I. (2013) *Celebrity Humanitarianism: Ideology of Global Charity*. New York: Routledge.
Lemvig, M. (2008) Maries Skiftetøj. *Nødhjælpen DK*. Available from: https://www. noedhjaelp.dk/nyheder/magasinet/maries-skiftetoej (accessed 30 May 2015).
Mostafanezhad, M. (2013) Getting in touch with your inner Angelina: Celebrity humanitarianism and the cultural politics of gendered generosity in volunteer tourism, *Third World Quarterly*, 34 (3), pp. 485–499.
Nelson, R. H. (2014) Calvinism without God: American environmentalism as implicit calvinism, *Implicit Religion*, 17 (3), pp. 249–273.
Nielsen, A. A. (2008) Skift Stil – Trade Fair! *Nødhjælpen DK*. Available from: https:// www.noedhjaelp.dk/nyheder/seneste-nyt/skift-stil-trade-fair! (accessed 30 May 2015).
Olwig, M. F. and Christiansen, L. B. (2015) Irony and Politically Incorrect Humanitarianism: Danish Celebrity-Led Benefit Events. In: Richley, L. A. (ed.),

Celebrity Humanitarianism and North-South Relations: Politics, Place and Power. Oxford: Routledge, pp. 170–188.

Orange Karma (2013) *Orange Karma Sensational 2013 – The Finals.* Copenhagen. Available from: https://www.youtube.com/watch?v=R3DzFzW2EcQ (accessed 30 May 2015).

Pagh, L. (2013) Manden, der grundlagde Roskilde Festival. *Roskilde Festival.* Available from: http://issuu.com/roskilde-festival/docs/orange_press_rf2013_07-07-2013/8 (accessed 30 May 2015).

Platter, C. (2001) Novelistic Discourse in Aristophanes. In: Barta, P., Miller, P. and Platter, C. (eds), *Carnivalizing Difference. Bakhtin and the Other.* London: Routledge, pp. 51–78.

Richey, L. A. and Ponte, S. (2014) New actors and alliances in development, *Third World Quarterly*, 35 (1).

Robbins, P. (2004) *Political Ecology: A Critical Introduction.* New York: Blackwell.

Roskilde Festival (2015a) About Roskilde Festival. *Roskilde Festival.* Available from: http://www.roskilde-festival.dk/more/about-roskilde-festival (accessed 30 May 2015).

Roskilde Festival (2015b) Camp of the Year @ Roskilde Festival. *Facebook.* Available from: https://www.facebook.com/pages/Camp-of-the-Year-Roskilde-Festival/403822206298420?sk=info&tab=page_info (accessed 30 May 2015).

Roskilde Festival (2015c) Fun Facts. *Roskilde Festival.* Available from: http://www.roskilde-festival.dk/more/press/fun-facts (accessed 30 May 2015).

Roskilde Festival (2015d) Roskilde Festival has won prestigious green award. *Roskilde Festival.* Available from: http://www.roskilde-festival.dk/news/2015/roskilde-festival-has-won-prestigious-green-award (accessed 30 May 2015).

Roskilde Festival (2015e) We're all non-profit. *Roskilde Festival.* Available from: http://www.roskilde-festival.dk/more/non-profit (accessed 30 May 2015).

Roskilde Lokalavis (2015) Hædrer orange madkamp. *Roskilde Lokalavis*, Roskilde, 21st January. Available from: http://roskilde.lokalavisen.dk/haedrer-orange-madkamp-/20150121/artikler/701219949/1618 (accessed 21 July 2015).

Steiner-Aeschliman, S. (1999) The Integrated Protestant Ethic and the Spirit of Environmentalism. Seattle, WA: Washington State University. Available from: http://www.bookpump.com/dps/pdf-b/1120400b.pdf (accessed 30 May 2015).

Tiessen, R. and Heron, B. (2012) Volunteering in the developing world: the perceived impacts of Canadian youth, *Development in Practice*, 22 (1), pp. 44–56.

Turner, V. (1967) *The Forest of Symbols: Aspects of Ndembu Ritual.* Ithaca, NY: Cornell University Press.

Weber, M. (2002) *The Protestant Ethic and the Spirit of Capitalism: and Other Writings.* London: Penguin.

Part II
Conservation and control

Introduction to Conservation and control

Eric J. Shelton

When may the word *ecologies* be used? What about the term *ecosystems*? Paul Ricoeur might ask: What is this *nature* of which we speak? And when should the word 'problem' be used to describe a particular environmental situation? Political ecology is a useful framework from which to examine contexts in which problems might be 'seen as one of control over access, aesthetics, and landscape production' (Robbins, 2012, p. 177). The chapters in this section describe tourism as more than a purely economic analysis of the normal tourism business model of *adding value* and *extracting revenue*. Rather, each chapter examines the 'production of protection' within the context of the conservation and control thesis, whereby 'control of resources and landscapes has been wrested from local producers or producer groups (by class, gender or ethnicity) through the implementation of efforts to preserve "sustainability," "community" or "nature"' (ibid., pp. 178–179). In the process of seeking to establish control, local officials and global stakeholders who claim to be acting in the best interest of 'the environment' and its preservation often disable local livelihoods and systems of production, obtaining 'consent of the governed ... through social technologies' (ibid., pp. 178–179). This consent of the governed is maintained through constant reorganization of the social institutions that appear to grant it, thus perpetuating a hegemonic governmentality. Tourism practices are regularly embedded in such processes of conservation and control. These practices are also readily entangled in the social construction of a dehumanized wilderness. The conservation and control thesis eschews the polygons of scientific ecology as the basis of conservation territories, finding them 'ecologically and socially problematic, and inadequate either to meet the goals of preservation either of wildlife or of livelihoods' (ibid., p. 179).

Frequently, the idea of a socially constructed wilderness takes the form of 'nature without people' or, the production of protection, embedded within the neoliberal managerial lexicon (an 'outcome') and involving the expenditure of measurable 'conservation effort' (an output). Brockington *et al.* (2008) illustrate how twenty-first century conservation is enmeshed with neoliberal economic practices and ideologies. In this way, late-capitalist economic production and consumption systems contribute to the production of protection as an economic act. A collection of such acts, whether achieved through preservation or through use, contributes to the broader growth of the conservation economy (Groser, 2009).

The conservation economy is consistent with more general neoliberal economic reform and sits easily alongside other features of Knight and White's (2009) framing of 'conservation for a new generation'. Other features of such conservation are:

> (t)he creation of capitalist markets for natural resource exchange and consumption ... privatisation of resource control within these markets ... commodification of resources so they can be traded within markets ... withdrawal of direct government intervention from direct market transactions; and ... decentralization of resource governance to local authorities and non-state actors such as non-governmental organizations (NGOs).
> (Fletcher, 2010, p. 172)

Tourism, particularly tourism in protected areas, fits neatly into this neoliberal order, offering improved biodiversity outcomes but Robbins offers instead a paradox:

> If environmental degradation is often associated with the marginalization of poor subsistence communities and working people, it might be logical to assume that conservation and preservation of environmental systems, resources, and landscapes are commensurate with community sustainability and the protection of livelihoods. This has proven far from true
> (Robbins, 2012, p. 176)

Sometimes, when environmental degradation is advanced, and existing livelihoods have all but disappeared, the task of amelioration or restoration can appear too difficult to engage with. The chapters in this section all resist the temptation to disengage with such problems as being too hard to consider.

The section begins with the Pacific Islands, where Forrest Young considers Rapa Nui (popularly known as Easter Island) and the ways in which the development of its UNESCO world heritage site and national (Chilean) park have become sites of intense and long-standing political challenges. Chapter 6 illustrates how the promotion of sustainable tourism has in many ways politically and economically marginalized indigenous Rapa Nui thus reproducing ongoing colonial relations of power. The role of governmentality and biopolitics in the struggle over access to land and natural resources in Rapa Nui, have as a result, become further inflamed as indigenous rights to self-determination among Rapa Nui material, social and symbolic worlds are challenged.

Continuing with an investigation of the colonial trope of national/global control over indigenous/local self-determination, Zhang sets Chapter 7 in Shangri-La County, southwest China. In a critique of the introduction and subsequent management of ecotourism in Shangri-La, Zhang illustrates how China's environmental issues have been widely discussed in the global context, through their relations with the 'modernization' of China. Zhang's chapter contributes to ongoing work in political ecology that politicizes Escobar's notion of hybrid natures, and nature-society relations. Finally, Zhang sheds critical light on the role of the widespread nationalist discourse surrounding tourism within protected areas.

Chapter 8 picks up on the idea of ecotourism in a post-industrial Japanese landscape, and how it is represented and performed, recalling Oelschlaeger's claim that 'the idea of wilderness in postmodern context is . . . a search for meaning - for a new creation story or mythology' (1991, p. 321). Cunningham situates the production of this landscape within a Japan that is past its peak, economically and socially. Such a situation evokes Worster's (2008, p. 266) suggestion that 'reliable indicators of whether nations become active in preserving wild places are the state of personal freedom, the degree of social equality, and the sanctity of human rights'. Cunningham's work thus begs the question: does the act of *creating* wild places act similarly?

In Chapter 9, Wearing and Wearing argue that the 'rights of nature' is best effected through an 'environmental citizenship' of all living beings. They remind us that as the fight for the rights of nonhuman nature evolved, a similar process was occurring in the tourism industry, which led to the idea of ecotourism, first perceived as a means to save nature through utilitarian value-use rather than use-value. Certainly, ethics, as one aspect of the production of protection has received growing attention over the last three decades. The nature of proper ethical relationships between humans and the nonhuman world has been of particular interest to tourism scholars. Wearing and Wearing outline several approaches to ethics and endeavour to support a Kantian ethics, which suggests that ethics must precede the development of environmental protection practices. In doing so, they reveal that political ecologies must first and foremost have solid ethical premises (and promises) on which to base assumptions and recommendations for action.

The intellectual traditions informing the chapters in this section are wide-ranging, spanning ecocentric and anthropocentric analytic positions, here harnessed together by a combined ecological/political understanding of the 'natural' world. The authors share an understanding of the dangers of ecological complexity being trampled by political expediency, often under the guise of sustainable tourism development. This caution makes the authors wary of uncritical application of the 'wise use' environmental paradigm. These practitioners of political ecology are aware of the need not only to be critical, but also never to offer simple formulations of complex situations.

References

Brockington, D., Duffy, R. and Igoe, J. (2008). *Nature Unbound: Conservation, capitalism and the future of protected areas*. London: Earthscan.

Fletcher, R. (2010). Neoliberal Environmentality: Towards a poststructural political ecology of the conservation debate, *Conservation and Society*, 8 (3), p. 172.

Groser, T. (2009). The Conservation Economy, *Research Cluster for Natural Resources Law Newsletter Two: 2009*, pp. 2–3.

Knight, R. and White, C. (eds). (2009). *Conservation for a New Generation:Redefining natural resources management*. Washington: Island Press.

Oelschlaeger, M. (1991). *The Idea of Wilderness: From Prehistory to the Age of Ecology*. New Haven: Yale University Press.

Robbins, P. (2012). *Political Ecology: A critical introduction (2nd Edition)*. Chichester: Wiley-Blackwell.

Worster, D. (2008). *A Passion for nature: The Life of John Muir*. Oxford: Oxford University Press.

6 Unsettling the moral economy of tourism on Chile's Easter Island

Forrest Wade Young

> *Iconic and mysterious Easter Island statues are all that remain of the clans that once lived there.*
>
> *Los Angeles Times Travel*, December 29, 2013

The development of 'Chile's Easter Island' into a UNESCO World Heritage Site and Chilean national park receiving an estimated 80,000 tourists each year has been increasingly targeted in the international media as a global space of degraded, unsustainable landscape threatening cultural heritage, fishing stocks, public health, and water supply (Long, 2014). Neoliberal discourse highlighting strategies for both international and Chilean stakeholders to manage sustainable tourism of the park has emerged as the primary solution to development problems. In response to the crisis, and working in conjunction with the state, the Coca-Cola Corporation, the National Geographic Society and the Cousteau Society, among other international groups, have begun to initiate conservation programmes ranging from building houses made of recycled bottles and cans to the development of reforestation projects and marine reserves (Young, 2014). Echoing problems of conservation park designation worldwide, the normalization and promotion of sustainable tourism on 'Chile's Easter Island' involves the management of a limited 'way of seeing, understanding, and (re)producing the world . . . a way of seeing and being in the world . . . seen as just, moral, and right' (West *et al.*, 2006, p. 252). In the context of the intensifying struggle of the Indigenous Rapa Nui Nation[1] against 'placelessness' (Escobar, 2001, p. 140) in Chile's Easter Island, any priority given to the problematization of conservation, as with any genealogy of landscape (Robbins, 2004, p. 124), requires interrogation as a political ecological strategy of governmentality and biopolitics. Consistent with reimagining political ecologies in terms of the discursive turn, multi-scalar analysis of park development is shown to reterritorialize the world of Rapa Nui in terms of 'generification' (West, 2005) that facilitates biopolitical control of Rapa Nui as a population to move around a park of things, rather than as a nation with peoplehood and Indigenous rights to self-determination of their material, social and symbolic cultural world. Entangling political ecological research in Indigenous studies, Pacific Island anthropology and post-structural critical theory, analysis of place-based Rapa Nui 'counter-work'[2] discloses 'buried epistemologies'

(Robbins, 2004, p. 125; Willems-Braun, 1996) that contest respatialization of their Polynesian island world into a sustainable Chilean national park and World Heritage Site. Following a historical overview of the local, national and global political ecology of the development of tourism in Chile's Easter Island, analysis of community narratives articulated in the Indigenous language discloses how the management of the island as a tourist habitat undermines Rapa Nui sacred sites and subjectivities. In the context of recent developments in other Pacific Island worlds, the production of sustainable tourism in Chile's Easter Island illuminates forces of contemporary neoliberal globalization as articulating not only a political economy of accumulation by dispossession' (Harvey, 2004) but also the production of a 'moral economy,[3] serving tourist imaginaries at the expense of Indigenous social movements for self-determination.

At the entrance of the Musée du Quai Branly in Paris, a stolen head of the world-famous *moai* statues of Easter Island greets consumers and introduces them to the museum's affective orders of things. Museum-goers, much like *Los Angeles Times* world travellers, are likely to fail to understand that *moai* severed from their Indigenous cultural context are actually entangled in a globally mediated 'spectacle' that displaces a multiplicity of historical and ongoing social relationships (Debord, 1994, p. 12). Tourism has seen steady annual growth from approximately 2,000 visitors in the mid-1970s, to 20,000 in the 1990s and over 40,000 in the first decade of the twenty-first century. Since the inception of small-scale modern tourism in Chile's Easter Island, tourists have desired to learn about the *moai* and archaeological sites more than about the contemporary Rapa Nui people, who now number over 2,500 (McCall, 2008, pp. 49–55). In the tradition of Edward Said's 'Orient' (1994) and Gerald Vizenor's 'Indian' (1998, p. 50) critical tourist studies theorist Beverly Haun (2008) has demonstrated that Western tourist imaginations of Chile's Easter Island are dialogical with a thick archive of texts, photos and paintings dating to the initial events of eighteenth-century Western imperialism on the island. Tourist representations of *moai* as metonyms for a lost island world, for example, trace to *A View of the Monuments on Easter Island*, a nineteenth-century painting by William Hodge during Captain Cook's encounter with Rapa Nui and the first Western representation to depict *moai* independently of Rapa Nui people. The painting grounds a Kantian aesthetic discourse of the '*moai* as sublime' (Haun, 2008, pp. 145–147) objects for artistic appreciation. This representation displaced the previously dominant discourse of Rapa Nui as an island of 'savages' that circulated in the Western public sphere following the violent imperial arrival of the Dutch West Indies Corporation to Rapa Nui on Easter Sunday 1722, a date that led to its Occidental naming, along with the murder of at least ten unarmed Rapa Nui by the Dutch upon landfall. By separating the Rapa Nui people from the *moai*, a stage was cleared for Westerners to begin to culturally forget the Western violence that occurred during Roggeveen's encounter with the island, and instead freely imagine a range of mysteries regarding the statues' construction. Haun (2008, p. 167) sees this as a social theatre that continues to be performed today, both in discourses of *moai* evolution (by scientists) as well as in discourses of *moai* creation by outer-space aliens (by new age mystics).

As tourism anthropologist Geoffrey White (2014) noted, for the Solomon Islands in particular and in the Pacific Islands more generally, modern tourism in Chile's Easter Island initially developed from a complex of global forces with little accountability to the Rapa Nui people. Though for different reasons, the US military was a factor as it was in the case of the Solomon Islands. Modern international tourism to Chile's Easter Island emerged in the context of infrastructure developed for a secret US Air Force base built in 1966 (McCall, 1995, pp. 1–2). Lindblad Travel, a luxury tour operator based in the US, organized the first tour packages in 1967 on chartered LAN-Chile propeller aircraft that landed at the US military airfield (McCall, 2008, p. 47). Accommodating approximately 40 tourists in tents, the tours visited key archaeological sites in Dodge pick-up trucks imported from the US (Fisher, 2005, p. 221). There was a social hierarchy within the operation; Linblad provided international guides while Rapa Nui participated as drivers (McCall, 2008, p. 47). When the socialist Allende government came to power in 1970, the Air Force departed the island and the Chilean State Hotel Corporation (HONSA) constructed a prefabricated 120-bed hotel to replace the tents as part of a development plan that proposed a self-sufficient economy balancing cooperative-based tourism with agriculture and fisheries (Porteous, 1981, p. 223). Following the US-supported coup d'état of President Allende, Chilean dictator Augusto Pinochet began the implementation of a neoliberal plan of privatization of state resources, eventually selling the HONSA hotel to private investors. Under Pinochet, tourism rather than socialistic self-sufficient agricultural development became the focus of island plans (Porteous, 1981, p. 223). After an initial decline of tourist numbers following the coup, tourism ultimately expanded, principally through the emergence of American and European tour packages to Rapa Nui marketed to people travelling between Pape'ete, Tahiti and Santiago, Chile (Porteous, 1981, pp. 205–207).

Studies of the Pacific Island region have tended to view tourism since the 1970s in Rapa Nui as a comparative success story, depicting the industry as 'largely under local control' (Harrison, 2003, p. 18). This neoliberal development narrative highlights the predominance of Rapa Nui-owned, tourism-related businesses such as accommodation facilities (Stanton, 2003, p. 116) and a general 'rise in wealth' associated with new opportunities for economic growth (McCall, 2008, p. 51). However, it is worth noting that most Rapa Nui are employed in entry-level jobs while professional offices are occupied almost exclusively by Chileans (Stanton, 2003, p. 117), and that, as is common with tourism in the Asia-Pacific region broadly (Connell and Rugendyke, 2008, pp. 17–20), wealth accumulation in Rapa Nui has been unevenly distributed and associated with rising costs of living (Stanton, 2000, p. 147). Increased tourism has also been entangled in accelerating ethnic conflict between Rapa Nui and Chileans (Stanton, 2000, p. 150) during which loss of language and culture has intensified and Rapa Nui are now threatened with becoming a minority on their own island as Chileans and expatriates have increasingly settled on the island (Porteous and Shephard-Toomey, 2005, p. 10).

Most importantly, however, the sense of local control of tourism in Rapa Nui fails to account for the fact that central to the island's tourism since the 1960s has been a political and legal system that Chile has imposed upon Rapa Nui. Though the national park system did not operate until tourism began in the 1960s, it was created in the early twentieth century – along with the marginal place of Rapa Nui within it. The organization of the park system provides only 13 percent of the island for Rapa Nui use under 'dominion title' (*titulo dominio*), while the state and private companies manage the rest of the island as a state farm, national park and for various public services (International Work Group for Indigenous Affairs, 2012, p. 10). Following late nineteenth-century violations of a 1888 document signed between Rapa Nui and Chile, known as the 'Agreement of Wills' (*Acuerdo de Voluntades*), Rapa Nui were kept behind barbed wire in conditions of 'semi-slavery' and effectively rendered 'stateless' until 1967 when they acquired a modicum of civil rights for the first time (International Work Group for Indigenous Affairs, 2012, pp. 5–8). Chile, conveniently neglecting its own version of the Agreement of Wills, which emphasized Indigenous chiefs were to keep their titles of governance (Fisher, 2005, p. 142), and made no reference to the introduction of land-tenure systems (Pereyra-Uhrle, 2005, p. 135), which leased the entirety of the island to a Chilean entrepreneur in 1895. The businessman, Enrique Merlet, partnered in 1903 with Williamson and Balfour Company, a transnational firm with offices in the United States, Great Britain and Chile (Fisher, 2005, pp. 157–158). Under a subsidiary firm rather bluntly named the Easter Island Exploitation Company, Rapa Nui became a 'company state' that was 'organized and maintained by aliens . . . in the company headquarters' (Porteous, 1981, p. 45) to produce wool for the global market. No doubt in part because, as a recent truth commission has publicized, Chile never officially ratified the Agreement of Wills (Gobierno de Chile, 2008, p. 565), Chilean title to the island lands lacked an explicit legal footing. In 1933, out of fear of the increasingly powerful company state securing its own title to island lands, Chile registered the entirety of island lands in terms of the state (Vergara 1939, pp. 37–61) and consequently designated the island as a national park in 1935 without consulting or notifying Rapa Nui (Delsing, 2009, p. 127; International Work Group for Indigenous Affairs, 2012, p. 8). Like the development of many national parks and other protected areas worldwide (West *et al.*, 2006, p. 60) Chile appears to have strategically constructed the park in order to dampen the capacity of 'forces generated by the encompassing political economy' (Cole and Wolf, 1999, p. xvi) to strengthen power and take over the island officially from the state.

Rapa Nui have resisted their marginalization in this highly political ecology from the inception of the company state and through its expansion in 1995 under UNESCO's 'declaration of the whole island as an open-air World Heritage museum' (Haun, 2008, p. 210). This was a designation the Rapa Nui people were not consulted on and was a change they contested out of fear of having their rights to self-determination compromised (Fisher, 2005, pp. 244–245). Ongoing tensions in Rapa Nui escalated into bloodshed between 2010 and 2011 amid the expansion of global tourism, including the development of the five-star Hangaroa Eco Village and Spa, the only hotel on the island fully privately owned by a non-Rapa Nui person. The violence also occurred within a context of perceived transnational Asia-Pacific Economic Cooperation (APEC) development plans without Rapa Nui consultation, increased Chilean settler migration to the island and a sense of intensifying poverty and ghettoization of Rapa Nui life (c.f. Young 2012b). Escalating from protests surrounding the appointment of Petero Edmunds Paoa as Governor in March 2010, protests that would later lead to his resignation amid charges of corruption, for more than six months Rapa Nui *hua`ai* (extended families/clans) reclaimed plots of land from the state and multifarious private stakeholders, and occupied a number of state institutions and the Hangaroa Eco Village and Spa. In December 2010, in defiance of a request for precautionary measures against the use of armed forces by the Chilean state filed at the Inter-American Commission on Human Rights (IACHR) by the Washington DC-based Indian Law Resource Center, Chilean special forces burned Rapa Nui national flags and exercised state force to end the Rapa Nui reclamations (Young, 2012b, pp. 194–196). During the events 21 Rapa Nui were reported injured; actions of 'government thuggery' (McCall, 2010) which the International Work Group for Indigenous Affairs (IWGIA, 2012, p. 32) found to be in extensive violation of international law covering the rights of Indigenous peoples. Rapa Nui were shot at with rubber pellets, beaten with batons and choked. A total of 17 Rapa Nui had to be hospitalized, including two men who were flown to Chile for more intensive care. One of them, Richard Tepano, was reported to have lost an eye from a pellet shot at close range. On 6 February 2011, just two days before the IACHR granted the precautionary measures on behalf of the majority of Rapa Nui *hua`ai* represented by the Indian Law Resource Center (Crippa, 2014, p. 247), Chilean state forces continued aggression with the criminalization and violent removal of members of the Hitorangi family from the Hangaroa Eco Village and Spa (IWGIA, 2012, p. 24).

As the Chilean Supreme Court supported the Chilean-owned Hangaroa Eco Village and Spa against the Hitorangi title to lands occupied by the hotel, and a Court of Appeals rejected an initial ruling of excessive violence by the Chilean police Captain Albornoz against Rapa Nui (Young, 2013, pp. 179–180), conflict has continued and international lawyers see Chile in Rapa Nui as imposing 'colonial rule onto a non-self-governing territory' (Crippa, 2014, p. 247). This has been made explicit in recent struggles of the Rapa Nui against state plans to privatize lands of the state-owned farm in Vaitea, an area for which Rapa Nui have resisted previous development plans when the state attempted to develop the area into a five-star golf resort in conjunction with a nearby project to construct a *moai*-shaped air traffic control tower (Shephard-Toomey, 2001, pp. 111–112). In January 2014, Rafael Tuki Tepano, the Rapa Nui leader of the Corporación Nacional de Pueblos Indígenas (CONADI), organized local, national and international political efforts against a state plan to distribute Vaitea's 1,052 hectares of land among 264 families in individual plots of 2.5 hectares each. Calling the plan 'colonial' and arguing that it violated international legal conventions on consultation, Tuki Tepano wrote a report co-signed by other leading Rapa Nui grassroots organizations, Parlamento Rapa Nui, Makenu Re`o Rapa Nui, Asamblea de Clanes and Autoridades Tradicionales de Rapa Nui, in which it was emphasized that Rapa Nui rights had been violated by the plan's development under consent rules articulated in International Labour Organization Convention 169. Tepano then organized a plebiscite against the plan, which was being coordinated by an armed Chilean police force against Rapa Nui will (Young, 2015a, p. 286).

Rapa Nui people today are enveloped not only within the Derridean 'textual violence' of an imperial archive that has rendered them absent from 'the scene of their own culture' (Haun 2008, p. 4), but equally within a complex web of global capitalism, Cold War geopolitics, Latin American neoliberalism and state violence against Rapa Nui social movements for self-determination. Chile's Easter Island appears in this context not only as a 'geographic imaginary' (Haun, 2008, p. 6), but also as a 'violent cartography' (Shapiro, 1997) of settler colonialism that disciplines Rapa Nui bodies into a limited subject position of a Foucauldian 'apparatus' (*dispositif*) modulated by a multiplicity of statecraft, international archaeology, global conservation programmes and tourism (Young, 2012a, p. 20). We can begin to distinguish the structural and symbolic violence of respatializing Rapa Nui into an 'open-air World Heritage museum' at the scale of contemporarily-lived subjectivity by reflecting upon a Rapa Nui discursive practice that contemporary Indigenous anthropologists of Oceania theorize as critical to Polynesian affective belonging, epistemology, identity and social ontology: namely, 'genealogical work'[4] (Tengan, *et al.*, 2010, p. 140). In 2007–2008 (Young, 2011), monologues and conversations with members of the majority of Rapa Nui *hua`ai* regarding the meaning of the Chilean land-tenure system were recorded. Analysis of the following narrative from a Rapa Nui woman in her forties about the land-tenure system discloses some of the ways in which Rapa Nui lived 'sense of place' (Feld and Basso, 1996, p. 11) stands in conflict with the 'avatars of place' (Escobar, 2001, p. 152) articulated through the powerful capillaries of the transnational production of Chile's Easter Island:

Tō`oku mana`u me`e nei he Titulo Dominio. `Ina he riva mātou mo to`o i te me`e nei. He aha ta`e riva te me`e nei te Titulo Dominio? Te ha`aura`a: E avai mai ā te Tire, te hau Tire, i te henua ki mātou. A mātou, mātou henua i rava`a ai: ta`e o te hau Tire; me`e hakaara mai a Hotu Matu`a. Mo mātou te kāiŋa. I te hora tuai, te matahiapo o te ariki to`o mai, he vahi te kāiŋa; vahi i roto i tā`ana poki era erima. I te vahiiŋa i vahi ai he oho mai ki te hora nei hai mata. Te tātou mata o nei o Rapa Nui, e ai rō `ā te rāua parehe. I te hora nei o aŋarina te ha`aura`a: au i tuha`a mai ai i tō`oku kona. Kona ke nō atu. Ta`e o`oku te aro era. E ai rō `ā te mata; o rāua ra aro. I te hora nei, ta`e te hua`ai i noho ai i roto te rāua kona era o te rāua mata tahito. Hakanohonoho, hakahapehape `ā te tuha`aiŋa o te henua tuai. He aha kī era pe nei ēe? Me`e ta`e au ra me`e. E ai rō `ā te kuhane i `ruŋa o te henua. Te me`e nei he kuhane, mātou tupuna. `Oira te nu`u o nei ta`e au te tuha`aiŋa o te hora nei o te henua. I roto te Rapa Nui, e ai rō `ā te nu`u e ma`u nō `ā te mana`uiŋa tahito era. He hakatere, he oho mai mo te rāua matu`a mai te rāua tupuna. Te ha`aura`a ki te matu`a, ki te ŋa poki. He hakatere, he oho mai pe mu`a. Tō`oku mana`u me`e nei, me`e hauha`a mo hāpī ki te ŋa poki. Me`e hauha`a rahi mo hakama`a ki te ŋa poki, mo ma`a i a rāua i te me`e nei, mo mo`a takoa. Tā`ato`a ahu, e ai rō `ā tō`ona mata, e ai rō `ā tō`ona hua`ai. O rāua te ahu, te moai.

My thought of this thing 'dominion title': It is not good for us to take this thing. Why is it not good, this 'dominion title'? The meaning of this thing: Chile, the Chilean government, gives land to us. We, our land, was already acquired. [But] not from the Chilean government; it is established by a genealogy traced to Hotu Matu'a. The land is for us. In ancient times, the eldest son of the chief [Hotu Matu'a] acquired [land], divided the land; divided the land among his five children. The way of distributing land divisions to this day was to be done by 'tribe' (*mata*). All of the tribes of Rapa Nui have a piece of land. At this time today, here is the significance: I have divided my land area. It is a strange land area without purposeful meaning. It is not my land 'section' (*aro*). It is of a [different] tribe; this section is of them. Currently, extended families do not reside upon the land areas of the ancestral tribes. They are made to reside on lands that ancestral divisions render false. What is said about this now? It is something unsuitable. There is a spirit upon the island. These spirits are our ancestors. Thus, groups of people do not consider the current divisions of the island suitable. Within Rapa Nui, there are groups of people who carry the thoughts of the ancestors. They would govern by coming to their elders and ancestors. This is significant to parents and children; to govern this way in the future. My thought is that this, this is valuable to teach to the children. It is of great value to give this knowledge to the children, for them to know these things, for them to respect these things. All of our 'ceremonial centers' (*ahu*), belong to particular 'tribes' (*mata*), and particular extended families (*hua`ai*). The *moai* and *ahu* belong to them.

As in the well-known poem of Sāmoan literary theorist and writer Albert Wendt (2000), inside the Rapa Nui woman 'live the dead', in particular the most famous ancestor of Rapa Nui, founding chief Hotu Matu`a, who, with his sister, founding royal Ava Rei Pua, are often in the foreground of contemporary genealogical work in Rapa Nui. The recorded genealogies are not articulated in terms of the Western literary genre of 'legends' popularized for tourist consumption (Englert, 2006) but, in the form of Indigenous 'ecologically based oral narratives' that distinguished Tongan anthropologist and artist Epili Hau`ofa argues are critical to dwell upon to begin to 'clear the stage' of colonial history and 'bring in new characters' (Hau`ofa, 2008, p. 64) in Indigenous histories of Oceania. Hotu Matu`a does not merely have symbolic value in these narratives; in terms of the contemporary 'political ecology of things', he is very much an 'actant' (Bennett, 2010, p. 8).

Actants – human or nonhuman, material or immanent – 'do things' in political ecologies (Bennett 2010, p. viii). The acts of Hotu Matu`a in 'dividing' (*vahi*) the island *kāiŋa* by 'tribe' (*mata*) deeply resonate with the narrator and render state policies problematic. Similar to the ways in which Hawaiian genealogy establishes cosmological relationships between 'Native Hawaiians' (Kānaka Maoli), animals, deities, families, fish, lands and spirits in Hawai`i in conflict with US neoliberal conservation and development projects (Goldberg-Hiller and Silva, 2011, p. 436), the acts of Hotu Matu`a are understood by the narrator to braid together a multiplicity of 'ceremonial centres' of *ahu* and *moai*, 'tribes' (*mata*), 'spirits' (*kuhane*) and 'ancestors' (*tupuna*) and 'children' (*poki*) into 'a single cord' (Tengan *et al.*, 2010, p. 141) grounded within *kāiŋa*. In terms of this 'broader ecology of affect' (White, 2006, p. 51) *moai* are clearly not to be understood as sublime objects for consumption by an 'aesthetic community' of tourists (Ranciére, 2008). *Moai* are significant to Rapa Nui ancient and contemporary families as 'spiritual tombstones' that 'protect the land and the blood matrix to which each clan belongs' (Hitorangi, 2013). In other words, as with many national memorials, *moai* 'construct relations between people and places' (White, 2006, p. 52). A consideration of official Rapa Nui translations of *kāiŋa* into the Spanish '*útero*' (womb, uterus) (Huki, 1988, p. 10), which signifies an 'umbilical connection to land' (Hito, 2004, p. 26), rather than the more common, crude translation of the term as simply 'land', further distils the Rapa Nui sense of familial place grounded by the acts of Hotu Matu`a.

Kāiŋa is understood by Rapa Nui to be a gift of Hotu Matu`a that shelters familial memorials and wombs of future progeny. It is thus clear why the narrator, like many Rapa Nui, challenges the Chilean state policy to administer Rapa Nui in terms of dominion titles of *kāiŋa*. Her conception of island land divisions by dominion title as 'very strange' (*Kona ke nō atu*) and ultimately 'unsuitable' (*Me`e ta`e au ra me`e*) are sensible because dominion titles are neither enacted by Hotu Matu`a nor linked to the 'spirits' (*kuhane*) and 'ancestors' (*tupuna*) that constitute *kāiŋa*'s genealogical value within the families. Polynesian genealogical work ties culture, land and people together into an *oikonomia*, or a complex social organization of '*oikos* (the home)' (Agamben, 2009, p. 9) that Chile's Easter Island, much like many state conservation parks (United Nations, 2009, p. 91), tries to divide through various forms of violence. Just as language loss often weakens the 'interwovenness' (Woodbury, 1998, p. 257) of a cultural world, so too does stripping a people of its affective sensibilities of *kāiŋa* and *moai*. By unbraiding the cord that weaves Rapa Nui into an *oikonomia*, the state creates stronger conditions for 'the conducting of conduct' (Foucault, 2008, p. 186) of Rapa Nui within Chilean biopolitical strategies. Agamben (1993, p. 156) famously disrupted Derrida's grammatological philosophy of infinitely deferred traces of meaning by a 'putting together' of language and social topology. By severing Rapa Nui from their familial places of *kāiŋa* and *moai*, the state harnesses Rapa Nui back into a deferral of semantic traces. This is what critical French historian Pierre Nora might term the ultimately placeless 'archival memories' (*lieux de memoire*) of state histories: memories that ' . . . have no referent in reality; or rather, they are their own referent: pure, exclusively self-referential signs. . . . a site of excess closed upon itself, concentrated in its own name' (Nora, 1989, pp. 23–24).

Archival memories contrast with 'real environments of collective memory' (*mileux de memoire*) associated with ancient worlds and contemporary communities grounded in relatively traditional practices, rituals and social life (Nora, 1989, p. 7). In the narrative below, a Rapa Nui elder woman in her fifties reflects upon some of the everyday struggles of families to retain their *mileux de memoire* against forces that conduct Rapa Nui towards archival memories of the state and its national park. She relates Rapa Nui resistance to a state plan to move a sacred stone named Pū ʻOhiro from a location near a remote dirt road that park management considered vulnerable to damage, in order to better conserve it as a tourist site in Chile's Easter Island. Conflict over the proper place of Pū ʻOhiro discloses some of the micropolitics circumscribing the everyday care of Rapa Nui material culture, livelihood and subjectivity:

... A Taha Roa kona eʻa o tupuna ki ruŋa o te vaka. ʻI ira e ai ro ʻana etahi mana tōʻona iŋoa ko Pū ʻOhiro, he maʻea hakauŋa i te roŋo ki te taŋata henua. Ra kona era ko Taha Roa kona eʻa hī i muʻa ʻana o te hora tahito era. Te tupuna he eʻa era ki ruŋa i te vaka hī mai i te kai mo haŋai o te taŋata; he kahi ʻana; he ika ʻana. ʻI ira i tapaʻo ai i te mana ko Pū ʻOhiro mo te roŋo o te henua. ʻI puhi era i a Pū ʻOhiro, he tuʻu ata ki ruŋa i te Motu Nui ko te aro era o te Hau Moana; ki tuʻa era o Poike ki Kava Kava Kioʻe ki Motu Maratiri; ki ʻAna te Pahu te roa o te puhi ka oho era i te roŋo. ʻA Pū ʻOhiro o te Mata era Ko te Ure o Mokomai . . . . I aŋarina e ai ro ʻana etahi taureʻareʻa tōʻona iŋoa ko ʻG'. Taʻe oʻona te mata ko Ure o Mokomai. Ko tuʻu mai ʻana te taŋata paʻari ko ʻT' tōʻona iŋoa, he taŋata tere vaikava . . . He ariki o Haŋa Piko. Tōʻona taŋi: pehē taŋata tere vai kava ʻina ko haka makenu i te maʻea ko Pū ʻOhiro mai tōʻona kona tupuna ki te rua aro. Peira o kua rāua i vānaŋa ai e Haŋa Piko te taŋata tere vai kava mo oho mai mo kī a aʻaku mo vānaŋa ararua ko ʻG' mo taʻe toʻo i te mana ke ki te rua aro o rāua te manaʻu nei. Ka hakare a Pū ʻOhiro tōʻona kaiŋa ko Taha Roa o ira a ia. O ira tōʻona kuhane o ira tōʻona varua. ʻE mo hakamau ka noho a ʻG', he poki āpī, ka rararama ka ʻuiʻui ki te taŋata paʻari o te kaiŋa nei ko te Pito o te Henua. Ka ʻui ki tōʻona nua ki tōʻona tupuna tetere mau o tōʻona mata. Ko ai a ia? ʻO aro hē a ia? Pehē makupuna? Pehē hinarere? ʻO koro era ko ʻK' ʻe ko nua era ko ʻM'. He raʻe ana ite: Ko ai tōʻona kuhane? Ko ai tōʻona varua? ʻAi ka manaʻu rō peinei ē; ina a ʻG' ko tuʻu ki te aro ke, mata ke a Ure o Mokomai. Ka ʻui raʻe ki kua Hitoraŋi ʻe ki te Tupa Hotu Rikiriki.

... Taha Roa is a land area from which the ancestors embarked upon in canoes. There [at Taha Roa] is something with 'spiritual power' (mana) named Pū ʻOhiro, it is a stone used to send messages to people of the island. That area Taha Roa is an area that was fished in front of in ancient times. Ancestors went out from there upon canoes to fish for food to feed the people: fish for tuna and other fish. There, the spiritual power of Pū ʻOhiro for [sending] messages to the island is significant. The sounds blown at Pū ʻOhiro, constituted messages that traveled far; they arrived from the islet Motu Nui to the kin region of Hau Moana; to behind Poike at Kava Kava Kio`e to islet Motu Maratiri; and to Pahu cave. Pū `Ohiro belongs to the clan area named Ure o Mokomai . . . Today there is a [Rapa Nui] youth named 'G'. Ure of Mokomai is not the clan area of this youth. The elder Rapa Nui named 'T', an open-ocean fisherman . . . traversed here . . . He is a leader at Haŋa Piko [the principal fishing dock of Rapa Nui]. His cry: open-ocean fishermen cannot allow the Pū `Ohiro stone to go from its ancestral place to a second kin area. Hence, their group, the open-ocean fishermen at Haŋa Piko discussed to go and to tell me to discuss with 'G' that their thought [the fishermen] is to not take the spiritual power [of Pū `Ohiro] to a second kin area. Pū `Ohiro must remain in its territorial land of Taha Roa. There is its ancestral spirit, its cosmic being. And it would be correct that 'G', a new child, wait, and search, ask the elders of this land Te Pito o te Henua. The youth should question the mother and grandparents. Who is she? Of what kin area? How will it be for the 'grandchildren' (*makupuna*)? How will it be for the 'great grandchildren' (*hinarere*)? The grandfather and the grandmother should know: Whose is its ancestral spirit? Whose is its spiritual being? This is the thought; it is not for 'G' to come to Ure of Mokomai, a different kin and clan area. The first thing is to ask the groups of Hitoraŋi and Tupa Hotu Rikiriki.

Competing knowledge practices and regimes of authority are articulated in this political ecological encounter. Rapa Nui families are shown to mobilize what can be seen as 'the production of biopower from below' (Hardt, 1999, p. 198) against state biopower from above and its strategies of control via 'heritage-power' (Castañeda, 2009, p. 112). Assuming it can be archaeologically understood as simply a portable artefact of a tourist site that the state can officially move within the national park at its discretion, the basic idea of the state proposal was to move the stone to a location where it could be better conserved and accessed by actors in the tourism industry. While on the one hand the story can be read as a simple illustration of 'heritage governmentality' (Castañeda, 2009, p. 117), the struggle over the consciousness of 'G' in the conflict also discloses how the dispute is entangled in governmentality of Rapa Nui society and subjectivity. The narrator and 'T' are in a fight not only against archaeology as a 'technology of government' that attempts to 'legislate' (Smith, 2009, p. 121) Pū `Ohiro into cultural heritage for enterprises of the state, but as an instrumental relay of 'relations of subjection' that 'fabricate subjects' (Foucault, 1980, p. 15).

Pū `Ohiro is entangled not simply in a representational politics of its significance, but within the biopolitics of maintaining an 'affective atmosphere' (Anderson, 2009) where Rapa Nui people can build relationships and a society enculturated in terms of kinship. 'G' is employed by the park, but is actually of Rapa Nui descent. To stop the movement of the stone, 'T' and the narrator feel compelled to deactivate the colonial mind of 'G' and teach her the fabric of kinship relations that anchor Pū `Ohiro to a specific familial place. Like the story of Hotu Matu`a, the genealogical work of this narrative situates the sacred stone Pū `Ohiro within a complex 'spiritual ecology' (Sponsel, 2011, p. 37). The narrator analyzes 'G' in terms of Rapa Nui kinship, rather than in terms of the offices and titles constructed by the discourse of Chilean bureaucracy where 'G' is of higher standing than both the narrator and 'T'. In doing so, she reduces the status of 'G' to the kinship status of an 'adolescent' (*poki āpī*) who is to wait and seek the opinion of 'elder Rapa Nui' (*taŋata pa`ari*), specifically, 'G''s mother ('*tō`ona nua*') and other elders responsible for managing the clan of 'G' ('*tō`ona tupuna tetere mau o tō`ona mata*'). Their concern is not how to conserve the exchange-value of Pū `Ohiro in the interests of the aesthetic communities of archaeology, the Chilean state and international tourism. Rather, they are preoccupied with preserving Pū `Ohiro for future 'grandchildren' (*makupuna*) and 'great grand-children' (*hinarere*), as well as contemporary Rapa Nui fishermen, who find affective spiritual and symbolic use-value of the stone within the *oikonomia* of Taha Roa.

Questions of the ecological sustainability of tourism are clearly real and serious for Rapa Nui, but they are not new. For example, Petra Campbell (2006), a founder of the International Help Fund Australia, independently initiated in 2004 a study of tourism sustainability in Rapa Nui that challenged the Chilean state to address what she perceives to be a looming 'environmental catastrophe'. Campbell predicted future aquifer contamination due to the state's failure to develop modern infrastructure for treatment of toxic raw sewage and other waste overflowing in poorly-provisioned landfills, continued land degradation from long-term erosion and mismanagement, marine resource destruction and a range of energy failures. While her results did stimulate Chile to initiate its own studies into the potential problems, Campbell was more successful in garnering some support from French and Australian government agencies (though not from Chile) for implementing her proposed multi-million dollar sustainability plan (Campbell, 2008, pp. 52–53). If it is the case that the Chilean state is now truly going to engage potential ecological collapse in terms of real strategies one might reasonably ask why Chile is suddenly interested in sustainable tourism now, many years after the dire warnings surfaced. For a diversity of reasons, environmental crises have acquired 'a sort of moral high ground to stand on when making decisions about local peoples and the global environment' (West, 2006, p. 32). In other words, global biopolitical priorities tend to defer the ethical and ontological goals of Indigenous self-determination projects (Byrd, 2013, p. 221). Internationally, the Chilean state lost moral ground as numerous human rights organizations and representatives of various nation states condemned its treatment of Rapa Nui during the height of the 2010–2011 conflicts. For example, US Congressman Eni Faleomavaega of American Sāmoa condemned Chile's actions on the congressional floor (US Congress, 2011), and United Nations special Rapporteur on Indigenous Affairs, James Anaya, called for UN-led investigations into excessive use of force, along with Amnesty International (Young, 2012b, p. 197). On the one hand, by highlighting its engagement with the global priority of reducing ecological crises, Chile may be hoping to divert attention from its violence against Rapa Nui and thereby amplify its moral status on the global stage as a progressive green state.

In the context of Chile's history in Rapa Nui, a state plan of sustainable tourism in Chile's Easter Island, absent of any concurrent plan for restoring Rapa Nui territorial rights and self-determination as has been internationally recommended within the Indigenous human rights community, can be read as a plan for a civil order that is 'fundamentally a battle-order' (Foucault, 1980, p. 16). Santi Hitorangi, a leader of his family against the hotel, the Chilean state and the globalization of Rapa Nui, emphasized at the United Nations Church Center, attended by approximately 100 attendees of the UN Permanent Forum on Indigenous Issues in 2014, that Rapa Nui 'remains a colony of Chile' and that 'Chile keeps Rapa Nui by guns' (Young, 2015a, p. 289). The 2010–2011 conflict, as well as the more recent armed plebiscites for Vaitea lands, serve to substantiate Hitorangi's assessment. This chapter also reveals that tourism has become a powerful form of control of Rapa Nui. Settler colonialism, as Patrick Wolfe (2008, p. 103) stated, is 'a structure rather than an event' and it applies an incessant 'logic of elimination' towards the 'summary liquidation of Indigenous people'. The presence of the *dispositif* of Chile's Easter Island works daily against the *oikonomia* of Rapa Nui subjectivity in bodies, lands and minds. If, following Edward Said (1983, p. 8), one considers that, for those who live it, culture is first and foremost about 'belonging to or in a place, being at home in a place', then it is clear that Rapa Nui will never be comfortably home in Chile's Easter Island. As is commonplace in the biopolitics of settler colonialism (Rifkin, 2014, pp. 150–151), Chile reproduces Rapa Nui as a 'population' to be individuated, manipulated and assimilated into its own settlement plans, rather than as a nation of peoplehood with a right to determine its own way of life. For the Chilean state, the actants of Rapa Nui living material culture are merely things to conserve for tourist exchange-value. Rapa Nui deactivated 'G' and saved Pu`Ohiro from displacement, but they will constantly have to battle the next 'G' for heritage-power over the next cord of Rapa Nui patrimony, and a new state plan to privatize the *kāiŋa* of their revered Hotu Matu`a. George Marcus (2011) has argued that 'crisis talk' normalizes and legitimates the situations that precede a crisis. Foregrounding ecology in Rapa Nui as 'the crisis' for the Chilean state to address disavows the crisis of the unstable place of th Rapa Nui Nation in Chile's Easter Island.

Tourism remains a powerful 'mode of governmentality' (Burtner and Castañeda, 2010, p. 2). In Hawai`i, the moral economy of the rising sense of the sustainability of theme park tourism in Pearl Harbor has been importantly questioned (White, 1997). Yet critical attention to the absence of Pu`uloa, the Indigenous name for Pearl Harbor, and Kanaka Maoli from this political ecology highlights a different kind of problematization: namely the erasure of Indigenous history, people and politics from tourism of Pu`uloa and US national memory (White 2014). Similarly, given the blatant absence of 'Indigenous Fijians' (Itaukei) from any marketing imagery of 'Fiji Water' (*wai ni tuka*), a development project with a history of Indigenous conflict (Kaplan, 2011, p. 230), one wonders about the efficacy of any simple political ecological analysis promoting the sustainability of the transnational commodification of Fiji Water. Antonio Negri (1999, pp. 87–88) tells us that 'the sublime has become normal' within the 'economy of desire' dominant today, and that it is 'on the basis of affect the enemy must be destroyed'. What is one to make of the transnational governmentality of increasingly common 'hollow ecologies' (West, 2012, p. 236) of Pacific Islands worlds of sublime *moai, wai ni tuka* and Pu'uloa emptied of their Indigenous atmospheres of affect? As elaborations of the nineteenth-century Western plan represent Egypt within 'the Europe of department stores and world exhibitions' (Mitchell 1991, p. 15), are these simply pathetic signs of a *dispositif* that 'long ago lost its soul' and 'carnivalized itself' (Baudrillard, 2010, pp. 9–10)? Or are they colonial signs of a 'continuation of war by other means' (Foucault, 2003, p. 15) that, like the Musée du Quai Branly itself, affects a sense of world cultural diversity while at the same time excluding the non-Western world from its regime of 'true' humanity (Dias, 2008, p. 302)? Whatever the answer to this question, Rapa Nui and other Indigenous peoples of the world, have increasingly moved beyond waiting for 'the politics of recognition' in order to engage in a 'politics of refusal' (Simpson, 2014, pp. 11–12) of the moral economy of what colonial states consider officially settled. The Rapa Nui will never be at home in Chile's Easter Island, but neither will the political ecology of tourism in Chile's Easter Island ever be at home in Rapa Nui. As writing of this chapter concluded, on 26 March 2015 leaders of the Rapa Nui Nation shut down the Chilean-administered national park of Easter Island, blocking access roads with felled trees and reterritorializing entrances signs to 'tourist sites' as places denoting 'ancestral property'. Chile has called in the riot police (Young, 2015b).

Notes

1 The proper noun 'Rapa Nui' is the Polynesian name for 'Easter Island' and also the name of its Indigenous language and people. In international politics (United State Congress, 2011) and law (Crippa, 2014) Rapa Nui is also referred to a nation.
2 The political ecology of 'counter-work' emphasizes local knowledge, ontologies, and values produced in conflict with global development programmes that enact 'alternative modernities' and 'decolonial configurations of nature, culture, and economy' (Escobar, 2008, p. 10).
3 I use the notion of 'moral economy' in terms of recent anthropological work (see Fisher, 2014) that extends the seminal writings on the concept from Max Weber, Karl Polanyi, Michel Foucault and James Scott, among others.
4 Unlike the Nietzschean genealogy developed by Foucault that applies genealogy as an alternative to epistemology and to dissipate rootedness and identity (Foucault 1977, p. 162), Polynesian genealogical work is an epistemological practice for rooting identity in one's ancestral relationships, as can be induced from examples in the chapter.

References

Agamben, G. (1993) *Stanzas: Word and Phantasm in Western Culture*, Trans. R. L. Martinez. Minneapolis, MN: University of Minnesota Press.
Agamben, G. (2009) *What is an Apparatus? And Other Essays*. Stanford, CA: Stanford University Press.
Anderson, B. (2009) Affective Atmospheres, *Emotion, Space, and Society*, 2, pp. 77–81.

Baudrillard, J. (2010) *Carnival and Cannibal*. London: Seagull Books.
Bennett, J. (2010) *Vibrant Matter: A Political Ecology of Things*. Durham, NC: Duke University Press.
Burtner, J. and Castañeda, Q. (2010) Tourism as a Force of World Peace: The Politics of Tourism, Tourism as Governmentality, and the Tourism Boycott of Guatemala, *The Journal of Tourism and Peace Research*, 1 (2), pp. 1–21.
Byrd, J. (2013) *The Transit of Empire: Indigenous Critiques of Colonialism*. Minneapolis, MN: University of Minnesota Press.
Campbell, P. (2006) Easter Island: On the Verge of a Second Environmental Catastrophe, *Rapa Nui Journal*, 20 (1), pp. 67–70.
Campbell, P. (2008) Easter Island: A Pathway to Sustainable Development, *Rapa Nui Journal*, 22 (1), pp. 48–53.
Castañeda, Q. (2009) Notes on the Work of Heritage in the Age of Archaeological Reproduction. In: Morteson, L. and Hallowell, J. (eds) *Archaeologists and Ethnographies*. Gainsville, FL: University of Florida Press, pp. 109–119.
Cole, J W. and Wolf, E. R. (1999) *The Hidden Frontier: Ecology and Ethnicity in an Alpine Valley*. Berkeley, CA: University of California Press.
Connell, J. and Rugendyke, B. (2008) Tourism and Local People in the Asia-Pacific Region. In: Connell, J. and Rugendyke, B. (eds) *Tourism at the Grassroots: Villagers and visitors in the Asia-Pacific*. New York: Routledge, pp. 1–40.
Crippa, L. A. (2014) Te Pito Te Henua: The Inspiring Rapa Nui Nation's Efforts to Rebuild its Government and Regain Control of its Territory, *Griffith Journal of Law & Human Dignity*, 2 (2), pp. 247–264.
Debord, G. (1994) *The Society of the Spectacle*. New York: Zone Books.
Delsing, M. R (2009) *Articulating Rapa Nui: Polynesian Cultural Politics in a Latin American Nation-State*. Unpublished PhD Thesis. University of California-Santa Cruz.
Dias, N. (2008) Double Erasures: Rewriting the Past at the Musée du Quai Branly, *Social Anthropology*, 16 (3), pp. 300–311.
Englert, P. S. (2006) *Legends of Easter Island*. Hanga Roa: Rapa Nui Press.
Escobar, A. (2001) Culture Sits in Places: Reflections on Globalism and Subaltern Strategies of Localization, *Political Geography*, 20, pp. 139–174.
Escobar, A. (2008) *Territories of Difference: Place, Movements, Life, Redes*. Durham, NC: Duke University Press.
Feld, S. and Basso, K. H. (eds) (1996) *Senses of Place*. Santa Fe, NM: School of American Research Press.
Fisher, E. F. (ed.) (2014) *Cash on the Table: Markets, Values, and Moral Economies*. Santa Fe, NM: School for Advanced Research Press.
Fisher, S. R. (2005) *Island at the End of the World: The Turbulent History of Easter Island*. London: Reaktion Books.
Foucault, M. (1977) Nietzsche, Genealogy, History. In: Bouchard, D. F. (ed.) *Language, Counter-memory, Practice: Selected Essays and Interviews*. Ithaca, NY: Cornell University Press, pp. 139–164.
Foucault, M. (1980) War in the Filigree of Peace, *The Oxford Literary Review*, 4 (2), pp. 15–19.
Foucault, M. (2003) *Society Must be Defended: Lectures at the College de France 1975–1976*. New York: Picador.
Foucault, M. (2008) *The Birth of Biopolitics: Lectures at the Collège de France 1978–1979*. New York: Picador.

Gobierno de Chile (2008) *Informe de la Comisión de Verdad Histórica y Nuevo Trato con los Pueblos Indígenas*. Santiago: Gobierno de Chile.

Goldberg-Hiller, J. and Silva, N. K. (2011) Sharks and Pigs: Animating Hawaiian Sovereignty against the Anthropological Machine, *The South Atlantic Quarterly*, 110 (2), pp. 429–446.

Hardt, M, (1999) Affective Labor, *Boundary 2*, 26 (2), pp. 89–100.

Harrison, D. (2003) Themes in Pacific Island Tourism. In: Harrison, D. (ed.) *Pacific Island Tourism*. New York: Cognizant Communication Corporation, pp. 1–23.

Harvey, D. (2004) The New Imperialism: Accumulation by Dispossession, *Socialist Register*, 40, pp. 64–87.

Haun, B. (2008) *Inventing 'Easter Island'*. Toronto: University of Toronto Press.

Hito, S. (2004) Vaai Hanga Kainga. Giving Care to the Motherland: Conflicting Narratives of Rapanui, *Journal of Intercultural Studies*, 25 (1), pp. 21–34.

Hitorangi, M. (2013) The Unknown Truth Behind the Moais. *IC Magazine*. Available at: https://intercontinentalcry.org/the-unkown-truth-behind-the-moais [Accessed 30 June 2014].

Huki, P. (1988) Prefecio. In: El Consejo de Jefes de Rapa Nui, Hotus, A. (eds) *Te Mau Hatu 'O Rapa Nui: 'Los Soberanos de Rapa Nui' Pasado, Presente y Futuro*. Santiago: El Centro de Estudios Politicos Latinamericanos Simon Bolivar, pp. 9–12.

International Work Group for Indigenous Affairs (2012) *The Human Rights of the Rapa Nui People on Easter Island: IWGIA Report 15*. Copenhagen: International Work Group for Indigenous Affairs.

Kaplan, M. (2011) Alienation and Appropriation. Fijian Water and the Romance in Fiji and New York. In: Hermann, E. (ed.) *Changing Contexts, Shifting Meanings: Transformations of Cultural Traditions in Oceania*. Honolulu, HI: University of Hawai'i Press, pp. 221–334.

Long, G. (2014) Trouble in Paradise for Chile's Easter Island. British Broadcasting Corporation, 17 April. Available at: http://www.bbc.com/news/world-latin-america-26951566 [Accessed 30 June 2014].

Los Angeles Times Travel, December 29, 2013. Available at: http://articles.latimes.com/keyword/easter-island/featured/4 [Accessed 22 February 2015].

Marcus, G. (2011) The Stakes of 'Crisis', *American Anthropological Association Annual Meeting*, Montreal, Canada.

McCall, G. (1995) Japan, Rapanui, and Chile's Uncertain Sovereignty, *Rapa Nui Journal*, 9 (1), pp. 1–7.

McCall, G. (2008) Another (unintended) legacy of Captain Cook? The Evolution of Rapanui (Easter Island) Tourism. In: Connell, J. and Rugendyke, B. (eds) *Tourism at the Grassroots: Villagers and visitors in the Asia-Pacific*. New York: Routledge, pp. 41–57.

McCall, G. (2010) Thuggery on Rapa Nui: Chilean Parliamentarian Reactions. *Pacific Scoop. On-line News*. Wellington, New Zealand. Available at: http://pacific.scoop.co.nz/2010/12/thuggery-on-rapanui-chilean-parliamentarian-reactions/ [Accessed 22 February 2015].

Mitchell, T. (1991) *Colonizing Egypt*. Berkeley, CA: University of California Press.

Negri, A. (1999) Value and Affect, *Boundary 2*, 26(2), pp. 77–88.

Nora, P. (1989) Between Memory and History: Les Lieux de Memoire, *Representations*, 26, pp. 7–25.

Pereyra-Uhrle, M. (2005) Easter Island Land Law, *Yearbook of the New Zealand Association for Comparative Law*, 11, pp. 133–142.

Porteous, J. D. (1981) *The Modernization of Easter Island*. Victoria, British Columbia: University of Victoria Press.

Porteous, J. D. and Shephard-Toomey, T. (2005) Resistance and Land Control on Rapa Nui, *Rapa Nui Journal*, 19 (1), pp. 10–12.

Rancière, J. (2008) Aesthetic Separation, Aesthetic Community: Scenes from the Aesthetic Regime of Art, *Art & Research: A Journal of Ideas, Contexts, and Methods*, 2 (1), pp. 1–15.

Rifkin, M. (2014) Making People into Populations: The Racial Limits of Tribal Sovereignty. In: Simpson, A. and Smith, A. (eds) *Theorizing Native Studies*. Durham: Duke University Press, pp. 149–187.

Robbins, P. (2004) *Political Ecology*. Oxford: Blackwell Publishing.

Said, E. W. (1983) *The World, the Text, and the Critic*. Cambridge, MA: Harvard University Press.

Said, E. W. (1994) *Orientalism*. New York: Vintage Books.

Shapiro, M. J. (1997) *Violent Cartographies: Mapping Cultures of War*. Minneapolis, MN: University of Minnesota Press.

Shephard-Toomey, T. (2001) The Development of Rapa Nui (Easter Island), Chile 1967–2001, *Rapa Nui Journal*, 15 (2), pp. 110–113.

Simpson, A. (2014) *Mohawk Interruptus: Political Life across the Borders of Settler States*. Durham, NC: Duke University Press.

Smith, L. (2009) Theorizing Heritage: Legislators, Interpreters, and Facilitators. In: Morteson, L. and Hallowell. J. (eds) *Archaeologists and Ethnographies*. Gainsville, FL: University of Florida Press, pp. 120–130.

Sponsel, L. E. (2011) The Religion and Environment Interface: Spiritual Ecology in Ecological Anthropology. In: Kopnina, H. and Shoreman, E. (eds) *Environmental Anthropology Today*. New York: Routledge, pp. 37–55.

Stanton, M. (2000) I am not a Chileno! Rapa Nui Identity. In: Spickard, P. and Burroughs, W. (eds) *We are a People: Narrative and Multiplicity in Constructing Ethnic Identity*. Philadelphia, PA: Temple University, pp. 142–150.

Stanton, M. (2003) Economics and Tourism Development on Easter Island. In: Harrison, D. (ed.) *Pacific Island Tourism*. New York: Cognizant Communication Corporation, pp. 110–124.

Tengan, T. K., Ka`ili T. O. and Fonoti, R. T. (2010) Genealogies: Articulating Indigenous Anthropology In/Of Oceania, *Pacific Studies*, 33 (2/3), pp. 139–167.

United Nations (2009) *State of the World's Indigenous Peoples*. New York: Department of Economic and Social Affairs, United Nations.

United States Congress (2011) *House Congressional Record*. February 16, 2011, pp. H947–H948.

Vergara, V. M. (1939) *La Isla de Pascua: Dominación y Dominio*. Memoria de Prueba, Facultad de Ciencias Jurídicas y Sociales de la Universidad de Chile. Santiago, Chile.

Vizenor, G. (1998) *Fugitive Poses: Native American Indian Scenes of Absence and Presence*. Lincoln, NE: University of Nebraska Press.

Wendt, A. (2000) Inside Us the Dead. In: Borofsky, R. (ed.) *Remembrance of Pacific Pasts: an Invitation to Remake History*. Honolulu, HI: University of Hawai`i Press, pp. 35–42.

West, P. (2005) Translation, Value, and Space: Theorizing an Ethnographic and Engaged Environmental Anthropology, *American Anthropologist*, 107 (4), pp. 632–634.

West, P. (2006) *Conservation is Our Government Now: The Politics of Ecology in Papua New Guinea*. Durham, NC: Duke University Press.

West, P., Igoe, J. and Brockington, D. (2006) Parks and Peoples: The Social Impact of Protected Areas, *Annual Review of Anthropology*, 35, pp. 251–277.

White, G. (1997) On Not Being a Theme Park, *American Anthropological Association Annual Meeting*, Washington DC.

White, G. (2006) Landscapes of Power: National memorials and the domestication of affect, *City & Society*, 18 (1), pp. 50–61.

White, G. (2014) (De)Mystifying Tourism Studies, *American Anthropological Association Annual Meeting*, Washington DC.

Willems-Braun, B. (1996) Buried Epistemologies: The Politics of Nature In (Post) Colonial British Columbia, *Annals of the Association of American Geographers*, 87 (1), pp. 3–31.

Wolfe, P. (2008) Structure & Event: Settler colonialism, time, and the question of genocide. In: Moses, A. (ed.) *Empire, Colony, Genocide: Conquest, Occupation, and Subaltern Resistance in World History*. New York: Berghahn Books, pp. 102–132.

Woodbury, A. C. (1998) Documenting Rhetorical, Aesthetic, and Expressive Loss in Language Shift. In: Grenoble, L. and Whaley, L. (eds) *Endangered Languages: Language loss and community response*. Cambridge: Cambridge University Press, pp. 234–258.

Young, F. W. (2011) *Unwriting 'Easter Island': Listening to Rapa Nui*. Unpublished PhD Thesis, University of Hawai'i at Manoa.

Young, F. W. (2012a) 'I Hē Koe? Placing Rapa Nui, *The Contemporary Pacific*, 24 (1), pp. 1–30.

Young, F. W. (2012b) Rapa Nui: Polynesia in Review, Issues and Events, 1 July 2010 to 30 June 2011, *The Contemporary Pacific*, 24 (1), pp. 190–199.

Young, F. W. (2013) Rapa Nui: Polynesia in Review, Issues and Events, 1 July 2011 to 30 June 2012, *The Contemporary Pacific*, 25 (1), pp. 172–183.

Young, F. W. (2014) Rapa Nui: Polynesia in Review, Issues and Events, 1 July 2012 to 30 June 2013, *The Contemporary Pacific*, 26 (1), pp. 214–225.

Young, F. W. (2015a) Rapa Nui: Polynesia in Review, Issues and Events 1 July 2013 to 30 June 2014, *The Contemporary Pacific*, 27 (1), pp. 281–293.

Young, F. W. (2015b) Indigenous Rapa Nui Shut Down Easter Island's 'Tourist Sites'. Pacific Islands Report. Available at: http://pidp.eastwestcenter.org/pireport/2015/April/04–22-07.htm [Accessed 30 June 2014].

7 Rethinking ecotourism in environmental discourse in Shangri-La

An antiessentialist political ecology perspective

Jundan (Jasmine) Zhang

Introduction

If ever you have been to Shangri-La, you will understand that it is one of those places where there are widespread expectations that tourism will improve the local economy. In 2013 I made my fifth trip to Shangri-La. Much had changed since I first visited in 2007. Upon walking out from the gate of Shangri-La airport early one morning in April, I was surrounded by dozens of taxi drivers. 'Zhuoma[1], do you need a cab?' 'Zhuoma, where do you want to go?' The voices of at least 20 drivers were directed at me, all young men who had clearly been waiting for some time to get their first fare of the day. They drove their private jeeps rather than hiring a registered taxi. A brief moment of indecision was stretched out longer by my mixed feelings, thoughts and questions about who these men were and where were they were from. The crowd became impatient.

'Just pick one, Zhuoma!' 'It doesn't matter! We are all the same! Same price!' 'Stop thinking, just pick one!' Compelled by their imploring me to 'pick' one, I did. 'OK, I'll go with you', I said to one young man wearing *chuba*, the traditional Tibetan clothes. The rest of the crowd quickly moved on. The young driver I selected to ride with took my backpack and started chatting: 'Zhuoma, your bag is quite heavy . . . we have many tourists like you nowadays . . . Here, take this'. He handed over his name card. 'We have the most beautiful views in the world. Welcome to Shangri-La!'

Fog arose from a meadow in the nippy morning air as we headed toward the center of town, located a 20-minute drive from the airport. I had a quick look at the driver's business card. It was professional, but I was most impressed by the back of the card (Figure 7.1), which presented scenic photos of locally well-known places around Shangri-La. I asked about his taxi business. He told me that he was from a Tibetan village near the airport. Four years ago he began working as a taxi driver and, occasionally, a tour guide. 'What about your farmland?' I asked, 'We rent it out. People rent it to build guesthouses. It's just easier that way'.

Figure 7.1 Images of Shangri-La.
Photo by the author.

Although the transformation in land use and off-farm labor has been occurring in China since the 1978 economic reforms (Chen and Davis, 1998), my conversation with the young driver reflects a relatively recent phenomenon of ecotourism in remote parts of China. Growing numbers of relatively remote areas like Shangri-La are marketed as tourism destinations, whereby they essentially 'sell' their nature and culture (Gao et al., 2009).[2] The meanings of cultural, social and geopolitical landscapes have dramatically changed as a result of this economic transition. From previous visits I knew that the places represented on the driver's business card were sacred sites that once were strictly guarded, but in recent years have increasingly been packaged into sightseeing opportunities for nature-oriented tourists. Along with other disciplines, cultural geographers and environmental anthropologists have made several calls for an integrative framework to study the interrelations among economic, sociocultural and environmental issues (Crumley, 2002; Mitchell, 2000). Political ecology emerged from this context both as a theoretical framework and epistemological approach by which to examine nature-society relations, with an emphasis on power relations in human-human and human-nonhuman interactions (Robbins, 2012). Within tourism studies, political ecology has been employed as a critical tool in examining the relationship between tourism and environmental governance and management (Cole, 2012; Stonich, 1998).

Drawing on my fieldwork between 2007 and 2013 in Shangri-La and its nearby areas,[3] this chapter examines Shangri-La's tourism development and environmental changes with respect to theoretical debates about how we might employ political ecology to delimit and redirect our inquiries surrounding nature and society. This will be accomplished in three parts. First, I situate Shangri-La's tourism development within relevant historical environmental changes. Closely related with development policy after the establishment of the People's Republic, China's forestry and land policies have undergone various phases addressing different foci at each period. Tourism, and in particular ecotourism in central Shangri-La, came into play in this context and has since been publicly identified as the sustainable development industry that is to replace the timber industry. Viewing Shangri-La's tourism development through a historical lens not only draws our attention to the discourses of development and sustainability, but also the dynamic interactions and frictions generated in ecotourism and conservation. For instance, from a political ecology perspective, it is also important to understand how the taxi driver as an individual has dealt with the recent changes in his living environment. The second section therefore introduces Escobar's (1999) antiessentialist political ecology, a framework that describes biological and historical articulations of nature in three loosely interconnected regimes. I argue that ecotourism needs to be situated in light of Escobar's notion of 'hybrid natures', through which beliefs about 'capitalist, organic and techno natures' can be held by individuals, in tension, simultaneously. In this way, I extend discussions around the conservation and control of natural resources in political ecology (Robbins, 2012) to questions such as: What is nature? Who has access to nature? and How may this access change over time, in the context of tourism development? Finally, I conclude by suggesting that analyses of ecotourism can shift focus from the dichotomy of culture and nature to an 'environmental discourse' that is at once cultural, biological and political. Rethinking ecotourism in Shangri-La through an antiessentialist political ecology helps us to understand that (1) ecotourism in Shangri-La results from transforming relationships between humans and 'nature'; (2) ecotourism in Shangri-La, rather than a form of commoditized nature, is an embodiment of hybrid natures; and (3) consequently, tourism can contribute to emerging work in the political ecology of tourism by continuously unsettling and re-questioning predetermined concepts such as 'nature' and 'culture'.

Ecotourism in Shangri-La

Ecological and economic background

Since 2001, when the central government announced that Zhongdian County in Diqing Tibetan Autonomous Prefecture in Yunnan Province is the 'real location' of the mythical Shangri-La in *Lost Horizon*, tourism, as expected, has grown into a major economic generator in the region (Allen, 2001; Glyn and Alistair, 2012; Hillman, 2003a, b; Hu, 2005; LePage, 1996; Llamas and Belk, 2011). The namechange was not simply fortuitous, and needs to be examined in a socioeconomic, political and ecological context. It is therefore useful to look into the ecological and economic background of Shangri-La. As we will see from the historical articulation of Shangri-La's ecological and sociocultural changes, 'nature' has played a central role in Shangri-La's tourism development and leads it in the direction of ecotourism.

Shangri-La lies in the core areas of The Three Parallel River Protected Area, a natural World Heritage Site (UNESCO, 2003). The three big rivers, Mekong, Salween and Yangtze, run parallel to each other and Shangri-La sits on the upper part of the Yangtze River before it turns East and, together with Yellow River, nurtures the vast plain of Eastern China. Altitude within the prefecture varies from 1,500 metres in deep gorges to over 6,100 metres in the great ranges and snowy mountains. The diverse topography and elevation create numerous microclimates within the region, producing an enormous diversity of fauna and flora. A consequence of the topography of Diqing Prefecture is that arable land is very limited, amounting to only 6 percent of the overall agricultural area. The importance of forestry as a primary economic activity in Diqing Prefecture is therefore apparent. Forestry policy, which has undergone some transformations in the last few decades, along with land reform in rural development, has shaped Diqing Prefecture's economic development profoundly (Wang *et al.*, 2004).

In the early years after the establishment of the Peoples' Republic of China (PRC) in 1949, forest policy focused on tree planting and timber harvesting. The latter was particularly emphasized in Southwest China, where 43.9 percent of the total timber reserves of China are situated (State Forestry Administration, 2000). In the 1950s' land reform and the period of the Great Leap Forward, logging expanded extensively using fuel blast furnaces, leading to massive forest destruction. In the 1960s, forests in Northwest Yunnan were opened to forestry industries and logging grew rapidly, supplying raw materials for national economic development and the recovery from the Great Leap Forward and the Great Famine. Ecosystems and wildlife habitat attached to these primeval forests were severely damaged (Richardson, 2000). In 1978, the Ministry of Forestry (later known as the SFA) was formed and with the land-tenure reform in place the same year, peasants were contracted to take responsibility for afforesting wastelands and bare land. This allowed the peasants to access timber and secure additional land for tree planting and intercropping. However, it remains a source of criticism that the forests of Southwest China have never been officially managed on a sustainable basis due to lack of funding and insufficient management (Harkness, 1998; Richardson, 1990). The unstable, ever-changing forestry regulations have overall formed a format of power circulation that peasants and villagers are not able to access (Xu and Ribot, 2004) triggering what is now considered to be 'illegal' logging in various villages in Northwest Yunnan. Especially in places such as Shangri-La, where ethnic groups have their own forest management systems and rituals and customs as well as where utilizing forestry resources are critical livelihood practices, it is (unsurprisingly) difficult to regulate so-called illegal logging.

Zackey (2007) suggests that such illegal logging is a form of political resistance. While I did come across Tibetan villagers who continued to log in prohibited areas and expressed their own opinions toward the logging ban, I also suggest scholars take a more nuanced perspective to interpret such 'political resistance', as it is not only taking form through the existing Tibetan/Han tension, but also other tensions such as peripheral/central and individual difference within the local community.

China's forestry reforms since the 1950s led to the large reduction of actual natural and mature forests in some regions, for instance, forest cover in Yunnan had fallen from 55 percent in the 1950s to 30 percent in 1975 (Rozelle *et al.*, 1998; Winkler, 1998). A recent report shows that only 9 percent of the primeval forest in Yunnan remains (Greenpeace, 2012). Ecosystems in this area, in which forests play a vital role, thus faced a perilous situation. Snow disasters and floods occurred frequently in the 1990s. Among them, the Yangtze River has flooded every two years since the 1980s and the situation worsened significantly between 1994 and 1996 (Studley, 1999). In 1998, the Yangtze River experienced the worst flooding since 1954 in the lowlands as well as the Tibetan plateau, claiming more than 3,650 lives and causing more than US$ 30 billion in damage (Studley, 1999).

Green economy and ecotourism in Shangri-La

It is widely accepted that the 1998 Yangtze flood marked the turning point for the central government's forest management strategy, from one of exploitation to conservation (Richardson, 2000; Studley, 1999; Wang *et al.*, 2004; Willson, 2006; Zhao and Shao, 2002). Several national programs, economic plans and development strategies were launched after the flood. First, in 1999 and 2000, the National Forest Conservation Program (NFCP – normally referred to as the 'logging ban') and the Sloping Land Conversion Program (SLCP) consolidated China's emergence as an *environmental state* and together constitute one of the largest environmental rehabilitation efforts in the world (Yin *et al.*, 2005). Alongside NFCP and SLCP, the number of nature reserves has grown slowly between the 1950s and the 1980s as the forestry and land reform policies fluctuated but has increased abruptly since the early 1990s (Zinda, 2012). Meanwhile, beginning in the 1980s, nature reserves and forest parks have been encouraged as sites to develop nature-based tourism. Up until the end of the 1990s almost all nature reserves were turned into tourism destinations (Ni, 1999). With this strong encouragement from the state there began a tremendous market-directed turn that conservation should couple with ecotourism for income (Han and Ren, 2001; Mu *et al.*, 2007). Indeed, in areas such as Shangri-La, where the natural environment has always been less a subject of protection than a resource to utilize, ecotourism following the logging ban has naturally become the vehicle for resolving the conflicts between regional development and environmental degradation. In a government report of Shangri-La's tourism development, the need to adopt a sustainable development strategy and green economy is clearly articulated:

> From a historical perspective, Diqing Prefecture has for a long time had an unenviable status and even today its social and economic development is still behind other regions. The majority of the populace is still in poverty. Following the international theme of development, Shangri-La ideally should, in the future, eliminate poverty. We must convert ecological and cultural resources into economic benefits for our people . . . We must at the same time balance the relationship between conservation and development, industrialize the ecological construction and ecologicalize the industrial construction. In so doing we will build Shangri-La's *Green Economy*, a *Shangri-La Path* that will lead to sustainable development . . .
>
> (Qi, 2007, my translation)

The path of the green economy is not a path that Shangri-La's economy takes alone. After the formation of the World Commission on Environment and Development and the establishment of the concept of sustainable development in 1987 (Brundtland, 1987), it was proposed that humanity must work together to manage a modern economy that is more sustainable. Based on the concepts of sustainable development and environmental economics, the basic concern of any green economy is how to maintain a non-declining natural wealth while developing economic wealth (Pearce *et al.*, 1989). One of the arguments of the early advocates of green economy is that environmental assets are either undervalued or not valued at all, which leads to inefficient consumption of natural resources as well as environmental degradation (Le Blanc, 2011). Therefore, being able to 'convert ecological and cultural resource into economic benefits for our people' is crucial. Another government report on Shangri-La's tourism development elaborates on how such re-evaluation of environmental (and sociocultural) assets can be made through the Shangri-La Strategy:

> *Shangri-La Strategy* is a sustainable development strategy with strong regional characteristics, which includes the inspirations from the fictional story (in Hilton's book), our Tibetan culture, and modern technology. The main idea of this strategy is to use more sustainable and cleaner resources in order to develop cleaner industries, such as tourism, hydropower and biological products. Within the principles of conserving resources, environment and ecology, we hope the regional economy will grow more rapidly.
>
> (Qi *et al.*, 2000 my translation)

It is important here to note that fiction, Tibetan culture and modern technology are identified as the three primary drivers for the development of Shangri-La's regional economy. The question that emerges from this development strategy is whether these drivers are each capable of facilitating the growth of 'natural wealth' in order to become valuable as various 'environmental assets'. If tourism is one strategy intended to complete the transfiguration, is it because 'the natural and cultural heritage is an unlimited treasure for Shangri-La for development of its tourism industry, a huge amount of invisible capital that blessed the people of

Shangri-La' (Ni and Cao, 2003)? It is this question that I focus on below. Shangri-La's nature reserves and protected areas have become increasingly open to tourists since the late 1990s and, as a result, ecotourism has become an significant part of the green economy and sustainable development initiatives in the area (Diqing Prefecture Tourism Bureau, 2008). Furthermore, it is widely assumed that conservation and tourism can be combined with other leisure activities and thus generate profits. One materialization of this belief is national parks, which were introduced as a new model of conservation and ecotourism (Zinda, 2014). Pudacuo National Park in Shangri-La, which claims to be the first national park in Mainland China, has received attention from scholars since its opening in 2007 (Wang et al., 2012; Zhou and Grumbine, 2011; Zinda, 2014). More recently, 'Shangrilazation' (Hillman, 2009), referring to the transformations Shangri-La's sociocultural and natural landscape have experienced since its tourism development, have been evaluated from a political ecology perspective (Coggins and Yeh, 2014).

An antiessentialist political ecology

As an interdisciplinary field, political ecology has provided a conceptual space in which scholars have been able to integrate various concerns about development and environment. As noted in the introduction to this collection, political ecology frameworks include examinations of society and nature as consequences of the exchange of power and influence. This focus on the political aspects of ecological change departs from an allegedly apolitical ecology approach. Early definitions of political ecology focused on the context of political economy, then combined the ecological concerns (Blaikie and Brookfield, 1987; Greenberg and Park, 1994; Peet and Watts, 1996). This move conceptualized the relationship between society and natural resources in the development of the material landscape as a process involving the production of nature (Smith, 2008). The majority of current studies of Shangri-La's changing landscape have analyzed environmental governance, ranging from national and international commodification and production of ecological knowledge, to new trends of market logics of ecological modernization (Coggins and Yeh, 2014). Often, using this perspective, tourism is characterized as facilitating the production of such environmental governance, in part, through conservation efforts.

This chapter takes a different – and I hope a more inclusive and observant – perspective, not only by viewing the society-environment, conservation-development relationship, but also in rethinking the nature of tourism/ecotourism. This requires us to first see that tourism is more than some sort of container (in the form of an industry) of cultural or social exchanges, or set of capitalist relations. Tourism is a mediator and a process itself, one that never has been and never will be contained within something circumscribed and fixed. This situation is partly because at its core tourism is about human-nonhuman-spatial interaction, which can never really be 'misused' or 'invented' by one party. Second, an antiessentialist political ecology perspective encourages us to renew 'the question of nature'.

Escobar (1999) proposes that in order to foster a more nuanced account of nature-society relations we need to recognize again the post-structuralist elements in political ecology, meaning the analyses of knowledge, institutions, development and ideologies. Nevertheless, employing a purely top-down governance approach may miss the point; nature is historically produced and experienced through discursive practices (Escobar, 1996). I argue that we need to rethink tourism, particularly ecotourism, within the local-global and nature-culture dialectics, rather than taking it merely as an outcome of green governmentality within a neoliberal economic, social and environmental system.

The core issue of the 'question of nature' lies in how to strive for and achieve a balanced or dynamic position that recognizes both the idea that nature is socially constructed and is a product of culture (Guneratne, 2010), as well as the belief that nature exists as an external and biological ordering (Walker, 2005). Holding on to either side of these polarized oppositions does not help to open up dialog between 'meanings' of nature and 'laws' of nature (Rappaport, 1990). It is in this context that I find useful, on antiessentialist grounds, Escobar's (1999) innovative and contentious framework of nature regimes, and thus use this taxonomy to structure this section. Based on his definition of political ecology as 'the study of the manifold articulations of history and biology and the cultural mediations (of) such articulations . . . ' Escobar (1999, p. 3) emphasizes the process of how the biophysical and the historical articulations are translated into, implicated with and produced through each other. By avoiding the common categories of nature, environment or culture, this approach attempts to investigate the historical constitution of 'nature' as 'a complexity of positions and determinations without any true and unchanging essence, always open and incomplete' (p. 3). The three nature regimes developed through such understanding therefore are not essentializing the rational connection between a certain cultural framework and a certain articulation of biology, but rather are relying on 'reassemblages and recombinations of organisms and practices' (p. 5). In the following sections I will situate ecotourism in Shangri-La operating against these regimes. Through presenting how ecotourism in Shangri-La is fit, as well as unfit, for each of the regimes, I address the necessity of understanding ecotourism in 'hybrid natures' instead of locating it in either of these regimes alone.

Capitalist nature

Capitalist nature, Escobar (1999) suggests, is the most conspicuous and well-understood regime of nature. Emerging from post-Renaissance Europe of the late eighteenth century, nature was portrayed as a passive object under the realistic and distant gaze prevalent in landscape art. This gaze was essential for the later development of modern science in the Western world, for it set the rational attitude toward nature, as well as inaugurated the idea of 'man' as an anthropological structure and the foundation of all possible knowledge. Although it was argued that the anthropocentric history of the world started much earlier than the eighteenth century (Russell, 1946), this Enlightenment era and its most influential thinking did provide the final push for creating a modern society where the separation of nature and society is regarded as its basic feature.⁴ In one way, developing Shangri-La into a tourism destination could be regarded as a variation on painting a realist canvas, except that locating the imagery of a paradise or utopia takes in 'real' time and space, and what is subject to the gaze of tourists does not remain focused on one piece of scenery but rather encompasses people and their entire living environment.

Capitalist nature regimes take forms of modern phenomena, such as rational management and organization of knowledge, modern development, planning and demographics, as well as postmodern phenomena, where capital and labor per se are not at stake but rather the code of production is (Escobar, 2011). For the former, nature is examined quintessentially as a commodity, a theme to which Marxist theorists have actively contributed. For instance, Li (2014) describes how the emergence of capitalist relations governed by competition and profit turns everything farmers engage with into commodities. For the latter form, Foucault's (1991) term 'governmentality' is often used to analyze the system of modern society in order to select, simplify, naturalize and normalize certain knowledges over others in order to achieve the result of commodification. Luke (1999) deploys the term 'green governmentality' to explain interlacing sustainable growth and governmental discourses in an effort to enhance national productivity and state power.

In this regime, nature is no longer an external domain but is a stock of capital (Escobar, 2011). At the same time, ideologies and discourses within these historical commodification events redesignate and redirect how nature can be experienced and utilized. A green economy therefore is situated in a capitalist nature regime because it signifies how capitalism is entering an ecological phase (O'Connor, 1993). The environmental changes in Shangri-La are directly related to capitalizing nature as a primary resource in the era of '(d)evelopment is imperative', at the same time indirectly ushered into postmodernity by a broader governance plan that is not only material but also ideological (Bramwell, 2011).

However, the master narrative of any capitalist nature regime can be confronted by the more subtle and dynamic human-nonhuman-human relations. For example, when talking with Tibetan villagers who seemed to have fully embraced the idea that developing tourism is the most ecological and efficient way of sustaining themselves, I was also often confronted with comments such as 'You city people must have too much money so you would spend money on walking in the mountains and looking at farmland', hinting that there are other understandings of nature, capital and leisure. These understandings cannot be reduced and interpreted into capital relations. Therefore we must look into the past and query whether it is legitimate to claim that attaching monetary value to flora and fauna is uniquely a capitalist or Western approach. These questions lead us to the second regime in Escobar's (1999) framework: that of *organic nature*.

Organic nature

Although Escobar (1999) admits his reluctance to use the word 'organic' for its problematic association with other essentialist terms such as 'purity', 'timelessness' and 'wholeness', he nevertheless adopts the word to underscore that how each individual is born as an organism and becomes connected and situated in a society therefore has much in common with how any single organism is created and merged into a larger, organic system. Often represented in the forms of 'local knowledge' or 'indigenous ecological knowledge', the organic nature regime features a 'local model' of nature-society relations different from the capitalistic nature regime, in the way that in this regime nature and society are not separated ontologically (Escobar, 1999). As Escobar summarizes, a local model of nature may exhibit features such as

> specific categorizations of human, social, and biological entities, boundary settings, and systematic classifications of animals, spirits, and plants. It may also contain mechanisms for maintaining good order and balance in the biophysical, human, and spiritual circuits or a circular view of biological and socioeconomic life ultimately grounded in Providence, gods, or goddesses. There may also be a theory of how all beings in the universe are raised or nurtured out of similar principles, since in many nonmodern cultures the entire universe is conceived as a living being with no strict separation between human and nature, individual and community, community and the gods.
>
> (Escobar, 1999, p. 8)

Shangri-La is populated by various ethnic groups, predominantly Tibetan, each with their own ways of interacting within their communities and with the 'natural environment'. Tibetan Buddhism, for example, is still commonly practiced in people's daily lives, especially in rural areas. However there is a lack of knowledge in terms of what is actually occurring in parallel with the seemingly coherent belief system. For instance, a brief browse through the photos and travel blogs shared by travellers who have been to Shangri-La shows that the place is presented as a package of tourism, unspoiled nature and Tibetan Buddhism. This representation brings forward the question, noted by Hillman (2010) in his work, of why the integration of Tibetan Buddhism into the process of developing ecotourism is more officially permitted in Shangri-La. We have learned in capitalist nature regimes of Shangri-La's geopolitical importance for the political economy of China. Therefore the local model interplayed and negotiated with the more universal model is in this case, the regime of capitalist nature. A few lines from the interpretation texts normally presented to visitors on the tour bus in Pudacuo National Park illustrate how an organic nature regime can intersect with a capitalist nature regime:

> Dear visitors, you must be amazed by how well preserved the nature is in our national park . . . today city people can only see such well-conserved and original nature in Pudacuo National Park, in Shangri-La . . . Perhaps you wonder, why? All thanks to Tibetan people who live on this land, their cultural understanding of nature and their caring behaviours towards nature . . . Nature in Shangri-La has nurtured an animistic cosmology and an ancient religion, the

'Bon Religion'.⁵ In this belief system Tibetan people believe that every item in the universe has its own spirit and guard, and humans should always respect them and not disturb them. Because everyone and everything shares the same spiritual source, only by respecting each other can harmony be achieved. In words of today's ecological value, humans will receive nature's revenge and punishment only if we objectify the law of nature and, by contrast, we must obey the objective law of nature in order to receive gifts from nature . . . and this is exactly why nature in Pudacuo National Park is protected so well.

(Pudacuo National Park, 2008)

It is noteworthy that while presenting the 'unique belief system' of Tibetan people and its link to unspoiled nature, the narrative seems to naturally connect to 'today's ecological value' as if it is axiomatically linked. I suggest the reason of such a claim can be put loosely in two streams of understanding: one stream of thinking believes that there is certainly an environmental economy existing within the 'organic nature regime' (cf. Anderson *et al.*, 2005; Netting, 1993; Pan and Yang, 2000); whereas the other stream posits that incorporation between capitalist and organic nature regimes is as a result of interference in the postmodern era (for instance, the introduction of tourism – see Makley, 2014; Martin, 2012). Reflecting on fieldwork in Shangri-La, my position falls somewhere in between. Villagers who initiate ecotourism development take up the task of operating their business in a typical package of transport, accommodation and tour guiding, as well as interpreting their spiritual and cultural landscape into something accessible. This is certainly not an easy task and it sometimes conflicts with other villagers' opinions of how things should be done. However, the significance of this insight is that individuals who are experiencing multiple modes of human–environmental relations do not necessarily need to be converted or transformed into a single fixed mode. While we search for alternative models of dealing with human–environmental relations (Coggins and Hutchinson, 2006), it is also important to see the diversity existing in the seemingly monolithic Tibetan perspective (Hakkenberg, 2008; Yeh, 2014). Thus, possibilities exist for people to be simultaneously premodern, modern and postmodern, or perhaps should not be categorized by these descriptors in the first place (Zhang, 2014). Escobar (1999) clearly explains that the natures of local communities cannot be 'reduced to the inferior manifestations of capitalist nature, nor can they be said to be produced only according to capitalist laws' (p. 7). I fully endorse this point and argue cases from Shangri-La constantly challenge the foundation of judging what is a systematic and coherent Tibetan cosmology/spirituality and what is in the capitalist arena.

Technonature

While capitalist and organic nature regimes mix and mingle with one another, one palpable channel missed in the picture is the methods of communication between the two. How does a villager living deep in the mountain make his ecotourism product known to tourists? How can energy and electricity needs be met in remote places when trying to accommodate the demands of increasing numbers of ecotourists? Answers to these questions are closely linked with the era of technonature: an epoch Escobar (1999) believes is marked by technology as pure antiessentialism. With contemporary technoscience there is a prevailing sense that everything can be recombined, modified and reinvented. Although it may seem to be absolutely on the other side of the organic nature regime, technonature shares the feature of organic nature that it is context-specific and highly localized. For instance, helping people in the Tibetan village of Shangri-La to develop their own recycling system would need to contextualize a certain technology as well as fit into their existing cultural models.

Rather than commodifying nature in the market product sense, technonature erodes the value of the here and now and reinvents a sense of 'virtuality' beyond the limit of concrete space and time, resulting in the increasingly reduced boundaries between the organic and the technological – as in machinery – to become the 'cyborg' (Haraway, 1991). In one way, technonature already has unsettled both organic nature and capitalistic nature regimes and opened up new space for new possibilities of dealing with nature, or the question of nature. In order to imagine the new possibilities for, and new meanings of, diversity, we need to address Escobar's question: 'Do technonatures make possible a new experience of the natural that could facilitate the re-creation of a (different) continuity between the social and the natural?' (Escobar, 1999, p. 11). In many ways, technonature urges us to reimagine 'another nature', thus nature does not need to follow either capitalist or organic regimes. Indeed, the reign of technonature began earlier than we realize. In fact, both development and conservation are results of the rise of technoscience, for it serves as the tool for developing technology and industry, as well as understanding the cost of such development. Endless recombination of nature, culture, body and commodity is the dominant logic in the new technoscience and requires us to recognize the ontological dynamics involved in producing manifold forms of being in the new era (Escobar, 1999). Ecotourism, I argue, has emerged as both the result of such recombinations as well as part of the drive for an ongoing process of engagement. The struggles between the capitalist nature regime and the organic nature regime have provided a creative space for individuals who wish to explore 'alternative nature' or 'environment after nature'.

Hybrid natures

The three nature regimes summarized by Escobar (1999) allow us to view ecotourism in different contexts from an antiessentialist political ecology point of view, for none of these regimes is closed-ended. Instead, they have overlapped with, reacted according to and reproduced each other. Capitalist, organic and technonatures are different from one another because humans who engage with these three regimes are differently positioned, demand different things from their biological surroundings and therefore create different historical articulations into the biological articulations of 'nature'. However, as humans interact and society is a 'floating territory' (Bauman, 2000), it is often the case that one embraces different natures at the same time. Tension, friction and affect generated from the very cohabitation of these nature regimes thus open one up to the possibilities of hybridization, a constant struggle that may never be resolved.

The notion of hybrid natures (Escobar, 1999) thus reminds us to be always multidimensional. Viewing the historical accounts of the Shangri-La area's deforestation and reforestation only from a political economy aspect blinds us from seeing other existing regimes, for example environmental governance in premodern China (Elvin, 1998), as well as the role of Tibetans' ecological knowledge and traditions in shaping the environment (Anderson *et al.*, 2005; Coggins and Hutchinson, 2006). At the same time, projecting local knowledge and beliefs onto human–nature relationships also withholds the fact that any locality is operating in collaboration and negotiation with universality, thus never stays the same and cannot be subjected to simple suppression, or the allocation of inferior status. Although capitalist nature regimes in Escobar's (1999) framework is about the transformation of nature into capital (in different forms and through different ways) it also seeks collaboration with organic nature and technonature to further expand and sustain its system, and vice versa. As mentioned in each sector, ecotourism in Shangri-La manifests how this conflation is not only always happening but also is inevitable.

To summarize, 'nature' has always had the potential to be defined by different parties in different ways. The challenge we face today is to understand the incorporation of multiple constructions of nature. It is from this process of constant conflating, shaking, consolidating, breaking and recombining that new possibilities and understandings of ecological subjectivities emerge. People are embedded in hybrid nature through interacting with one another, and further interaction would lead to the emergence of new regimes of nature.

Conclusion: nature as environmental discourse

Ecotourism development in Shangri-La is the result of green economic forces and green governmentality that are historically situated in Shangri-La's environmental changes in the context of China's contemporary land and forestry policy. However, from an antiessentialist political ecology perspective (Escobar, 1999), nature cannot be read only as being subjected to governmentality and commodification. Rather, it should also be understood as a vibrant environmental discourse that various regimes conflate and collaborate on and which are, at times, frictional and contradictory. Such environmental discourse allows us to see tourism as a social practice caught in a constant process of making/unmaking/remaking that may bring new ideas about nature into the current debates concerning nature and culture, conservation and development. The hybrid natures – the recombination and hybridization of capitalist nature, organic nature and technonature regimes – stress that we ought not to regard the ideas of nature as coherently and systematically dependent on categories such as geographic locations or cultural background. Places or cultures are never isolated from each other or from the transversal ideas in individuals who experience and exhibit constant changes.

Tourism should be recognized as contributing to the debates on human–environmental relations in that it is a process that can mediate between how individuals practice, present and negotiate their interactions with 'nature', and with their own means to interpret and process 'environmental discourse'. The question of how, from the past to the present, humans have dealt with 'nature' is held within each human/nature encounter, especially those encounters in tourism. The legitimacy of developing ecotourism is claimed to be its modern approach, characterized by its separation of nature and society, and as being a regime based on the idea that both the human and natural worlds can be organized and subjected to rational, totalizing control (Pálsson, 2006). But this view is starting to change; Fletcher (2014, p. 3) views ecotourism as 'a cultural or discursive process, embodying a particular constellation of beliefs, norms and values that inform the activity's practice'. Hollinshead (2009), meanwhile, adopts the term 'worldmaking' from art studies to tourism studies to describe the function of tourism as a powerful medium through which to revalue stories, images and memories about people and places. Ecotourism development in the Shangri-La area constitutes more complex elements than merely economically driven forces, ecological resource utility changes and local resistance. Through tourism we are directed to see that, rather than producing coherent and culturally specific ways of relating to 'nature', people's relationships and interactions are constantly in tension within categorical systems.

This chapter contributes to the political ecology of tourism by introducing the idea of 'environmental discourse' through which 'nature' is always experienced as a certain hybridization of the three nature regimes and how tourism can be reviewed in the idea of 'environmental discourse'. Environmental discourse applies itself to individuals' everyday lives and marks an individual's identity by how one locates oneself within society and nature. Moreover, environmental discourse is made up of multiple and irreducible relations between different regimes of nature; individuals are the primary resource of such relations and hence environmental discourse is ever present but without origin. This is to say, by our very being in one or several regimes of nature, and by handling the tensions between different regimes of nature, we impose various sorts of power onto 'nature'. Individuals' models of dealing with relations with biological and societal others are intrinsically coincided and negotiated. Through coexisting among these couplings and negotiations, we are always contributing to environmental discourse, consciously or unconsciously. This is a process we are only just beginning to comprehend, that culture needs elucidations from nature, as much as nature needs explanations from culture (Ingold and Pálsson, 2013). Both culture and nature evolve through interplay with each other. Environmental discourse therefore draws our attention to events, phenomena and interactions where individuals are constantly carrying various ideological, philosophical and poetic ideas. Situating tourism, especially ecotourism, in environmental discourse helps us to gain a perspective that forces us to rethink linear and logical sustainable development paradigms.

Notes

1 Zhuoma is the Tibetan word used to address young women.
2 The term 'nature' is employed as the problematic concept that always is entangled with various hegemonies and situated within philosophical debates of dualism, such as human/nature divide (Soper, 1995).

3 Besides the central town of Shangri-La, I spent time in Luorong Village, Pudacuo National Park and Niru Village.
4 Latour (1993) argues otherwise.
5 Bon Religion is a religion that existed before Buddhism was brought to Tibetan areas from Nepal, India and Mainland China. Today's Tibetan Buddhism is arguably an outcome of the intermingling of both Bon Religion and Buddhism.

References

Allen, C. (2001) *The Search for Shangri-La: a journey into Tibetan history*. London: Abacus.
Anderson, D. M., Salick, J., Moseley, R. K. and Xiaokun, O. (2005) Conserving the Sacred medicine mountains: a vegetation analysis of Tibetan sacred sites in Northwest Yunnan. *Biodiversity & Conservation*, 14 (13), pp. 3065–3091. Bauman, Z. (2000) On writing: on writing sociology. *Theory, Culture & Society*, 17 (1), pp. 79–90.
Blaikie, P. and Brookfield, H. (1987) *Land Degradation and Society*. London: Methuen.
Bramwell, B. (2011) Governance, the state and sustainable tourism: a political economy approach. *Journal of Sustainable Tourism*, 19 (4–5), pp. 459–477.
Brundtland, G. H. (1987) *Report of the World Commission on environment and development: Our Common Future*. United Nations. Available at: http://www.un-documents.net/our-common-future.pdf [Accessed: 10 May, 2014].
Chen, F. and Davis, J. (1998) Land reform in rural China since the mid-1980s. *Land reform*, 2, pp. 123–137.
Coggins, C. and Hutchinson, T. (2006) The political ecology of geopiety: nature conservation in Tibetan communities of Northwest Yunnan. *Asian Geographer*, 25 (1–2), pp. 85–107.
Coggins, C. and Yeh, E. (2014) Introduction: Producing Shangri-La. In: Yeh, E. and Coggins, C. (eds), *Mapping Shangrila: Nature, Personhood, and Sovereignty in the Sino-Tibetan Borderlands*. Seattle, WA: University of Washington Press, pp. 3–18.
Cole, S. (2012) A political ecology of water equity and tourism: a case study from Bali. *Annals of Tourism Research*, 39 (2), pp. 1221–1241.
Crumley, C. L. (2002) *New Directions in Anthropology and Environment: Intersections*. Plymouth: Altamira Press.
Diqing Prefecture Tourism Bureau. (2008) The tourism industry development of Diqing Prefecture. Available at: http://dq.xxgk.yn.gov.cn/Z_M_014/Info_Detail.aspx?DocumentKeyID=ABB024B9140048089D7FDA8EFB07B21C [Accessed: 10 May, 2014].
Elvin, M. (1998) The environmental legacy of imperial China. *The China Quarterly*, 156, pp. 733–756.
Escobar, A. (1996) Construction nature: Elements for a post-structuralist political ecology. *Futures*, 28 (4), pp. 325–343.
Escobar, A. (1999) After nature: steps to an antiessentialist political ecology. *Current Anthropology*, 40 (1), pp. 1–30.
Escobar, A. (2011) *Encountering Development: The Making and Unmaking of the Third World*. Princeton, NJ: Princeton University Press.
Fletcher, R. (2014) *Romancing the Wild: Cultural Dimensions of Ecotourism*. Durham, NC: Duke University Press.
Foucault, Michel. (1991) Governmentality. In: Burchell, G., Gordon, C. and Miller, P. (eds), *The Foucault Effect: Studies in Governmentality*. Chicago, IL: University of Chicago Press, pp. 87–104.

Gao, S., Huang, S. and Huang, Y. (2009) Rural tourism development in China. *International Journal of Tourism Research*, 11 (5), pp. 439–450.

Glyn, A. and Alistair, W. (2012) Is this Shangri-La? The case for authenticity in the Chinese and Indian hospitality industry. *Journal of Brand Management*, 19 (5), pp. 405–413.

Greenpeace (2012) *The Yunnan Forests in Crisis: A Report on Investigation of Current Status of Yunnan Natural Forests*. Beijing: Greenpeace.

Greenberg, J. and Park, T. (1994) Political ecology. *Journal of Political Ecology*, 1 (1), pp. 1–12.

Guneratne, A. (ed.) (2010) *Culture and the Environment in the Himalaya*. London, New York: Routledge.

Hakkenberg, C. (2008) Biodiversity and sacred sites: vernacular conservation practices in Northwest Yunnan, China. *World Views: Environment, Culture, Religion*, 12 (1), pp. 74–90.

Han, N. and Ren, Z. (2001) Ecotourism in China's nature reserves: opportunities and challenges. *Journal of Sustainable Tourism*, 9 (3), pp. 228–242.

Haraway, D. (1991) The actors are cyborg, nature is coyote, and the geography is elsewhere: postscript to 'cyborgs at large'. *Technoculture*, 3, pp. 183–202.

Harkness, J. (1998) Recent trends in forestry and conservation of biodiversity in China. *China Quarterly*, 56, pp. 911–934.

Hillman, B. (2003a) Paradise under construction: minorities, myths and modernity in Northwest Yunnan. *Asian Ethnicity*, 4 (2), pp. 175–188.

Hillman, B. (2003b) *The poor in paradise: tourism development and rural poverty in China's Shangri-La*. Paper presented at the landscapes of diversity: proceedings of the III symposium on montane mainland South East Asia (MMSEA), Kunming, China.

Hillman, B. (2009) Ethnic tourism and ethnic politics in Tibetan China. *Harvard Asia Pacific Review*, 10 (1), pp. 3–6.

Hillman, B. (2010) China's many Tibets: Diqing as a model for 'development with Tibetan characteristics?' *Asian Ethnicity*, 11 (2), pp. 269–277.

Hollinshead, K. (2009) 'Tourism state' cultural production: the re-making of Nova Scotia. *Tourism Geographies*, 11 (4), pp. 526–545.

Hu, Y. (2005) *Shangri-la and the Tibet Imagination in Colonialism Context*. (M. A. H138818), Tsinghua University (People's Republic of China), Peoples Republic of China.

Ingold, T. and Pálsson, G. (2013) *Biosocial Becomings: Integrating Social and Biological Anthropology*. New York: Cambridge University Press.

Latour, B. (1993) *We Have Never Been Modern*. Cambridge, MA: Harvard University Press.

Le Blanc, D. (2011) Special issue on green economy and sustainable development. *Natural Resources Forum*, 35 (3), pp. 151–154.

LePage, V. (1996) *Shambhala: The Fascinating Truth behind the Myth of Shangri-la*. New York: Quest Books.

Li, T. (2014) *Land's End: Capitalist Relations on an Indigenous Frontier*. Durham, NC and London: Duke University Press.

Llamas, R. and Belk, R. (2011) Shangri-La: messing with a myth. *Journal of Macromarketing*, 31 (3), pp. 257–275.

Luke, T. W. (1999) Environmentality as Green Governmentality. In: Darier, E. (ed.), *Discourses of the Environment*. Oxford: Blackwell.

Makley, C. E. (2014) The Amoral Other: State-led Development and Mountain Deity Cults among Tibetans in Amdo Rebgong. In: Yeh, E. and Coggins, C. (eds), *Mapping*

Shangri-La: Contested Landscapes in the Sino-Tibetan Borderlands. Seattle, WA and London: University of Washington Press, pp. 229–254.

Martin, E. (2012) Manifestations of Tibetan Buddhism in Pudacuo National Park and its Effectiveness as an Environmental Education Tool. *Independent Study Project (ISP) Collection. Paper 1461*. Available at: http://digitalcollections.sit.edu/isp_collection/1461 [Accessed: 10 May, 2014].

Mitchell, D. (2000) *Cultural Geography: A Critical Introduction*. Oxford: Blackwell.

Mu, B., Yang, L. and Zhou, M. (2007) The progress of eco-tourism researches of nature reserves in China. *Journal of Fujian Forest Science and Technology*, 34 (4), pp. 241–247.

Netting, R. M. (1993) *Smallholders, Householders: Farm Families And The Ecology Of Intensive, Sustainable Agriculture*. Stanford, CA: Stanford University Press.

Ni, Q. (1999) A summary of the study of China's ecotourism in recent years [J]. *Tourism Tribune*, 14 (3), pp. 40–45.

Ni, R. and Cao, L. (2003) Exploring tourism industry in Shangri-La 香格里拉旅游文化产业初探. *Creation 创造*(1), 3.

O'Connor, M. (1993) On the misadventures of capitalist nature. *Capitalism Nature Socialism*, 4 (3), pp. 7–40.

Pálsson, G. (2006) Nature and Society in the Age of Postmodernity. In: Biersack, A. and Greenberg, J. B. (eds), *Reimagining Political Ecology*. Durham, NC: Duke University Press, pp. 70–96.

Pan, F. and Yang, G. (2000) Shangri-La and development of Religious ecotourism 香格里拉与宗教生态旅游开发. *The Ideological Front*, 1, pp. 82–85.

Pearce, D. W., Markandya, A. and Barbier, E. (1989) *Blueprint for a Green Economy* (Vol. 1). London: Earthscan.

Peet, R. and Watts, M. (1996) Liberation Ecology: Development, Sustainability, and Environment in the Age of Market Triumphalism. In: Peet, R. and Watts, M. (eds), *Literation Ecologies: Environment, Development, Social Movements*. New York: Routledge, pp. 1–45.

Pudacuo National Park, (2008) Interpretation Texts for Bus Narrators in Pudacuo National Park, internal publication. Shangri-La, YN: Pudacuo National Park.

Qi, Z. (2007) Shangri-La welcomes you. Shangri-La, Yunnan: Yunnan Province Diqing Tibetan Autonomous Prefecture Tourism Bureau.

Qi, Z., Cheng, S. and Shen, L. (2000) A Shangri-La Strategy for sustainable development and its practice in the Qinghai-Tibet plateau. *Resource Science*, 22 (4), pp. 83–85.

Rappaport, R. (1990) Ecosystems, Populations, and People. In: Moran, E. (ed.), *The Ecosystem Approach in Anthropology*. Ann Arbor, MI: University of Michigan Press, pp. 41–73.

Richardson, S. D. (1990) *Forests and Forestry in China: Changing Patterns of Resource Management*. Washington, DC: Island Press.

Richardson, S. D. (2000) *Forestry, People and Places: Selected Writings from Five Decades*. Rotorua, New Zealand: Business Media Services.

Robbins, P. (2012) *Political Ecology: A Critical Introduction*. Chichester, West Sussex; Malden, MA: J. Wiley & Sons.

Rozelle, S., Albers, H., Guo, L. and Benziger, V. (1998) Forest resources under economic reform in China. *China Information*, 13, pp. 106–125.

Russell, B. (1946) *History of Western Philosophy and its Connection with Political and Social Circumstances from the Earliest Times to the Present Day*. London: Allen & Unwin.

Smith, N. (2008) *Uneven Development: Nature, Capital, and the Production of Space* (3rd edn). Athens, GA: The University of Georgia Press.

Soper, K. (1995) *What is Nature?* Oxford: Blackwell.

State Forestry Administration (2000) *Forest Resources Statistics of China (1994–1998)*. Beijing: Department of Forest Resources Management, SFA.

Stonich, S. C. (1998) Political ecology of tourism. *Annals of Tourism Research*, 25 (1), pp. 25–54.

Studley, J. (1999) Forests and environmental degradation in SW China. *International Forestry Review*, 1 (4), pp. 260–264.

UNESCO (2003) World Heritage Nomination – IUCN technical evaluation: Three Parallel Rivers of Yunnan Protected Areas (China) ID N1083: UNESCO.

Walker, P. A. (2005) Political ecology: where is the ecology? *Progress in Human Geography*, 29 (1), pp. 73–82.

Wang, G., Innes, J., Wu, S., Krzyzanowski, J., Yin, Y., Dai, S., . . . Liu, S. (2012) National park development in China: conservation or commercialization? *AMBIO: A Journal of the Human Environment*, 41 (3), pp. 247–261.

Wang, S., Cornelis van Kooten, G. and Wilson, B. (2004) Mosaic of reform: forest policy in post-1978 China. *Forest Policy and Economics*, 6 (1), pp. 71–83.

Willson, A. (2006) Forest conversion and land use change in rural Northwest Yunnan, China. *Mountain Research and Development*, 26 (3), pp. 227–236.

Winkler, D. (1998) Deforestation in Eastern Tibet: Human Impact Past and Present. In: Clarke, G. (ed.), *Development, Society, and Environment in Tibet*. Band Austria: OADW, pp. 78–95.

Xu, J. and Ribot, J. (2004) Decentralisation and accountability in forest management: a case from Yunnan, Southwest China. *The European Journal of Development Research*, 16 (1), pp. 153–173.

Yeh, E. (2014) Reverse Environmentalism: Contemporary Articulations of Tibetan Culture, Buddhism and Environmental Protection. In: Miller, J., Yu, D. S. and van der Veer, P. (eds), *Religion and Ecological Sustainability in China*. New York: Taylor & Francis, pp. 194–219.

Yin, R., Xu, J., Li, Z. and Liu, C. (2005) China's ecological rehabilitation: the unprecedented efforts and dramatic impacts of reforestation and slope pretection in Western China. *China Environment Series*, 7, pp. 17–32.

Zackey, J. (2007) Peasant perspectives on deforestation in Southwest China. *Mountain Research and Development*, 27 (2), pp. 153–161.

Zhang, J. J. (2014) *No Gods, No Shangri-La': Rethinking tourism in an environmental discourse*. Paper presented at the RC 50 International Tourism Committee, 'Science and Power relations in Tourism Studies' session, The World Congress of Sociology, Yokohama, Japan.

Zhao, G. and Shao, G. (2002) Logging restrictions in China: a turning point for forest sustainability. *Journal of Forestry*, 100 (4), pp. 34–37.

Zhou, D. Q. and Grumbine, R. E. (2011) National parks in China: experiments with protecting nature and human livelihoods in Yunnan province, Peoples' Republic of China (PRC). *Biological Conservation*, 144 (5), pp. 1314–1321.

Zinda, J. A. (2012) Hazards of collaboration: local state co-optation of a new protected-area model in Southwest China. *Society & Natural Resources*, 25 (4), pp. 384–399.

Zinda, J. A. (2014) Making National Parks in Yunnan: Shifts and Struggles Within the Ecological State. In: Yeh, E. and Coggins, C. (eds), *Mapping Shangrila: Nature, Personhood, and Sovereignty in the Sino-Tibetan Borderlands*. Seattle: University of Washington Press, pp. 105–128.

8 (Re)creating forest natures
Assemblage and political ecologies of ecotourism in Japan's central highlands

Eric J. Cunningham

Introduction

I am walking through a forest in the Kiso region of central Japan on a chill October morning. My companion is a retired Forestry Agency employee who now periodically offers tours of the area for paying guests. His many years of experience in the region's forests have instilled in him a broad and detailed knowledge of their ecologies, biological characteristics, geological makeups, and histories. The pages of my field notebook from the day are littered with the names and identifying characteristics of plants and trees, as well as notations about rock formations. We pass through a stand of towering *hinoki* cypresses (*Chamaecyparis obtusa*), some 200 to 300 years old, and then climb onto the vanishing bed of a railroad. My companion tells me this was once part of a larger rail network built in the early part of the twentieth century to haul timber to nearby lumber mills. It allowed the government to intensively and extensively extract wood from the area until the middle part of the century. Further along we encounter a large swathe of planted Japanese larch (*Larix kaempferi*) standing in straight, uniform rows. Plantation-style forests like this one are common today in the Kiso region. Because larch grows quickly and can be used for timber it was planted extensively after the Second World War to supply materials for reconstruction. However, Japan's domestic timber market evaporated after regulations on imports of foreign woods were relaxed in the 1960s. As a result, most of the planted larch was never harvested. The widespread forest conversion that has resulted from government-sponsored 'natural resource development' in the Kiso region has had deleterious social and ecological effects. However, these now ubiquitous plantation-style forests, my companion tells me, are perceived as natural and are aesthetically pleasing to the majority of people who visit the region, so their anthropogenic origins go largely unrecognized. He laments that '[t]ourists these days just want to see these boring forests'.

This lamentation denotes a disconnect that is at the heart of many ecotourist experiences in Japan. There is a misalignment between the natural landscapes that ecotourists desire to see and experience and the actual conditions of environments (Fletcher and Neves, 2012; West and Carrier, 2004). The forest that I describe above, like many in the Kiso region, could accurately be characterized as a post-industrial landscape. Historical practices of industrialized forestry, followed by post-industrial transitions towards conservation, have transformed the area's ecologies and left its human communities in precarious states. As I detail further below, this forest history exemplifies two central themes addressed by political ecologists, identified by Robbins (2012, p. 21) as 'the degradation and marginalization thesis' and 'the conservation and control thesis'. The first thesis refers to the transition of ecologically benign production systems to overexploitation through state development and integration into regional, national, and/or global markets. The second thesis is concerned with the ways in which resources and landscapes are wrested away from local inhabitants in the name of preservation, sustainability, and/or conservation.

Yet, despite landscape markings denoting this exploitative history, the Kiso region's scenic beauty, as well as its remoteness, has made it a popular ecotourism destination. The common images and discourses of ecotourism in the Kiso region, however, do not depict these landscapes as places transformed through a political-economy of natural resource exploitation. Rather, landscapes are pasted together in a decoupage of images and discourses depicting pristine natures, traditional cultural lifeways, and harmonious ecological relations. These images and discourses of nature comprise a 'tourist gaze' that frames the experiences of ecotourists in the region (Urry, 2002). Meanwhile, the 'backstage environment' (Campbell et al., 2008, p. 214) – the actual conditions of social and ecological life, as well as the contingent histories that brought them into being – remains largely unseen. Teasing apart what is 'seen' and what remains 'unseen' within ecotourist experiences calls for a political ecology approach and an analysis that is historically situated, place-based, and multi-scalar, which is the approach taken in this chapter.

My primary argument is that images and discourses are critical to ecotourism practices in the Kiso region because of the roles they play in mediating disconnections between ecotourist fantasies and the actualities of post-industrial social and ecological life. In other words, I suggest that images and discourses hold the power to transform lived reality into something that enables consumption, or is itself consumed. In Japan, the power to transform marginalized rural landscapes into spaces of ecotourist consumption is allowing for novel forms of capital accumulation that may not be as 'green' or as 'socially responsible' as they are presented to be. In fact, for local residents of the Kiso region, ecotourism entails a new set of vectors by which social and ecological life is being rearranged to meet the needs of a broader political economy.

Situating ecotourist practices in the Kiso region within the historical contingencies of political-economic relations of natural resource use and rural transformation allows for greater contextualization of these practices within a broad constellation of connections involving various human and nonhuman actors, institutions, and cultural beliefs. Such contextualization is a key strategy of political ecology approaches, one that views 'ecological systems as power-laden rather than politically inert' (Robbins, 2012, p. 13). An additional aim of this chapter then, is to situate localized ecotourism practices in Japan within power-laden contexts at broader temporal and spatial scales in order to critically interrogate the roles they play in shaping local ecologies.

Recently, scholars have begun to address broad sets of social and ecological components through the concept of *assemblage* (Anderson and McFarlane, 2011; Dewsbury, 2011; Li, 2007). Similar to concepts of network, assemblage describes a vast set of connections and relations between 'heterogeneous elements that may be human and nonhuman, organic and inorganic, technical and natural' (Anderson and McFarlane, 2011, p. 124). Assemblage, however, differs from network in at least two distinct ways. First, assemblage denotes activity rather than stasis. As Li (2007, p. 264) puts it, '[a]ssemblage flags agency, the hard work required to draw heterogeneous elements together, forge connections between them and sustain these connections in the face of tension'. Second, whereas network tends to reference hierarchical organization, with certain elements always governing others, assemblage references heterarchical organization, meaning that cause-and-effect relations among elements can shift through time or across space. The concept of assemblage, 'enables us to remain deliberately open as to the form of the unity, its durability, the types of relations [. . .] involved' (Anderson and McFarlane, 2011, p. 124), and thus allows us to consider ecotourism not only as a set of practices that take place *in* the world, but as a set of practices linked to forms and movements that help to *create* the world. Employing assemblage to develop a political ecology perspective offers analytical space in which to interrogate normative claims about ecotourism as a set of practices that is benign or beneficial to local ecologies or, in the words of the Japan Ecotourism Society (JES), a set of practices that 'aims to activate local economies by providing tourists with opportunities to experience local attractions accompanied by competent interpreters while preserving local resources such as natural environment, the culture and historical heritages [sic]'.[1] Employing a political ecology perspective to thinking about assemblage qualities of ecotourism, and their situatedness within a larger political economy, raises interesting questions concerning what it is that ecotourists desire to encounter; what they actually do encounter; and what effects these encounters have on humans, nonhumans, and other ecosystem elements at different scales.

Images and discourses of nature

A central aim of ecotourist practice is to bring paying customers into contact with some kind of 'other', be it other-than-human nature, other-than-ordinary people, or other-than-typical lifestyles. Disconnection is thus a central premise that is consistently reiterated within ecotourism assemblage – humans are disconnected from nature; the present is disconnected from the past; and modernity is disconnected from tradition. The premise of disconnection, however, is rarely overtly conveyed, but rather feeds into broader ecotourist imaginaries that are upheld and transmitted through images and discourses (Salazar, 2012, p. 864). By 'images and discourses' I am referring to Japan's mediascape in its entirety, which includes at least television, movies, magazines, public advertisements, pamphlets, brochures, emails, social networking applications, websites, and other forms of internet content. Like other (post)industrial nations, in Japan, images and discourses dominate landscapes at multiple scales. Nearly any physical surface has been, or is liable to be, co-opted for the dissemination of images and discourses. For example, the interiors of trains are littered both with moving and still images that advertise a variety of products. At the same time, these public images and discourses must battle for attention against portable electronic devices, which are ubiquitous in Japan and deliver a variety of digital content. Within this mass of images and discourses, nature and rural areas are overwhelmingly presented as pristine and bucolic spaces in which Japan's urban population can find relief from the stresses and anxieties of everyday life. Within the realm of images and discourses, Japan's natural and rural environments are constructed as spaces disconnected from the conditions of everyday modern life.

Images and discourses hold the power to present themselves as mere representations of reality, rather than as integral parts of those realities. A number of scholars have made it clear, however, that environments, particularly those designated as sites of conservation, often are deliberately crafted to match natures fetishized through images and discourses (Brockington and Duffy, 2011; Duffy, 2002; West and Carrier, 2004). Guy Debord (1994) theorizes that in modern societies, relations between people are mediated by an accumulation of images and discourses (what he calls 'spectacle') so that '[l]ived reality is materially invaded by the contemplation of the spectacle' (Debord, 1994, Thesis 8). Igoe (2010, p. 376) expands on Debord's work to include mediations of relations between people and environments, and suggests that analyzing these mediations and their effects 'requires moving beyond simple dichotomies of representation and material reality'. In considering the role of images and discourses within ecotourism assemblage, I aim, following Igoe, to move beyond such dichotomies. I argue that images and discourses are key components of ecotourism in Otaki and the Kiso region, due not only to their capacity to frame ecotourist encounters, but also to their capacity to 'invade' lived reality and shape material environments through the active bodies that continually work to assemble ecotourism.

Focusing attention on the seemingly transient realm of images and discourses and considering how these are translated into material environmental realities moves us towards a political ecology of ecotourism in Japan by throwing light on three interconnected processes: (1) The fetishization of nature and rurality through the production, circulation, and consumption of images and discourses; (2) The work of crafting ecotourism landscapes, both through the manipulation of physical environments (cutting trees, removing trash, and so forth) and the disciplining of bodies (guiding where to walk or where to look, for example); and (3) The affirmation of images and discourses through direct encounter.

I suggest that ecotourist practices in the Kiso region involve more than bringing tourists to nature; they involve bringing both tourists and natures into being. In other words, the abilities of images and discourses to invade material realities lead us to consider the power that ecotourism assemblage holds to *produce* both ecologies and environmental subjects (Robbins, 2012, pp. 22–23). Considered in this framework, ecotourism is revealed to be fundamentally political, in that the various lines of articulation through which it is assembled are drawn within larger

(Re)creating forest natures 173

contexts of multi-scalar, political-economic relations in which benefits and harms are not evenly distributed, meaning that rural ecologies and human communities are often made 'losers', even as they are presented as 'winners' (Robbins, 2012, pp. 87–88).

Chapter structure and methods

In what follows, I develop my argument through empirical case studies of ecotourism in the Kiso region. I begin with an account of natural resource development and use in the region, in particular state-sponsored industrial forestry, in order to situate modern practices of ecotourism within the region's broader historical ecologies. Against this backdrop of natural resource (over)use, I next examine uses of images and discourses within emergent forms of ecotourism and suggest the ways in which they serve to frame ecotourist encounters in part by obfuscating the actualities of the Kiso region's historical ecologies. These empirical accounts are based on 25 months of ethnographic fieldwork, conducted between 2008 and 2015 in the village of Otaki, which included participant observation, interviewing, group discussions, and archival research. I then draw from these case studies to consider the roles that images and discourses play within assemblages of ecotourism in Japan today. This analysis is based on ongoing research of environment-related media in Japan consisting of surveys of digital (websites, videos, social network content, email, magazines) and print (brochures, magazines, pamphlets) media. All translations of Japanese materials are my own, unless otherwise noted. I conclude with a discussion of assemblage in terms of a politics of imagination in which I suggest that often what is at stake within political ecologies is the power to imagine, envision, and actualize ecologies and ways of being. Considered in a political ecology framework, therefore, ecotourism takes on significance not as a pathway towards sustainability, but rather as a field of practices situated at multiple scales and flush with all sorts of different political possibilities.

Historical ecologies of the Kiso region

The Kiso region is an area of approximately 600 square miles located in the central Japanese prefecture of Nagano, about 250 miles west of Tokyo. A steep river valley (the Kiso Valley) runs north to south through the region, with a number of smaller valleys running perpendicularly into it. Over 90 percent of the region is forested, with the majority designated national forest (*kokuyūrin*). Depopulation continues to be a serious problem, which has resulted in a series of municipal amalgamations. As of April 2015, the region consisted of three towns and three villages, with a total population of 28,631 individuals.[2] Like other rural regions in Japan, patterns of labor in the Kiso region have shifted in the post-war era. Fewer and fewer people work as farmers or foresters, which were dominant forms of livelihood until the 1970s. In 2010, among the working population over the age of 15, 8 percent were employed in primary industries, 30 percent in secondary industries, such as manufacturing, and 62 percent in tertiary industries mostly related to tourism.[3] The Kiso region thus typifies what anthropologist William W. Kelly has labeled 'regional Japan', 'the necessary reserves of metropolitan Japan' (1990, p. 209).

Intensive timber exploitation in the Kiso region began in the sixteenth century when elites began to seize forestlands there (Hirada, 1999; Totman, 1989). However, industrialized forestry began after the Meiji government took power in 1868 and set the new nation-state on a course of modernization. Meiji officials sought centralized control, which required simplifying and making legible a great variety of complex social and ecological phenomena. James C. Scott (1998, pp. 11–52) theorizes desire among state officials to deconstruct the complexities of real world phenomena in order to achieve an encompassing view and to isolate particular parts for efficiency. He notes that such myopia results in problematic distortions, which misrepresent a broad range of biological, spiritual, social, and cultural elements that comprise landscapes. State projects of simplification aim to create particular orderly arrangements out of worlds that are complex and messy. Employed and deployed within projects of simplification are a variety of tools, techniques, and institutions that serve to create authority and mystify those on the ground who cannot decipher the complex 'simplicity' embedded within. In their quest to create a totalizing vision of Japan and a naturalized 'domestic armature' (Wigen, 2010, p. 1) Meiji officials undertook a series of surveys, land consolidations, and ideological reconfigurations (Thomas, 2001; Tsutsui, 2003a; Waswo, 1988). Also, the Meiji government deployed a new vision for Japan's forests, which it began to take control of in 1874 through the imposition of land taxes (Ushiomi, 1968). The Meiji government's enclosure of what had up until that point been common-use forestlands disrupted local ecologies, which had long been structured by patterns of low-intensity anthropogenic disruption (McKean, 1982; Ushiomi, 1968). Through their incorporation into a national system of management, forests across Japan were drawn into an adaptive strategy situated within a broad web of political-economic forces and informed by modernist ideologies.

At the beginning of the twentieth century, forest ecologies in the Kiso region were further entangled in the political-economy of Japan's expanding empire and a desire for timber to fuel imperialist and militarist ambitions (Tsutsui, 2003b; Morris-Suzuki, 2013). In 1903, an Imperial Forest Administration Bureau (*teishitsu rinyakyoku* or, after 1908, *teishitsu rinya kanrikyoku*) was established in the town of Kiso-fukushima and three years later work began on a forest railroad system for hauling timber. This eventually included lines that stretched into every part of the Kiso region, with trains operating almost non-stop until the mid-1970s (Morishita, 1998). Between 1897 and 1935 profits from the sale of timber extracted from the Kiso region alone accounted for up to 80 percent of the total income from all imperial forests in Japan (Oura, 1992).

Industrial forestry remained intensive in the Kiso region into the post-war era as domestic timber markets boomed to meet growing needs for building materials to supply reconstruction efforts. However, in the 1960s, as demand for domestic timber began to outpace supply, the Japanese government quickly lifted trade barriers to allow for imports of raw logs (Iwamoto, 2002; Ota, 1999). With this, Japan took center stage in a political-economy of 'dual-decay' in which environmental quality is negatively affected in both importer and exporter locations (Seo and Taylor, 2003, p. 92). While appetites for wood in Japan have left forest ecologies in exporter locations in various states of degradation (Dauvergne, 1997), heavy investments in afforestation at home have resulted in material transformations (i.e. over-forestation) that have negatively impacted upland forest ecologies and human communities (Knight, 1997). In simple terms, through a history of state-sponsored industrial forestry, followed by heavy planting, focusing on coniferous timber varieties, Japan's domestic forests (particularly national forests) have been converted on a large scale from diverse, mixed ecologies to simple, plantation-style ecologies (Miyamoto and Sano, 2008).

Despite the actual ecological conditions of forests, Japan's Forestry Agency (*rinyachō*), which today governs and manages all national forests, and is a product of the US led Occupation Authority's restructuring of imperial institutions, touts them as 'national treasures' (*kuni no zaisan*) that function to benefit the citizenry (Ota, 1999). As John Knight (1997, pp. 711–712) notes, 'while Japanese forests are cited as parasitical on deforestation elsewhere, the official (and widespread) view in Japan itself is that Japanese reforestation is an expression of recovery from (earlier) deforestation'. Forests in Japan, including those in the Kiso region, have been ideologically reconfigured to fit the mandates of conservation and management for the good of the nation, rather than of intensive resource use and profit-making for the good of the imperial family. In pamphlets and online media, Japan's Forestry Agency liberally employs terms like 'citizens' forests' (*kokumin no mori*), signifying that forests in the Kiso region, like all national forests, belong to the nation. In brochures, pamphlets, email magazines, and on websites, forests in the Kiso region are presented to Japanese audiences as natural resources, accompanied by descriptions of beneficial purposes to which the Forestry Agency is (supposedly) putting them to use. The lists include a host of hybrid, specialized forests, including 'water and land protection forests', 'resource cycling forests', and 'forest and human coexistence forests'. Forests in this latter category – forests with which humans are said to 'coexist' – are the focus of recreation and tourism initiatives in Japan today and provide productive spaces within which to assemble ecotourism.

Imagining and assembling ecotourist natures

As a term and concept, iterations of ecotourism in Japan share similarities to those in Europe, North America, and elsewhere. The government of Japan's Ministry of the Environment defines ecotourism as, 'an arrangement with the aim of working together with local areas to convey to tourists the natural environment, history and culture, and characteristic charms of that area, so that their value and importance can be appreciated and connected to preservation'.[4] This definition resonates with that offered by Japan's largest ecotourism non-profit organization, the Japan Ecotourism Society (see http://www.ecotourism.gr.jp), as well as its international partner organization, The International Ecotourism Society (see http://www.ecotourism.org). Thus, in its conceptual framing, ecotourism in Japan tends to fit with the characterization of ecotourism offered by West and Carrier (2004, p. 483) as leisure travel with the twin aims of experiencing what is seen as the natural environment and interacting with people perceived as being exotic, and doing both in a way that minimizes negative impacts, or is perhaps even beneficial.

Though conceptualizations and practices of ecotourism in Japan are similar to those found around the globe, visions of nature within Japanese ecotourism differ somewhat from those that occupy the imaginaries of ecotourism in 'the West'. Salazar (2012, p. 864) defines *imaginaries* as forms, meanings, and values that emerge from individual and collective imaginings to create 'unspoken schemas of interpretation' premised on shared understandings. Compared to Western imaginaries of nature, premised on ideas of wilderness as opposed to civilization or culture (Fletcher, 2015; West and Carrier, 2004), the imaginaries that drive ecotourism in Japan differ in at least two ways. First, they are concerned with nature that is 'encultured'. As Catherine Knight (2010, p. 422) phrases it, this is nature 'in which nature and culture intersect, and [. . .] is reminiscent of a more idyllic rural lifestyle of the past, euchronia, when the Japanese "lived in harmony with nature"'. Second, Japanese imaginaries of nature are primarily (almost exclusively) concerned with domestic spaces. Animated films produced in Japan, such as Miyazaki Hayao's *Tonari no Totoro*, have helped make images and discourses of distinctly Japanese rural nature mainstream. The seemingly timeless rural nature depicted by Miyazaki is increasingly what ecotourists in Japan fantasize about and aspire to have contact with. Thus, landscapes and communities that appear – or can be made to appear – to embody such nature gain value in the emergent ecotourism economy, which means that local residents have a stake in working to make these imaginaries real.

Amid the post-war political-economic shifts noted above, forests in the Kiso region have been ideologically and materially transformed from spaces of production to spaces of consumption. They have been remade into kinds of commodities; 'deadened' objects to be contemplated (Igoe, 2013, p. 38). At the same time, traditional forms of production, such as forestry and agriculture, have all but evaporated from rural landscapes (Jussaume Jr., 2003; Kelly, 1990; Knight, 2000; Ota, 2002). Within this context, tourism has emerged as a central economic strategy for rural communities, one that has been vigorously pursued from national to local levels (Creighton, 1997; Funck and Cooper. 2013; Moon, 2002; Thompson, 2004). The majority of Japan's rural communities are now part of a 'vanishing' Japan (Ivy, 1995), with limited economic options and declining populations (Mock, 2014). While the vanishing of upland communities is lamented, this only valorizes them and their natural landscapes as 'recovered' national treasures, which remain outside of history, timeless, and precious. Situated within this context, ecotourism is assembled as a sort of 'tourism 2.0', promising to be healthier for the environment, more beneficial for local communities, and more enriching for tourist practitioners.

As components of ecotourism assemblage, rural spaces in Japan are imagined and produced as both 'exotic' and 'familiar'. Yano (2002, p. 15) describes

Japan's rural landscapes as 'ironic frontiers', distant and peripheral spaces that are enlisted to serve as center spaces of contemplation. Ecotourist practices in Japan are entangled not only with imaginaries of pristine natures and exotic peoples, but also with imaginings of the nation and its perceived connections to an indigenous, homogenous culture, ethnicity, and race. Ecotourism is premised on desires to 'discover' and consume 'things' Japanese (nature and culture) that are at once strange and familiar. I suggest that such imaginaries constrain the possibilities of life in the Kiso region, asking people and natures to remain static, pristine, and timeless though, as detailed above, the ecologies of the region are none of these things. Rather, they are dynamic, with contingent histories that have left human communities in precarious situations, with few economic options apart from tourism. Ecotourism imaginaries thus call forth subjects and landscapes that can cater to the desires of ecotourists. Local residents, eager to attract business, actively work with governmental and corporate actors, as well as tourism practitioners, to maintain ecotourism assemblage. In the process, these local actors (re)produce imaginaries of forests and traditional rural lifeways as assemblage components. The 'boring' plantation-style forest mentioned at the beginning of this chapter, for example, is (re)imagined as a 'Japanese style' forest: light, airy, and accommodating to ecotourist fantasies and practices.

Forest bathing – assembling real imaginaries

The concept of 'green tourism' (*gurīn tsūrizumu*) predates ecotourism in Japan by a few years, but references similar practices that are largely coterminous. The Ministry of Agriculture, Forestry and Fisheries (MAFF) introduced the term in 1992 and articulated a conception of rural space as a common good of the nation to be utilized by all citizens, not just local inhabitants. MAFF describes green tourism, 'as a bridge to connect the rural need for development and the urban need for leisure space' (Funck and Cooper, 2013, p. 188). As alluded to previously, the development of 'green' and 'eco' forms of tourism are dialectically related to material reconfigurations of rural spaces, especially forests. As landscapes have become economically unproductive in terms of converting biological processes into resources and commodities, they have been reconceptualized and reconfigured as spaces where nature is preserved so that it can benefit citizens. In other words, in post-war Japan, nature is preserved so that, rather than being extracted, it can be commodified and/or consumed *in situ* as a kind of fixed capital (Büscher, 2013: pp. 21–22).

'Forest bathing' (*shinrin yoku*) is one of the earliest forms of ecotourism in Japan. The term refers to a set of practices and institutions oriented around visiting forested areas in order to 'bathe' or 'bask' in their atmosphere. The history of forest bathing begins in the Kiso region and its renowned conifer forests. In 1970, part of the Akazawa national forest, located in the central part of the region, was designated by the Forestry Agency as Japan's first 'natural recreation forest' (*shizen kyūyō rin*). Twelve years later, in 1982, the concept of forest bathing was introduced through a conference held at the site. In the intervening years there has been increased interest among scholars, government officials, and the general public in the therapeutic benefits of forest bathing. In 2005 the Forestry Agency introduced the *Therapeutic Effects of Forests Plan* to encourage research concerning the health effects of forest bathing in order to develop a more scientific field of practice, known today as 'forest therapy' (*shinrin serapī*) (Tsunetsugu et al., 2010).

Forest bathing practices are firmly rooted in imaginings of forest natures as 'pristine' environments that have been crafted through years of caring human intervention. They are held up as alternatives to the artificiality of urban areas, in which the signs of human activity are overt and overbearing. Tsunetsugu *et al.* (2010) suggest that, because human evolution is closely linked to nature, living in modern 'artificial' societies induces various forms of stress. Forest natures are therefore conceived of as spaces of connection (*tsunagu koto*) and healing (*iyasu koto*) for those alienated by the disconnections of modernity. Visiting forested areas and 'taking in the atmosphere' is said to be an effective way of reducing psychological stress, increasing sensations of relaxation, and improving physiological well-being through visual, olfactory, and tactile stimulations, as well as the absorption of *phytoncides* (antimicrobial volatile organic compounds) (Tsunetsugu *et al.*, 2010). The website of the Forest Therapy Society (FTS), Japan's largest non-profit organization related to forest therapy, states that '[i]n our modern stressful society, which is represented by techno-stresses, the physiologically relaxing effects brought about by forests, wood materials, and the like are raising citizens' interests and hopes' (Forest Therapy Society, 2015). In Japan, a nation that some label as increasingly precarious in economic, social, and spiritual ways, connection and healing have become buzzwords of the times as people struggle to deal with the 'pain of life' (Allison, 2013). Japan's forests are framed as places of refuge, with forest bathing and other forms of ecotourism serving as modalities for those seeking escape from the disenchanted landscapes of modernity.

The 'enchanted' forests within which forest bathing practices take place are as much artifacts of intensive human intervention as they are pristine natures, though pamphlets, websites, videos, and books do not generally lead people to this recognition. Yet, visiting the Akazawa national recreation forest in the Kiso region affords tourists opportunities to encounter forests that *appear* to correlate to the pristine natures represented in media. Making the imaginary real – pulling together disparate components, images, and discourses, as well as organic and inorganic materials, creating correlations, and presenting a world that appears real, authentic, and authoritative – is both the practice and result of ecotourism assemblage in the Kiso region. In the case of forest bathing, the creation of what are called 'forest therapy bases' (*shinrin serapī kichi*) is a crucial practice of assemblage. According to FTS, there are 60 forest therapy bases throughout the country. These bases, their website explains, are areas within which to verify the benefits of forest therapy and also to maintain general standards for related social and natural institutions (Forest Therapy Society, 2015).

The creation of forest therapy bases, in which forest bathing can be experienced as meaningful social practice, requires more than simply setting aside and preserving forested areas. The Akazawa forest therapy base, for example, is a space in which imaginaries of forest nature are made real through the assembling of various components. Critical to this assemblage are two sets of interrelated practices, identified by Li (2007, pp. 264–265) as 'rendering technical' and 'authorizing knowledge'. Rendering technical refers to the extraction of conceptual representations from the messiness of life in order to formulate problems that call for intervention. In the Kiso region, rendering technical has been accomplished through the establishment of national forests and the conceptual conjuring of 'eco-functional' natures that can be 'disassembled, recombined, and subjected to the disciplinary design of expert management', which makes them, '[appear] as though [they] can be calibrated to optimize ecosystem health and economic growth' (Igoe, 2013, p. 38). As noted previously, one category of 'eco-functions' ascribed to forests is recreation and leisure, which takes form in 'human-forest coexistence forests'. I suggest the creation, through images and discourses, of these and other eco-functional forest natures in the Kiso region, has been an institutional response to post-war, political-economic shifts that accompanied Japan's declining domestic timber market. Declining budgets and losses of human labor within Japan's Forestry Agency prompted modifications both of ideologies and practices. This was an attempt to offset the loss of industrial forest production by transforming forests into 'natural capital' in the (eco)service of the nation (Büscher, 2013, p. 22). It is no surprise, therefore, that Japan's first natural recreation forest was established by the Forestry Agency a few years after foreign wood imports began to erode the country's domestic timber industry.

In the years since the birth of forest bathing as a concept and field of practice, efforts have been directed at 'authorizing knowledge' by specifying and constraining a domain of knowledge surrounding it. Today, an assemblage of published scholarly work, governmental and non-governmental organizations, practitioners, magazines, pamphlets, films, and websites swirls around the concepts of forest bathing and forest therapy. Within this assemblage, forest environments, particularly those designated as 'forest therapy bases' take on the form of 'real simulacra' (Desmond, 1999, p. 178), actualized embodiments of forest nature imaginaries that are stabilized through layers of authenticity comprised of scientific discourses, images, and physical markers that enable the forest to represent itself (Culler, 1981). Assembled in this way, the Akazawa forest becomes what it is presented as, and perceived to be, pristine forest nature. As such, it is equipped with the power to consolidate, encapsulate, and obfuscate the complexities of its own (re)production and contingent historical ecological origins. This ability, and its effects in the Kiso region, is an implicit part of ecotourism assemblage that is largely accomplished through the production, circulation, and consumption of images and discourses.

Popular outdoor sports – assembling ecotourism subjects

Within Japan's ecotourism assemblage, remote mountain landscapes and communities, such as those of the Kiso region, emerge as ideal spaces to encounter the pristine natures and traditional lifeways that occupy rural nature imaginaries. Outdoor sports, such as mountaineering, mountain biking and trail running in particular, continue to attract visitors to rural areas throughout Japan. Such leisure activities in Japan date at least to the Meiji period (1868–1912) when mountaineering and hiking emerged as elite, cosmopolitan, 'Western' pursuits (Wigen, 2005). Mountain sports focusing on the jagged peaks of the 'Japanese Alps' (*nihon arupusu*), which surround the Kiso region, continue to be popular forms of recreation and tourism, especially among middle-aged and elderly segments of the population. However, in recent years, outdoor recreation has gone mainstream. For example, in the late 2000s 'mountain girl' (*yama gāru*) emerged as a buzzword in Japan's popular media. A feature on the phenomenon published on Yahoo!, Japan's *X Brand* webpage, offers an illustrated depiction of a mountain girl as a gendered representation of outdoor recreation. The depiction focuses on her unique fashion, including her cute sport tights and trekking boots (Figure 8.1). Mountain girls, a caption reads, 'are women in their 20s to 30s who completely have their eyes on mountains'.[5]

In the late 1990s, well before the arrival of the mountain girls, outdoor sports performed in natural settings, such as running, biking, and swimming, began to gain popularity. In 1999, Jiro Takikawa, an avid trail runner, started Power Sports, Inc. with the goal of becoming the Japanese distributor for PowerBar, a US based producer of energy bars and other sports-related foodstuffs. Takikawa and Power Sports, Inc. achieved this goal in 2009, but the market for PowerBars and other outdoor sports products in Japan was limited. Thus, in order to create a market, Takikawa began assembling a network of governmental and non-governmental institutions, local residents, and athletes with 'powersports' at the center. In a greeting on the Power Sports, Inc. homepage, Takikawa states:

> Within the limited amount of budget, we had to raise awareness of the brand, and to accomplish that, we've come up with various outdoor sports events. Through such events, we enjoyed the close relationships with participants, and explained the effectiveness and importance of PowerBar products to them directly on site. At the same time, by bringing hundreds and thousands of participants to the rural towns in deep mountains or on beautiful island beaches, we have supported district areas throughout Japan, where local people are troubled with serious depopulation [sic].[6]

Among the various events noted by Takikawa is the 'Self Discovery Adventure', a series of races hosted by Otaki village. In addition to the outdoor sports races, the Self Discovery Adventure includes a variety of other activities directed at families and onlookers, such as kayaking and a concert featuring well-known DJs. Each spring and summer since 2008, hundreds of athletes and other participants have flooded into Otaki for two or three days at a time, filling local guest houses (*minshuku*), eating at the village's few restaurants, and shopping at one of its two grocery stores. In exchange for this influx of power sport ecotourists, the village

(Re)creating forest natures 181

Figure 8.1 Depiction of 'Mountain Girl' (yama gāru).

government provides use of a large park and assists with setting up race courses. In addition, many village residents contribute by volunteering their time and labor to make foods, sell t-shirts and other local goods, and provide entertainment. The stated aims of the race events resonate with images and discourses present throughout ecotourism assemblage: to provide opportunities for outdoor recreation and encounters with nature while preserving the integrity of local communities and environments. Meanwhile, capital accumulation through the production, circulation, and consumption of PowerBars and other outdoor sports-related products appears as background to these more 'central', idealized practices.

Despite the intentions of Takikawa and Power Sports, Inc., ecotourism assemblage in the Kiso region is not solely a corporate strategy, nor a governmental project; neither is it a resultant formation, nor something completed. Rather, assemblage marks collectivities and ongoing processes (Anderson and McFarlane, 2011) through which 'situated subjects [. . .] do the work of pulling together disparate elements' (Li, 2007, p. 265). At the same time, ecotourism assemblage is constantly hailing abstract ecotourist subjects (Althusser, 1971) which, when embodied, can be enlisted to do further assemblage work. For example, embodied ecotourist subjects, who are hailed through their desires to contact nature, serve as witnesses to the realness of those natures through personal experiences of encounter. Affective feelings of awe, sensations of relaxation, and even aching muscles, help to give shape to and enliven material manifestations of nature imaginaries, offering cohesion to the larger ecotourism assemblage. Ecotourists validate the reality of these natures by circulating 'authentic' images and discourses through Facebook posts, Twitter updates, photos shared on Instagram, blog entries, and even stories told to friends over beers. These further iterations of nature imaginaries offer testimony of the reality of ecotourism assemblage, building up a mimetic ontology and presenting a cohesive veneer that further enfolds ecologies.

Mountain girls, power sports enthusiasts, and other variations of ecotourist are not the only subjects that emerge from within outdoor and power sports iterations of ecotourism assemblage. Local residents also are hailed as the targeted objects of ecotourism. According to ecotourist ideologies, they, along with the environments in which they dwell, are to be the beneficiaries of ecotourist practices. However, here again we find a disconnect as local people, who lack livelihood alternatives, are enlisted as the service workers of emergent ecotourism economies. Labor in the Kiso region's ecotourism economy is precarious, with local residents often working in low-paying, seasonal, and insecure positions as cooks, housekeepers, guides, and so forth. What is more, residents offer also large amounts of unpaid labor as part of village revitalization efforts that are intimately connected to ecotourist ideologies and practices aimed at helping rural communities. In actuality, however, much of this unpaid labor is spent on activities and projects, such as thinning trees, cutting weeds, and maintaining trail systems, which are meant to craft local environments befitting ecotourist imaginaries. This is nothing particularly new in the Kiso region. Previous tourist 'booms' compelled similar reconfigurations of social and ecological life and left their marks on local landscapes, as the many empty ski hills and abandoned tennis courts throughout the region attest. Ecotourism in Japan promises something different; it promises, in the words of the Japan Ecotourism Society, '[s]ocial contribution for environment and culture through tourism [sic]'.[7] However, applying a political ecology framework and situating ecotourism within broader political-economic contexts raises critical questions concerning the basic premises of ecotourism in Japan and reveals the crucial role that images and discourses play in transforming rural landscapes into spaces of consumption that allow for novel forms of capital accumulation that may not be as 'green' or as 'socially responsible' as they initially appear to be.

Conclusion

In Japan, as elsewhere, ecotourism has gone largely unexamined in terms of its capacities to produce particular landscapes, subjects, and ecologies. The myriad images and discourses of nature that circulate through the assembled components of ecotourism serve to maintain its cohesiveness across spatial and temporal scales. On the one hand, images and discourses 'present a fictional universe in which many stories are possible and each feels like it fits with the others' (Igoe, 2010, p. 377). Yet, on the other hand, as Debord (1994, Theses 5–6) theorizes, images and discourses are not simply additional decorations or supplements to the real world. Rather, they are materially translated and entail 'a world vision which has become objectified'. In other words, though images and discourses often present fictions – an industrial plantation forest might appear in a pamphlet or on a website as a CO_2 sequestration forest – when they are assembled across scales with institutions, social practices, cultural values, scientific knowledge, biological organisms, and material environments, and when the assemblage is added to, subtracted from, and reworked to maintain cohesiveness through time, its fictions take on the weight of truths and the whole construct becomes real.

In this chapter I have argued that images and discourses are critical components of political ecologies of ecotourism in Japan. Drawing on case studies from the Kiso region, I suggested that images and discourses help mediate disconnections between ecotourist imaginaries and the actualities of post-industrial social and ecological life in which rural peoples are increasingly marginalized. Ecotourism assemblage in Japan increasingly exhibits the ability to transform marginalized rural landscapes into spaces of consumption, opening new horizons of capital accumulation, while suggesting an approach to doing so that is more just and environmentally friendly. Though the ideals of ecotourism may, in particular times and places, translate into lived realities, a political ecology of ecotourism in Japan suggests that in many locales we are witnessing simply novel iterations of longer term trajectories of disempowerment and decline in which political-economies of tourism play key roles.

We might, therefore, consider ecotourism assemblage in Japan as the latest iteration of a longer set of assembling processes, in which landscapes, subjects, and ecologies are continually reconfigured to meet the needs or desires of bodies (individual, but more often collective) with capacities to express power. Political ecologies are an effect of such assemblage processes. The difficulty in critically engaging with these processes relates to the quality of assemblages to attain cohesiveness by pulling together greater numbers of components through time and across space. Li (2007, p. 265) identifies this capacity as an effect of 'anti-politics', another of the 'crucial' practices of assemblage. Practices of anti-politics entail a narrowing of political terrain by 'reposing political questions as matters of technique; closing down debate about how and what to govern and the distributive effects of particular arrangements by reference to expertise; encouraging citizens to engage in debate while limiting the agenda'. Viewed from this angle, nature, as both an idea and a lived reality, plays a critical role in obfuscating the workings of power within political ecologies of ecotourism, in large part by defining the limits of what is imaginable, possible, and probable. In her study of concepts of nature in Japanese political ideology, Thomas (2001, pp. 2–3) notes '[w]hoever can define nature for a nation defines that nation's polity on a fundamental level. A nation's sense of nature [. . .] bespeaks its sense of collective and individual possibilities'. In its ability to signify a variety of phenomena, including human actions, material environments, and even the world as it is, practices of defining nature are perhaps the ultimate (anti)politics.

Given this 'trouble with nature' (Thomas, 2001) a key challenge for developing a political ecology of ecotourism is to uncover ways to engage anti-politics in order to expand the terrains of eco-politics. The cohesiveness of assemblages is always subject to change, and what assemblages are or might become is, at any given moment, never completely settled. Deleuze and Guattari describe assemblages as being comprised of lines of articulation and lines of flight (cited in Dewsbury, 2011), meaning that they are not static, but rather active, formations. In his ethnography of conservation practices in southwest China, Michael Hathaway uses the metaphor of winds to describe globalization, which, he explains, 'is not the self propelling movement of one form, logic, or modality but a place of articulation and human work that not only transforms what is often described as the global *but actually brings it into being*' (2013, p. 7 [emphasis in original]). In a similar way, ecotourist practices in Japan's Kiso region are involved in the bringing into being of various phenomena across scales, including nature, rural spaces, human subjects, the nation, and avenues of capital accumulation. As I have argued in this chapter, ecotourism is not, as it often appears in images and discourses to be, a panacea for the social and ecological ills of Japan's central highlands. In fact, more often than not the worlds brought into being through ecotourism are worlds already rooted in dominant modes of capitalist production and its attendant values and rationales. Ecotourist practices, therefore, could be said to add more 'lines of articulation' to a longer trajectory of assemblage. Yet, we should not underestimate that assemblages are also uncertain and that social, ecological, or conceptual spaces can, at any moment open and, within these spaces, winds may shift and novel lines of flight may emerge. Cleaving through these lines of flight and seeking ways to articulate alternative ways of being that are affirming of all life and productive of abundant and caring ecologies should be a key goal for further developing a political ecology of ecotourism both in Japan and beyond.

Notes

1 http://www.ecotourism.gr.jp/index.php/english [Accessed June 12, 2015].
2 Nagano Prefecture homepage, http://www3.pref.nagano.lg.jp/tokei/1_jinkou/jinkou.htm [Accessed May 25, 2015].

3 Ministry of Internal Affairs and Communication, Statistics Bureau, Japanese Government, http://www.e-stat.go.jp/SG1/estat/eStatTopPortal.do[Accessed May 25, 2015].
4 http://www.env.go.jp/nature/ecotourism/try-ecotourism/about/index.html [Accessed May 21, 2015].
5 http://xbrand.yahoo.co.jp/category/travel/5478/1.html [Accessed May 27, 2015].
6 http://www.powersports.co.jp/corpo/ [Accessed May 27, 2015].
7 http://www.eco-tourism.gr.jp/index.php/english/ [Accessed May 19, 2015].

References

Allison, A. (2013) *Precarious Japan*. Durham, NC: Duke University Press.
Althusser, L. (1971) Ideology and Ideological State Apparatuses. *Lenin and philosophy, and other essays*. London: New Left Books.
Anderson, B. and McFarlane, C. (2011) Assemblage and geography. *Area*, 43 (2), pp. 124–127.
Brockington, D. and Duffy, R. (2011) *Capitalism and Conservation*. Malden, MA: John Wiley & Sons.
Büscher, B. (2013) Nature on the move I: the value and circulation of liquid nature and the emergence of fictitious conservation. *New Proposals: Journal of Marxism and Interdisciplinary Inquiry*, 6 (1–2), pp. 20–36.
Campbell, L. M., Gray, N. J. and Meletis, Z. A. (2008) Political ecology perspectives on ecotourism to parks and protected areas. In: Hanna, K. S., Clark, D. A. and Slocombe, D. S. (eds) *Tranforming Parks and Protected Areas: Policy and governance in a changing world*. New York: Routledge.
Creighton, M. (1997) Consuming rural Japan: The marketing of tradition and nostalgia in the Japanese travel industry. *Ethnology*, 36 (3), pp. 239–254.
Culler, J. (1981) Semiotics of tourism. *The American Journal of Semiotics*, 1, pp. 127–140.
Dauvergne, P. (1997) *Shadows in the Forest: Japan and the Politics of Timber in Southeast Asia*. Cambridge, MA: The MIT Press.
Debord, G. (1994) *The Society of the Spectacle*. New York: Zone Books.
Desmond, J. C. (1999) *Staging Tourism: Bodies on Display from Waikiki to Sea World*. Chicago, IL: University of Chicago Press.
Dewsbury, J. D. (2011) The Deleuze-Guattarian assemblage: plastic habits. *Area*, 43 (2), pp. 148–153.
Duffy, R. (2002) *A Trip Too Far: Ecotourism, Politics, and Exploitation*. Sterling, VA: Earthscan.
Fletcher, R. (2015) Nature is a nice place to save but I wouldn't want to live there: environmental education and the ecotourist gaze. *Environmental Education Research*, 21 (3), pp. 338–350.
Fletcher, R. and Neves, K. (2012) Contradictions in tourism: the promise and pitfalls of ecotourism as a manifold capitalist fix. *Environment and Society: Advances in Research*, 3 (1), pp. 60–77.
Forest Therapy Society. (2015) *shinrin serapī sōgō saito (Forest Therapy Total Web)* [Online].
Forest Therapy Society. Available: http://www.fo-society.jp/quarter/ [Accessed May 15 2015].
Funck, C. and Cooper, M. (2013) *Japanese Tourism: Spaces, Places And Structures*. New York: Berghahn Books.

Hathaway, M. J. (2013) *Environmental Winds: Making the Global in Southwest China.* Berkeley, CA: University of California Press.

Hirada, T. (1999) Kiso-dani no rekishi (History of the Kiso Valley). Tokyo: Rindoren-kenkyusha.

Igoe, J. (2010) The spectacle of nature in the global economy of appearances: anthropological engagements with the spectacular mediations of transnational conservation. *Critique of Anthropology*, 30, pp. 375–397.

Igoe, J. (2013) Nature on the move II: contemplation becomes speculation. *New Proposals: Journal of Marxism and Interdisciplinary Inquiry*, 6 (1–2), pp. 37–49.

Ivy, M. (1995) *Discourses of the Vanishing: Modernity, Phantasm, Japan.* Chicago, IL: University of Chicago Press.

Iwamoto, J. (2002) The Development of Japanese Forestry. In: Iwai, Y. (ed.) *Forestry and the Forest Industry in Japan.* Vancouver: UBC Press.

Jussaume Jr., R. A. (2003) Part-time farming and the structure of agriculture in postwar Japan. In: Waswo, A. and Nishida, Y. (eds) *Farmers and Village Life in Twentieth-Century Japan.* New York: RoutledgeCurzon.

Kelly, W. W. (1990) Regional Japan: the price of prosperity and the benefits of dependency. *Daedalus*, 119, pp. 209–227.

Knight, C. (2010) The discourse of 'encultured nature' in Japan: the concept of Satoyama and its role in 21st-century nature conservation. *Asian Studies Review*, 34 (4), pp. 421–441.

Knight, J. (1997) A tale of two forests: reforestation discourse in Japan. *Journal of the Royal Anthropological Institute*, 3 (4), pp. 711–730.

Knight, J. (2000) From timber to tourism: recommoditizing the Japanese forest. *Development and Change*, 31 (1), pp. 341–359.

Li, T. M. (2007) Practices of assemblage and community forest management. *Economy and Society*, 36 (2), pp. 263–293.

McKean, M. (1982) The Japanese experience with scarcity: management of traditional common lands. *Environmental Review*, 6 (2), pp. 63–91.

Miyamoto, A. and Sano, M. (2008) The influence of forest management on landscape structure in the cool-temperate forest region of central Japan. *Landscape and Urban Planning*, 86, pp. 248–256.

Mock, J. (2014) Hidden behind Tokyo: Japan's rural periphery. *The Asia-Pacific Journal*, 12 (12), No. 3.

Moon, O. (2002) The countryside reinvented for urban tourists: rural transformation in the Japanese muraokoshi movement. In: Hendry, J. and Raveri, M. (eds) *Japan at Play.* New York: Routledge.

Morishita, T. (1998) *Omoide no kiso shinrintetsudō – Yama no kurashi wo sasaeta rokujūnen (Memories of the Kiso Forest Railroad–60 years that supported mountain livelihoods).* Matsumoto: Kyodoshuppansha.

Morris-Suzuki, T. (2013) The nature of empire: forest ecology, colonialism and survival politics in Japan's imperial order. *Japanese Studies*, 33 (3), pp. 225–242.

Ota, I. (1999) Declining situation of Japanese forestry today and its challenges toward the 21st Century. *Natural Resource Economic Review*, 5, pp. 103–124.

Ota, I. (2002) The shrinking profitability of small-scale forestry in Japan and some recent policy initiatives to reverse the trend. *Small-scale Forest Economics, Management and Policy*, 1 (1), pp. 25–37.

Oura, Y. (1992) Kokuyūrin ni okeru shinrin rekuriēshiyon jigyō no hatten: Ōtakimura eirinsho kannai kokuyūrin wo jirei toshite (The expansion of recreation forest practices in national forests: An example from national forests under the jurisdiction of the Otaki Forestry Office). Shinshu University, Graduate School of Agriculture.

Robbins, P. (2012) *Political Ecology: A Critical Introduction*. West Sussex: Wiley.

Salazar, N. B. (2012) Tourism imaginaries: a conceptual approach. *Annals of Tourism Research*, 39 (2), pp. 863–882.

Scott, J. C. (1998) *Seeing Like a State: How Certain Schemes to Improve the Human Condition Have Failed*. New Haven, CT, Yale University Press.

Seo, K. and Taylor, J. (2003) Forest resource trade between Japan and Southeast Asia: the structure of dual decay. *Ecological Economics*, 45 (1), pp. 91–104.

Thomas, J. A. (2001) *Reconfiguring Modernity: Concepts of Nature in Japanese Political Ideology*. Berkeley, CA: University of California Press.

Thompson, C. S. (2004) Host produced rural tourism: Towa's Tokyo antenna shop. *Annals of Tourism Research*, 31 (3), pp. 580–600.

Totman, C. (1989) *The Green Archipelago: Forestry in Preindustrial Japan*. Berkeley, CA: University of California Press.

Tsunetsugu, Y., Park, B.-J. and Miyazaki, Y. (2010) Trends in research related to 'Shinrin-yoku' (taking in the forest atmosphere or forest bathing) in Japan. *Environmental Health and Preventive Medicine*, 15 (1), pp. 27–37.

Tsutsui, M. (2003a) The Impact of the local improvement movement on farmers and rural communities. In: Waswo, A. and Nishida, Y. (eds) *Farmers and Village Life in Twentieth-Century Japan*. New York: RoutledgeCurzon.

Tsutsui, W. M. (2003b) Landscapes in the dark valley: toward an environmental history of wartime Japan. *Environmental History*, 8 (2), pp. 294–311.

Urry, J. (2002) *The Tourist Gaze*. London: Sage Publications.

Ushiomi, T. (1968) *Forestry and Mountain Village Communities in Japan: A Study of Human Relations*. Tokyo: Kokusai Bunka Shinkokai.

Waswo, A. (1988) The transformation of rural society, 1900–1950. In: Duus, P. (ed.) *The Cambridge History of Japan: Volume 6, The Twentieth Century*. Cambridge: Cambridge University Press.

West, P. and Carrier, J. G. (2004) Ecotourism and authenticity: getting away from it all? *Current Anthropology*, 45, pp. 483–498.

Wigen, K. (2005) Discovering the Japanese alps: Meiji mountaineering and the quest for geographical enlightenment. *Journal of Japanese Studies*, 31 (1), pp. 1–26.

Wigen, K. (2010) *A Malleable Map: Geographies of Restoration in Central Japan, 1600–1912*. Berkeley, CA: University of California Press.

Yano, C. R. (2002) *Tears of Longing: Nostalgia and the Nation in Japanese Popular Song*. Cambridge, MA: Harvard University Asia Center.

9 Ecotourism or eco-utilitarianism
Exploring the new debates in ecotourism[1]

Stephen J. Wearing and Michael Wearing

Introduction

In the 1990s, O'Brien and Penna developed an 'ecological citizenship approach' to the welfare state which remains a strong and trenchant critique of the profit motive as commodifying the use of nature, particularly in contemporary Western ecotourism development. This approach captured the internationally recognized definition of ecotourism as 'responsible travel' that conserves the environment and 'improves the welfare of local people' (The International Ecotourism Society, http://www.ecotourism.org). Understanding the social, economic and human rights not only of nation state citizens, but also in relation to environmental concerns such as animal well-being and care, is just as urgent an agenda for future life on this planet as it was several decades ago (Benton, 1993; Carson, 1962; Leopold, 1949; Linzey, 2009; Naess, 1973, 1989; Singer, 1975).

As Robbins (2011) indicates, we need to return to the big questions of human displacement, environmental degradation, sociocultural citizenship rights and local cultural understanding of land use and property rights in host communities. Utilizing a political ecology approach, we also need to engage more broadly with questions in societies ridden with social inequalities as to who benefits. And who loses from tourism industries? Hosts and nature seem to become the most marginalized and vulnerable to the economic and sociocultural risk of profiteering and free market economics in tourist and ecotourism planning. We would like to help stem this tide and create alternative political imaginings and community development for authentic modes of ecotourism. Thus, we adopt a political ecology framework that ' . . . proposes that environmental change and the bases of social citizenship are inextricably intertwined . . . the domination of environmental agendas by the interests of big business is currently stifling the creative political alternatives promoted through alternative political ecologies of modern society' (O'Brien and Penna, 1998, p. 182). In the 1970s and 1980s the deep ecologists began to use an ecosystems worldview and protection of biodiversity framework that encouraged a holistic approach to the spontaneous experience of nature. In their outline of different perspectives on the well-being of the living planet, the work of authors such as Rachel Carson, Arne Naess and Peter Singer helped initiate the rising moment of eco-politics in parliamentary democracies. In this period Western green parties and green social movements began to have substantive effects on democratic processes across Europe, Australia and New Zealand. Begun in the 1970s and emerging through several decades of green activism, the Tasmanian and New Zealand Greens are now recognized as the world's first two green parties. Also, non-profit activist organizations such as Greenpeace and the World Wide Fund for Nature (known in the US and Canada as the World Wildlife Fund) emerged as significant players in the environmental sector with important successes in protecting and preserving Australian and New Zealand flora and fauna.

Green parties – which are typically identified as a leftist political ideology – are now present in 17 European states and have won 46 (2009) and 50 (2014) seats in the European Union's parliament (Connelly *et al.*, 2012, pp. 102–110) and 56 (2014) seats for the broadly socialist 'green party' the Nordic Green Left. We follow Naess (1989) and Lovelock (2006) in our assertion that the forces of economic globalization require a concentrated political struggle and activism to sustain global ecosystems and deep ecology[2] for all (human and nonhuman) life (Devall and Yap, 1985). This has, for many 'greenies', involved a deep ecological consciousness and utilitarian moral action that uses gestalt thinking and deeper questioning about human interaction with nature to capture the sublime beauty of nature and the sanctity of all life and the sustainability of Gaia itself. In our conclusion to the chapter we will reassess how community-based ecotourism, grounded in some aspects of eco-utilitarianism and others alternative views, can be reinvigorated for local 'green tourist development' by the philosophies of scholars such as Naess (1973) and Leopold (1949) in using the theory of global and local ecosystems and deep ecology.

Political ecology as a theory and ideology is especially helpful in that it provides new and insightful ways for researchers to examine the human and environmental interactions that take place within local political economies and across a variety of stakeholders and institutions that significantly shape such social politics. Political ecology seeks to combine 'the concerns of ecology and a broadly defined political economy' (Blaikie and Brookfield, 1987, p. 17). This theoretical approach means we are concerned with not only how the state intervenes in the relations between capital and labour but also how nature, conservation and preservation are socially organized in civil society under the auspices of policy stakeholders to harm, be indifferent to or to protect the environment. We note that the expanding ecotourism literature is fragmented and has failed to connect with other topical areas such as local, regional and community development issues (Weaver and Lawton, 2007). This fragmentation is evident particularly in its links to the literature in political ecology, while it is also noted there is a lack of literature that addresses certain issues in ecotourism (Weaver and Lawton, 2007). We include deeper ethical issues to give more weight to this critique (Taylor, 2011). We are particularly concerned with how neoliberal conservation has influenced the entrepreneurship of community-based tourists as tools for the establishment of local economies that facilitate such development while undermining bottom-up wealth generation for host communities (Brenner *et al.*, 2010; Büscher *et al.*, 2012; Fletcher, 2012, 2014; Hunt *et al.*, 2014; Wight, 1993).

Is there a new social ethics and local political ecology agenda for communities that challenge such instrumental and market-based economising views of ecotourism? This chapter provides some discussion around the political ecologies of ecotourism as an eco-utilitarian activity that aligns with the actions of individuals, much the same as those resulting from Edward Abbey's *The Monkey Wrench Gang* (1975) and Peter Singer's *Animal Liberation* (1975). We consider also some of the implications of the dominance of 'neoliberal conservation' (Büscher et al., 2012) in ecotourism, seen as pro-market arguments that 'nature can be saved only through submission' to the power of capital and capital accumulation.

We develop counter-environmental arguments that move against the ecotourist industry's potential submission to market-based thinking. Neoliberal ideology embraces local practices and networks that support the commodification and privatization of nature. A strong counter position has opened up in line with green and green – left political parties and movements that encourages deeply thought-through environmental ethics and eco-practices in ecotourism and further afield (cf. Taylor, 2011). The arguments centre contextually around how, when and where, and under what governance, community-based and local ecotourism can be effectively developed. Beeton's (2006) important work on community development and community-based tourism provides an apt outline of how sustainable ecotourism can involve people at the local community level:

> The term ecotourism is contested and at times has been used simply as a marketing tool. Basically, ecotourism is nature based, educative and managed in a sustainable fashion. Sustainable management refers to the community as well as the natural environment: it requires operators to support the local community through employment, products, education and in other ways. Ecotourism also incorporates aspects of TBL *(triple bottom line or 3D reporting on ecological, social and financial dimensions of tourist endeavours)* sustainability including generating financial support for protection and management of natural areas, benefits for residents and resident support for conservationists.
>
> (Beeton, 2006, p. 92)

How will individuals and collectives act within the limitations of constraint and consent in local and national political contexts? The important strategic and tactical questions are how will such social agents refuse and challenge the hegemony of neoliberalism, economic dominance and business models that deny the deeply rooted basis of eco-citizenship. We add to the eco-utilitarian position of common good and common wealth grounded in nature, an understanding of 'the commons' and stewardship of nature that is imparted by human action, and consent to governing the planet – effectively as a 'land ethic' or eco-utilitarianism (Dryzek, 2006; Leopold, 1949; Lovelock, 2006; Ostrom, 1990).

Unfortunately, the macro-neoliberal context of capitalist economies and economic globalization means that there are serious ideological and economic constraints on the agency of local communities to protect nature and govern their own tourist trade. Beeton (2006), in particular, highlights the impacts of neoliberalism on lower-income rural communities and tourism. This critique raises questions about the fair distribution of benefits and burdens, not only to the communities involved, but also in terms of environmental impact on habitat and nature. Can local institutions and stakeholders guarantee environmental conservation, protection and sustainability in the tourism industry? To take this local community-versus-capital argument further we will discuss issues of ecotourism and protected areas. We use case studies of conservation of sea turtles and whale watching in countries such as Australia and Costa Rica, nature walking in Australia, and excluded indigenous communities in Finland and other locations (Campbell, 2007; Hunt *et al.*, 2014; Devine, 2014; Puhakka *et al.*, 2009; Xu *et al.*, 2014). Each case requires a consideration of multiple social, political, economic and technological dimensions and variables that impinge upon the local, social and environmental context for host communities. In recent years, other authors have managed to give considerable research weight to such studies and arguments (Hunt *et al.*, 2014; West and Carrier, 2004). Our arguments will question, in a general way, the current dominant neoliberal paradigm of ecotourism. We also pose deeper arguments than those who 'marketeer' ecotourism, and explore the possibilities of new ecotourism spaces supported by political and deep ecology thinking that is outside purely commercialized tourism (cf. Wearing and Wearing, 2015). Further we would argue that such alternative thinking adds social and cultural value to the sustainability of community-based ecotourism endeavours, local stakeholders' forms of management and negotiation of tourist sites and eco-conscious travellers.

What does a political ecology perspective offer in our analysis? The political ecology perspective encourages eco-citizenship among all people in a society to sustain and enhance nature and local host communities in conservation and protection of life and land use. For some, the areas of political ecology and sustainable development are viewed not as an entity that is utilized in sustaining the environment, but of one sustaining the capitalist 'business as usual' mentality that is entrenched within neoliberal economic policy. Such deeply conservative ideological views also encourage a conservation ethics that enables profit-making without restraint and truncates or minimizes the possible harmful and unjust effects of global industries such as tourism (Büscher *et al.*, 2012; Fletcher, 2012). The economic interest of neoliberal conservationist and ecotourist developers are preserved in this framework. These interests, while not of one ilk, necessarily include stakeholders such as medium-to-large tour operators and owners of tourist enterprises from outside local communities. Within this context, as Escobar explains, 'in the sustainable development discourse, nature is reinvented as environment so that capital, not nature and culture, may be sustained' (Escobar quoted in Watts and Peet, 1996, p. 10). Under these contemporary conditions, where the neoliberal economy is privileged over the environment, it is clearly highly unlikely that the environment and capitalism will ever be able to coexist in a sustainable manner (O'Connor, 1998). Indeed, as Harvey (2005, p. 175) notes, neoliberalization 'happens to be the era of the fastest mass extinction of species in the Earth's recent history'.

If the damage to ecosystems is as a direct result of neoliberal practices that promote the benefits of global market economies and capital accumulation as profit, then there is a need to explore where ecotourism fits into the political ecologies debate. The authors have always insisted that sustainability of ecotourism is possible in a parliamentary democracy and through localized politics but it is not a zero sum game (Wearing and Wearing, 2006, 2015). As with all commercial business enterprises undertaken in local communities by either for-profit or non-profit organizations there will be winners and losers. We want to encourage equity, fairness and cooperation at a local level in ecotourism enterprises rather than profit maximization and harm done to nature and/or local people. As we will argue here the social realities on the ground have demonstrated in this mission many ambiguities and potential contradictions across and within countries (cf. Brenner *et al.*, 2010; Fletcher, 2012).

In the dominant free market economies of the developed world policy implications are heavily influenced by the interplay of government regulation and market forces. Tourism is often promoted by government or industry without any overall strategy, without adequate attention to legislative frameworks, without consultation or inclusion of local communities and without effective protected area management plans. We will examine the key policy issues related to ecotourism, including a discussion of mechanisms to ensure that it does not exceed its sustainable base in moving toward understanding the provision of infrastructure for development and the policy and institutional prerequisites for planning and managing this activity.

How does local ecotourism's consumption of nature work?

> Locally ecotourism is viewed as the activity contributing most to improvements in residents' quality of life in the Osa Peninsula and to increased levels of financial and attitudinal support for parks and environmental conservation. Ecological ownership by local people is substantial and many local ecotourism workers plan to launch their own businesses.
>
> (Hunt *et al.*, 2014, p. 1)

Hunt *et al.*'s study of the Osa Peninsula on Costa Rica illuminates what is in many ways a best-practice community for ecotourism in a developing country. This is perhaps one case where practice does match the high-minded expectations and the green ideology of ecotourism. Nonetheless, when it comes to evaluating the sociocultural and economic costs and benefits of ecotourism to local host communities, several definitions set a high bar on ecological, sociocultural and financial grounds. According to the Japan Ecotourism Society (JES, 2015), ecotourism should: (a) utilize unique, local, natural, historical and cultural resources; (b) promote the conservation and preservation of local resources through appropriate management; and (c) activate local communities through responsible tourism and economic development that makes sustainable use of the natural and social resources. Ecotourism Australia (EA, 2015) defines ecotourism as 'ecologically sustainable tourism with a primary focus on experiencing natural areas and that fosters environmental and cultural understanding, appreciation and conservation'.

In 2001, the *Journal of Sustainable Tourism* published an article looking at the transition from whale hunting to whale watching in Tonga (Orams, 1999). Since then, the rapid growth of the whale-watching industry has 'industrialised the ocean' (Corkeron, 2004, p. 848). Viewed by the International Whaling Commission (IWC) in 1983 as an alternative 'use' for whales, today whale watching is recognized as a legitimate form of ecotourism (Orams, 1999), although is viewed by some as 'an acceptable form of benign exploitation' (Gillespie, 2003, p. 408). Leading up to 2001, the international whale-watching industry was valued at over US$1 billion (Hoyt, 2001), and attracted oven 9 million people annually. According to the International Fund for Animal Welfare (IFAW), by 2008 this number had grown to over 13 million people, participating in over 119 countries (O'Connor *et al.*, 2009).

The whale-watching industry also delivers some best-practice benefits and developments of local coastal communities, notably in Western countries such as Australia. Several rural towns in Australia, such as Huskisson and Eden on the New South Wales south coast, have benefited from the efforts of local whale- and dolphin-watching businesses. This is certainly an ecotourism area that combines building capacity for local economies and communities while making sustainable use of natural resources and environments. Patrick Ramage, the Director of the IFAW whale program, notes that whale-watching revenues have more than doubled since 1998, and that whale-watching operations around the world now include 3,330 operators, and employ an estimated 13,200 people, with the fastest growth seen in Asia (O'Connor *et al.*, 2009). Peter Garrett, the former Australian Environment Minister, reports the whale-watching industry generated US$ 2.1 billion of tourism revenue worldwide (Milne, 2009). Ramage adds, '[w]hile governments continue to debate the future of whaling, the bottom line is increasingly clear: Responsible whale watching is the most sustainable, environmentally-friendly and economically beneficial "use" of whales in the 21st century' (O'Connor *et al.*, 2009, p. 9).

The interesting social and ecological impact and evaluating question is: does such ecotourism deliver benefits to local communities in terms of income and jobs and infrastructure, the lack of which previously meant such towns and their surrounding communities suffered economically, with a good percentage of their populations remaining on low incomes relative to national average incomes for that country? This can be said to be true for rural and regional communities that have established ecotourism, both in developed and developing countries, where even more so in the developing countries the income disparities between rural and city populations are extreme.

As a best-practice model that moves away from neoliberal conservation and market-based ecotourism imperatives, whale-watching guidelines suggest it can be operated within the boundaries of sustainable practice. However, whale watching appears to have grown at an average rate of 3.7 percent per year, compared to global tourism growth of 4.2 percent per year over the same period (O'Connor *et al.*, 2009, p. 23), so it may not be possible to ensure that all stakeholders operate within the parameters of sustainable practice for whale watching. At a regional level, whale-watching average annual growth has exceeded tourism growth rates in five of seven regions reported: Asia (17 percent per year), Central America and the Caribbean (13 percent per year), South America (10 percent per year), Oceania and the Pacific Islands (10 percent per year) and Europe (7 percent) (O'Connor *et al.*, 2009, p. 23).

Currently, since it is viewed positively by tourists, it is logical from within a community-based ecotourism frame to pursue whale watching. For example, in 2007 tourists in the Dominican Republic were surveyed to determine whether the stance of a country toward whale conservation or whaling would affect their decision about whether to visit that country on holiday. The majority (77.1 percent) reported that if a Caribbean country supported the hunting or capture of whales or dolphins that they would be less likely to visit it. An even larger majority (81.1 percent) stated that if a country had a strong commitment to whale and dolphin conservation, they would be more likely to visit that country on vacation (Parsons and Draheim, 2009). Additionally, it is within the interests of the whale-watching industry to ensure that it remains within sustainable practices aligned to ecotourism. Higham and Lusseau (2008, p. 63) note the need for such sustainability, lest operators find themselves 'slaughtering the goose that lays the golden egg'. Effectively this would mean market values win out, with few standards in place for the quality of the ecotourist experience or more professional and responsible tourist operators and guides in these endeavours.

Further empirical research and evaluation is called for in order to investigate the values and views of tourists on the issue of whale watching, and the wider range of animal welfare issues. Also, as we briefly reflect upon at the end of this chapter, a reinvigoration of deep ecological ethics is required to maintain sociocultural, environmental and financial standards in ecotourism practice. The example of whale watching, like that of saving sea turtles, provides us with tools counter to neoliberal tourism and conservation, within arguments about how deep ecology, global ecosystems and the sustainability of the planet for life might work and may go beyond eco-utilitarianism in tourist endeavours. It is with this in mind that we argue that ecotourism creates a market value for the observation of animals through the commodification of wildlife and its habitats. At the same time it provides education through direct experience and kindles non-use, or intrinsic values, for the natural environment.

However, some have criticized on ethical grounds the commodification and consumption of animals through whaling and whale watching (Scarpaci *et al.*, 2008). Ecotourism provides us with an opportunity to provide both conservation and commercialization, where the direct human 'gaze' on wildlife is central to the experience (Ryan and Saward, 2004, p. 246). Given its alignment to alternative tourism (Wearing and Neil, 2009) it is believed that community-based ecotourism provides a mechanism to improve animal welfare and to conserve nature in general (Beeton, 2006). Further, these sustainability and deep ecological effects for the ecosystems and biodiversity of the environment have flow-on effects in the local economies of host communities. As recent evaluations of community-based ecotourism have demonstrated, an approach linked with 'bottom-up' community economic development and strategies on the part of planners ensures that a greater degree of success is likely to occur. Nonetheless, our arguments follow, both protected conservation areas, and social and economic protection for local community involvement in ecotourism, are necessary components of alternative approaches.

Protected areas discourse and community-based ecotourism

How can protected areas discourse challenge accepted thinking on market-driven ecotourism and neoliberal conservation? Nowhere are the conflicting views over ecotourism more evident than the current debate over the function and purpose of protected areas (Wearing and Neil, 2009). It is a conflict over two primary orientations; 'preservation' versus 'use', and tourism. As it is often delivered in protected areas, ecotourism embodies precisely this dilemma. Such an opposition is illustrated and reinforced through accepted institutional arrangements in which tourism and conservation goals are pursued by independent organizations. The current focus of the debate on tourism in protected parks is the extension of a long controversy, one that has existed since the conception of protected areas and similar reserves (see for example Runte, 1997).

The contemporary imperative for conservation advocates is then *how* to conserve rather than whether or not to conserve. This focus is on how to buy into a neoliberal conservation agenda and its corollary profit motivated economic development (Büscher *et al.*, 2012; Fletcher, 2012). Modes of community-based ecotourism, as a sustainable development strategy, are increasingly being turned to as part of a political philosophy for protected area managers and conservation agencies as a means of providing practical outcomes in the struggle to provide a basis for continued protection for these areas. The difference in effects between the anthropocentric and ecocentric worldviews is profound; they are separated by a chasm created by differences rooted in religion, beliefs and behaviour. Anthropocentrism assumes that humans are the most important part of the global ecosystem, and an ecosystem's function should be sustained primarily for human benefit. From this perspective, nature is a utility and of value because it is of use to human beings, even if only as a place for camping, recreation and tourism.

Further, what are the local beliefs and cultures that inform how ecotourism relates to people and nature? Xu *et al.* (2014) argue that in Chinese culture nature and its relationship to humans are different to those of the West and that local cultures in Chinese protected areas encourage 'harmony' in living with animals. This is illustrated by the relationship between ecotourism and protected areas in countries such as China and Costa Rica (Campbell, 2007) where there are significant socio-economic and cultural divides between how urban, national governments and local people interpret common property rights. Also, indigenous peoples such as Inuit and Australian Aboriginals make claims to the use of natural resources in protected areas for the hunting of whales and saltwater crocodiles, respectively, each of which are often on protected species lists. In these communities there is a case for different and de-colonized views being acceptable to meeting the needs of local people while sustaining the natural environment, including protected species.

In this discussion, protected areas that are a central focus for ecotourism are in effect an anthropocentric measure to save nature from humanity. Some would say they are islands in a resource-consumptive planet. Are they capable of saving wildlife and natural habitat, particularly as they overlap with human interest and the other, unprotected, 95 percent or more of the Earth's surface? We suggest that if we are to see the survival of the ideals of protected areas, they must have both an eco-utilitarian orientation based on the commons, for all life, and a deep ecology value based on the interdependence of all life on Earth within contemporary societies. Otherwise, the measures used over time to sustain these islands of biological value will not succeed. One salient example that highlights the politics of this issue has been the conservation of sea turtles in countries such as Costa Rica, where differences in scale at the local and national level can mask the realities of communities using turtles as a natural resource (Campbell, 2007, p. 328). Privileging the national over the local marginalizes local communities' ability to manage, harvest or otherwise use their natural resources, and local initiatives for conservation of turtles as resources and common property. In this case, conflicting claims by biologists and other scientists about the numbers of sea turtles can have direct and adverse effects on local communities' use of this resource. The specifics of this analysis make it difficult to justify an ecotourism industry based solely on turtle watching or assistance in turtle breeding.

In this view, we can identify ecocentric approaches, where the ecological well-being of the whole planet is emphasized, regardless of the direct benefits to the human population inhabiting it. Employing ecocentrism to prolong the integrity of natural ecosystems would be in sharp conflict with surrounding uses, and overpowering Western utilitarian ideology and neoliberal conservation. Ecocentrism is not a policy that can be applied with officiousness, nor is it politically palpable – it is a philosophy that requires a 'deep' thought process and a change in our exploitive attitude toward nature as a whole. In this chapter, we aim to show that, because the ideas behind ecocentrism are revolutionary and require a conceptual leap into the unknown in order to move forward, we must look to forms of what we term eco-utilitarian action. Ecocentrism cannot be applied as a practical measure to protect nature and areas of biological wealth and, if we are unable to save nature outside protected areas, very little will survive inside.

While a clear ideological basis for ecotourism is considered at best vague and difficult to reconcile as a reality in tourist practice, it provides us with links to the above discussions. Although it is sometimes suggested that the economic aspect of ecotourism has overridden the environmental considerations (Buckley, 2001; Wight, 1993) it can still move us toward our ecocentric goals. The many attempts to reconcile the economics of the capitalist market with environmental imperatives are increasing, and these attempts to reconcile economy and environment almost invariably leave unchanged the world's neoliberal political economy that took shape, in part, through the implementation of policies based on the Washington Consensus. Often, the dominance of the market model means that it is the environmental paradigm and the ecological and social values of local communities that are required to adjust (Hay, 2002, p. 201). Since at least as far back as the mid-1990s this neoliberal paradigm for change has gained ascendancy as the dominant model for economic and social development of nation states in general (Harvey, 2005), particularly involving the dominance of Western markets in income and wealth accumulation, and specifically in this case for ecotourist operations, both in developed and developing nations (Brenner *et al.*, 2010).

Do we need an industry-based neoliberal business model for ecotourism to be successful?

The development of ecotourist industries at a local level, based on neoliberal business models that support pure market and *laissez faire* ideologies is, of course, deeply flawed. Evidence in sustainable tourism research has shown time and again that the social and cultural impacts of such models are commonly harmful, both to nature and to the rights and needs of local citizens (Wearing and Wearing, 2006). Such development is tantamount to denying evidence from social scientists that neoliberal conservation and neoliberal tourism – 'profit for profit's sake' – harm local communities both culturally and socio-economically. This avoidance of clear evidence is not dissimilar to the rationale of climate change deniers who irrationally ignore the overwhelming evidence from the geo and physical sciences that the Earth is warming up and that warming is due to human activity (Hoffmann, 2011). The ecotourism industry makes extensive use of natural assets such as forests, reefs, beaches, mountains and parks, but what does it contribute to the management of these assets? The provision of ecotourism infrastructure, and the costs of managing the impact of tourism on host communities, is often borne by the environment, the local community itself and the government.

Local communities – especially indigenous or traditional custodians – are particularly vulnerable to the deleterious impacts of tourism development, as they directly experience the sociocultural impacts of tourism. In many cases indigenous cultures are used extensively to promote destinations to overseas markets, yet many indigenous people justifiably feel that the tourism industry has a poor track record, often disregarding locals' legitimate interests and rights while profiting from their cultural knowledge and heritage. In examining the sustainability impact of national parks in north-eastern Finland, researchers have argued that the inconsistent findings on social benefits to community, and the nature of tourism, mean that the ongoing local benefits and burdens need to be monitored and evaluated (Puhakka *et al.*, 2009). Thus, Nordic countries who, through a rigorously-managed system of national parks, place a strong emphasis on maintaining ecosystem integrity, find it difficult to gauge the impact of tourism in local areas.

Despite ecotourism's potential as a model for sustainable development we need to be aware of ecotourism's future direction, as well as its possible challenges and pitfalls. Political ecology offers critical frameworks that are needed to evaluate ecotourism, mindful that economic benefits from tourism often create insufficient incentives for local communities to support conservation. In the eyes of the local communities, benefits often are offset by the physical intrusion of tourists, greater income inequality within and between local communities, increased pollution, sequestering of profits by outsiders and rising local prices. Without this continual questioning of alternative evaluative frameworks we risk losing the promise of change that ecotourism offers. Traditional approaches often are resistant to new approaches to operational and institutional arrangements. Without adequate regulation of private sector activities and sound protected area management ecotourism development may have adverse impacts on the resource base upon which it depends. However, any viable tourism practice needs to address the imperatives of the market. Alternative approaches that may keep ecotourism at the cutting edge of change in society are in the areas of research, management, marketing and planning, which can provide new answers to perennial questions. Above, we have mentioned the Osla Peninsula in Costa Rica as one best-practice model for a community-based and potentially decommodified form of ecotourism (Hunt et al., 2014). Devine (2014) has pointed to the worrying trend in developing countries to militarize ecotourism. For example, she examines how Guatemala's military is being used to protect national parks such as the Maya Reserve, a practice that obscures the issue of poverty and deforestation and undermines environmental conservation and social justice issues. These examples help to highlight our general argument that deeper thought and inquiry is needed to protect species and environments, people and communities in local areas that embrace ecotourism, if we are to counteract neoliberal conservation measures and marketized tourism.

Social impacts and conservation ethics

> The relationship between tourism and the environmental movement has often been strained. Mass tourism has been viewed by many environmentalists as the extractive industries of forestry, fishing and mining. Environmentalists have searched for alternative forms of tourism which are more compatible with protecting the environment. Often however, tourism has been developed in the same destructive manner as have other more destructive industries.
> (Reid, 2003, p. 117)

Reid's (2003) comments here on the relationship between tourism and the environmentalist branding of 'ecotourism', written in the context of escalating economic globalization qua neoliberalism, ring true. In spite of the complexities of these issues, ecotourism is one of the few areas where the link between economic development and conservation of natural areas is clear and direct, and it is imperative that we consider this in explorations of its future role in contemporary society.

Despite conflicting interpretations and convenient deployment of the term 'ecotourism' within the tourism industry, one thing is certain: the increasing global interest and growth in ecotourism cannot be explained simply as another tendency in a long line of recreational trends that will inevitably fade away to be replaced by a new one. Instead, it reflects a fundamental shift in the way human beings view and engage with nature and tourism (Chafe, 2005). So, are we changing our fundamental views of nature?

Ecotourism suggests this is the case, as it was originally conceived of as an alternative to the increasing threat posed by mass tourism both to the culture and the physical environment of destination areas. The original emphasis of ecotourism was on low-key, unobtrusive tourism, which has minimal impacts on natural ecosystems. However, the term 'alternative tourism' is interpreted by various authors in sometimes disparate and openly contradictory ways. For some, alternative tourism is up-market package tours of rich people to exotic destinations, mostly wilderness areas, or young people carrying rucksacks wandering the globe with limited financial means, or travel that encourages contact and understanding between locals and tourists, as well as protecting the physical environment (e.g. Büscher *et al.*, 2012; Butler, 1990; Cohen, 1972; Curran *et al.*, 2009; Dowling *et al.*, 2002; Hunt *et al.*, 2014; 2002; Priporas and Kamenidou, 2003; Richards and Wilson, 2004).

While we find that, in an attempt to balance anthropocentrism and ecocentrism, O'Riordan (1989, pp. 85–6) suggests that 'environmentalism seeks to embrace both worldviews'. Gough, Scott and Stables (2000) suggest a blurring of the extremes between the ecocentric/anthropocentric polarities that is rendered necessary partly by an acceptance that all human worldviews must be in some sense anthropocentric. Nonetheless, aspects of environmentalism may be organized along the single ecocentricity/anthropocentricity dimension. This view is the one possibly best aligned to ecotourism, where its contribution provides benefits to the environment and communities, while it still inevitably impacts on that section of natural environment and on those communities. Ecotourism is viewed as an eco-utilitarian activity that is enabling for a population that is seeking to do something for the environment; it is not an overall solution but a mechanism that provides a link for an increasingly neoliberalized society, to one supportive of 'nature'.

The use of protected areas for ecotourism offers us some insights into the changing nature of 'nature', as it has been advocated as a way in which the political establishment attempts to keep nature alive and kicking for some time, while we slowly, under the auspices of EISs and the like, extract nature's ecological resources in a 'sustainable' manner. For example, there is still no unified agreement on the concept of 'sustainable development' (Vojnovic, 1995). It can be said that the concept connects abstract environmental issues with people's personal and commercial interests. In many cases the sustainability goal is being applied only to the economic part of the development process, and the ecological part is considered only as a background for the stage where the economy is developing (Fowke, 1996). Ecotourism has used its banner of sustainable development of and in protected areas to establish its credentials. However, just as with other development in protected areas, we see issues arise with regard to overuse while, at the same time, it offers opportunities to connect nature to an interested market and influences the individuals' in that market valuing of nature.

Ecotourism offers people the enjoyment of nature within their leisure (eco-utilitarianism), and it has been this non-biological form of conservation (eco-centrism) that has been the driving force behind initiatives such as the national parks movement and the protection of wild areas and, although we might like to pretend otherwise, the continued valuing of these areas now is reliant on what might be termed an eco-utilitarian valuing *through* activities and experiences such as ecotourism. Indeed, we would not need national parks if we did not have such an exploitive relationship with nature (McNeely and Miller, 1989). The potential for pressure from outside protected areas will always be great as long as Western society continues to value wildlife in terms of its human use. The Western paradox is that 'humanity depends upon that which it threatens' (Diamond, 1991) and therefore the ways it values those phenomena is of key interest in the future.

An eco-care ethic: holding environmental and social values in balance

> To the ecological field-worker, the equal right to live and blossom is an intuitively clear and obvious value axiom. Its restriction to humans is an anthropocentrism with detrimental effects upon the life quality of humans themselves. This quality depends in part upon the deep pleasure and satisfaction we receive from close partnership with other forms of life
>
> (Naess, 1973, pp. 96–97)

Thus far, our argument has hinged on the valuing and rights of both local citizenship and nature itself in developing and sustaining ecotourism. The important dimension missing is a mechanism for holding nature and local community interests in balance. For example, as the evidence shows in some African countries (Curran *et al.*, 2009), local jobs and income sources from farming land can be lost through conservation protection. This is also a real threat to jobs and workforce incomes from large mining and logging companies in countries such as Australia and throughout Southeast Asia. Do such local economic and income difficulties also plague ecotourism development, especially where local people who live on low incomes, are poorly educated, and reside in rural communities, do not have the training or formal education needed to work in the tourist sector and are replaced by transient workers from outside, such as fly-in fly-out staff akin to what goes on in the mining industry? For community-based ecotourism, the question of values is central to considerations of its often competing and incompatible conception of, and practices toward, the natural world (Belshaw, 2001; Weaver, 2002). If we want to respect nature, and if we consider the way in which we relate to nature as not morally neutral, we must stop seeing this simply as a set of instrumental values (resources) and be willing to recognize that nature has intrinsic values (Larrère and Larrère, 2007).

Godfrey-Smith (1980) identifies two primary ways in which value is assessed in Western society. If the value that something is said to hold is a means to a valued end then it is designated as being of 'instrumental value'. 'Intrinsic value', on the other hand, is value that exists in its own right, for its own sake. What is central here is the ethic that such ideas and values underpin. For example:

- An ethic of 'use' – this is the normative or dominant mode of how human beings relate to nature: where nature is viewed predominantly as a set of resources which humanity is free to employ for its own distinct ends. It is an instrumental and anthropocentric view embodied by neoliberal conservationist views as how to divide the spoils or profits of a conservation enterprise e.g. in creating industry with the badge ecotourism (Wight, 1993);
- An ethic 'of nature' – holds that nonhuman entities are of equal value with the human species. It is broadly intrinsic and ecocentric, and is one that aligns with deep ecological questioning of life as part of larger less knowable ecosystems than human endeavour and science. It accepts the political ecology and eco-citizenship approach of equally valuing both nature's and citizens' rights.

An ethic of use begins from a human locus and it is this univocal perspective that often is described as anthropocentrism. The ultimate grounding of value in the Western world is intrinsic, as human beings are placed as the source of all value and, by extension, are the measure of all things. Such a view allows nature no intrinsic value in itself and for itself as its value lies only in satisfying human needs and desires. However, it is unfair to make the philosopher John Locke the villain of this piece of anthropocentrism, which has a very long and deeply entrenched history. It has, for all intents and purposes, been the single deepest and most persistent assumption of all the dominant Western philosophical, social and political traditions at least since the time of the classical Greeks and the rise of the Judeo-Christian religions (White, 1967; Nash, 1989; Fox, 1990; Taylor, 2011). This legacy of anthropocentrism is fuelled today also by neoliberal conservation that places economic values and the profit motive as utmost in struggles over the environment, notably through the dominance of rich nation states, such as Australia, the UK and the US, support for economic globalization and globally marketable ecotourism (Büscher et al., 2012).

Nonetheless, the alternative ecosystems and gestalt of all-life models of environmental discourse have impacted local, national and global politics across developed and developing countries (Dryzek, 2006). Ecotourism, mobilized as a commercialized set of discourses, forgoes a more radical agenda to promote green change and deep ecology thinking. On the other hand, guides and modes of practice have moved on embracing local and specific cultural experiences while engaging in more physical and robust travel. This in itself requires a rational calculating tourist who, somewhat in contradiction, engages at a deeper imaginative and creative level with traditional cultures and peoples. In Tibet, for example, a much more difficult, challenging and rewarding 'eco experience' occurs in modes such as trekking. Again, those trekking-like tourists who cycle, leave a small ecological and ecotourist footprint that makes it difficult for local communities to benefit unless they embrace a local entrepreneurial spirit, launching small 'green' businesses and developments such as in Costa Rica (Hunt et al., 2014; McCue, 2010). Most of all, this approach will involve an ethics of care for all living things – and for the planet itself – that does not compromise on challenging property rights and the submission of communities to greater industry-focused and international capital and profit (Taylor, 2011).

Conclusion

> *The theory of ecosystems contains an important distinction between what is complicated without any Gestalt (wholistic) or unifying principles – we may think of finding our way through a chaotic city – and what is complex. A multiplicity of more or less lawful interacting factors may operate together to form a unity, a system . . . Such complexity makes thinking in terms of vast systems inevitable. It also makes for a keen, steady perception of the profound human ignorance of biospherical relationships and therefore of the effect of disturbances.*
>
> (Naess, 1973, p. 97)

Naess's gestalt thinking and eco-practice in terms of the global ecosystem preceded and yet in some ways predicted the difficulties and complexities that climate-change scientists have had in communicating the idea of global warming (Garvey, 2008; Hoffmann, 2011). In a similar way, social scientists have had the problem of warning of the sociocultural and environmental impact of tourism as it began to emerge as an economically, socially and environmentally significant industry in the twentieth century. Yet, there is no better understanding of how political and deep ecology, supported by scientific research, can be used to synthesize and challenge the current global order. In relation to tourism and ecotourism the ascendancy and dominance of neoliberalism as a global ideological trajectory is catastrophically affecting the governance of nation states and local communities. The rise of independent green parties in the West and green social movements may continue, with the support of planners, academics, policy makers and activists, and by local communities developing their economies and civil societies from the bottom up. In contributing to these struggles we have followed Naess's thinking in embracing the qualities of nature as building contentment and inner satisfaction for local communities, and by way of sustaining and developing human interaction with nature through ecotourism endeavours that do not submit to market criteria as a priority, or act single-mindedly. There is perhaps no better platform than a book on political ecology to challenge neoliberal conservation and tourism.

Notes

1 We dedicate this chapter to Professor Arne Naess (1912–2009) for his deep and brilliant thinking on ecology, and our father Reverend Leslie Wearing (1931–2012) for taking us both, as brothers in our childhood years, into rural communities and into 'nature'.

2 Deep Ecology is regarded as a radical environmental movement and philosophy that stems from a belief that humans are equal to other organisms within the global ecosystem.

References

Abbey, E. (1975) *The Monkey Wrench Gang*. Philadelphia, PA: Pippincott Williams & Wilkins.
Beeton, S. (2006) *Community Development through Tourism*. Collingwood, VIC: Landlinks.
Belshaw, C. (2001) *Environmental Philosophy*. Stocksfield: Acumen.
Benton, T. (1993) *Natural Relations: ecology, animal rights and social justice*. London: Verso.
Blaikie, P. and Brookfield, H. (1987) *Land Degradation and Society*. London: Methuen.
Brenner, N., Peck, J. and Theodore, N. (2010) Variegated neoliberization: geographies, modalities, pathway, *Global Networks*, 10 (2), pp. 182–222.
Buckley, R. (2001) 'Major Issues in Eco-labelling', In X. Font and R. Buckley (eds), *Tourism Eco-labelling: Certification and Promotion of Sustainable Management*. Oxford: CABI, pp. 19–26.
Büscher, B., Sullivan, S., Neves, K., Igoe, J. and Brockington, D. (2012) Towards a synthesized critique of neoliberal biodiversity conservation, *Capitalism Nature Socialism*, 25 (2), pp. 4–30.
Butler, R.W. (1990) Alternative tourism: pious hope or Trojan horse? *Journal of Travel Research*, 28 (3), pp. 40–45.
Campbell, L.M. (2007) Local conservation practice and global discourse: a political ecology of sea turtle conservation, *Annals of the Association of American geographers*, 97 (2), pp. 313–334.
Carson, R. (1962) *Silent Spring*. Boston: Houghton, Mifflin.
Chafe, Z. (2005) Consumer demand and operator support for socially and environmentally responsible tourism. CESD/TIES Working Paper Number 104.
Cohen, E. (1972) Toward a sociology of international tourism, *Social Research*, pp. 164–182.
Connelly, J., Smith, G., Benson, D. and Saunders, C. (2012) *Politics and the Environment: from theory to practice*, 3rd edn. London: Routledge.
Corkeron, P. J. (2004) Whale watching, iconography, and marine conservation, *Conservation Biology*, 18 (3), pp. 847–849.
Curran, B. *et al.* (2009) Are Central Africa's protected areas displacing hundreds of thousands of rural poor? *Conservation and Society*, 71 (1), pp. 30–45.
Devall, J. and Yap, N. (1985) *Deep Ecology*. Salt Lake City, UT: Peregine Smith.
Devine, J. (2014) Counterinsurgency eco-tourism in Guatemala's Maya Biosphere Reserve, *Environment and Planning*, 32, pp. 984–1001.
Diamond, J. (1991) *The Third Chimpanzee: the evolution and future of the human animal*. New York: Harper-Collins.
Dowling, R., Newsome, D. and Moore, S. (2002) *Natural Area Tourism: Ecology, Impacts and Management*. Clevedon: Channel View Publications.
Dryzek, J. (2006) *The Politics of the Earth: environmental discourses*, 2nd edn. Oxford: Oxford University Press.
Ecotourism Australia (2015) What is Ecotourism? Available at: http://www.ecotourism.org.au/eco-experiences/why-choose-ecotourism/. Accessed 2 October 2015.

Fletcher, R. (2012) Using the master's tools? Neoliberal conservation and the evasion of inequality, *Development and Change*, 43 (1), pp. 295–317.

Fletcher, R. (2014) *Romancing the Wild: cultural dimensions of Eco-tourism*. Durham, NC: Duke University Press.

Fowke, R. (1996) Sustainable development, cities and local government, *Australian Planner*, 33 (2), pp. 61–66.

Fox, W. (1990) *Towards a Transpersonal Ecology*. Boston: Shambhaia.

Garvey, J. (2008) *The Ethics of Climate Change: right and wrong in a warming world*. London: Continuum.

Gillespie, A. (2003) Legitimating a whale ethic, *Environmental Ethics*, 25 (4), pp. 395–410.

Godfrey-Smith, W. (1980) Travelling in time, *Analysis*, 40 (2), pp. 72–73.

Gough, S., Scott, W. and Stables, A. (2000) Beyond O'Riordan: balancing anthropocentrism and ecocentrism, *International Research in Geographical and Environmental Education*, 9 (1), pp. 36–47.

Harvey, D. (2005) *A Brief History of Neoliberalism*. Oxford: Oxford University Press.

Hay, P. (2002) *Main Currents in Western Environmental Thought*. Bloomington: Indiana University Press.

Higham, J.E. and Lusseau, D. (2008) Slaughtering the goose that lays the golden egg: are whaling and whale-watching mutually exclusive? *Current Issues in Tourism*, 11 (1), pp. 63–74.

Hoffmann, M.J. (2011) *Climate Governance at the Crossroads: experimenting with a Global Response after Kyoto*. Oxford: Oxford University Press.

Hoyt, E. (2001) Whale watching 2001. *International Fund for Animal Welfare, UNEP*.

Hunt, C.A., Durham, W.H., Driscott, L. and Honey, M. (2014) Can eco-tourism deliver real economic, social and environmental benefits? A study of the Osa Peninsula, Costa Rica, *Journal of Sustainable Tourism*, 23 (3), pp. 339–357.

Japan Ecotourism Society (2015) Available at: http://www.ecotourism.gr.jp/index.php/english/. Accessed 2 October 2015.

Larrère, R. and C. Larrère (2007) Should nature be respected? *Social Science Information*, 46 (1), pp. 9–34.

Leopold, A. (1949) *A Sand County Almanac*. New York: Oxford University Press.

Linzey, A. (2009) *Why Animal Suffering Matters*. Oxford: Oxford University Press.

Lovelock, J. (2006) *The Revenge of Gaia*. London: Allen and Unwin.

McCue, G. (2010) *Trekking Tibet: A Traveler's Guide*, 3rd edn. Seattle, WA: The Mountaineers Books.

McNeely, J. A. and Miller, K. R. (eds) (1989) *National Parks, Conservation, and Development: The Role of Protected Areas in Sustaining Society*. Washington DC: Smithsonian Institute Press.

Milne, G. (2009) Garrett axes globetrotting whale envoy. Available at: http://www.dailytelegraph.com.au/garrett-axes-globetrotting-whale-envoy/story-e6freuy9-1225715198271. Accessed 2 October 2015.

Naess, A. (1973) The shallow and the deep, long-range ecology movement: a summary, *Inquiry: an interdisciplinary Journal of Philosophy*, 16 (1), pp. 95–100.

Naess, A. (1989) *Ecology Community and Lifestyle* (trans.) David Rothenberg. Cambridge: Cambridge University Press.

Nash, R. (1989) *The Rights of Nature*. Sydney: Primavera Press.

O'Brien, M. and Penna, S. (1998) *Theorising Welfare: enlightenment and modern society*. London: Sage.

O'Connor, James (1998) *Natural Causes: Essays in Ecological Marxism*. New York: Guilford.
O'Connor, S., Campbell, R., Cortez, H. and Knowles, T. (2009) Whale Watching Worldwide: tourism numbers, expenditures and expanding economic benefits, a special report from the International Fund for Animal Welfare. Yarmouth MA, USA, prepared by Economists at Large, 228.
Orams, M. (1999) *Marine Tourism: Development, Impacts and Management*. London: Routledge.
O'Riordan, T. (1989) The challenge for environmentalism, *New Models in Geography*, 1, 77.
Ostrom, E. (1990) *Governing the Commons: the evolution of Institutions for Collective Action*. Cambridge: Cambridge University Press.
Parsons, E.C.M. and Draheim, M. (2009) A reason not to support whaling – a tourism impact case study from the Dominican Republic, *Current Issues in Tourism*, 12 (4), pp. 397–403.
Peet, R. and Watts, M. (2004) *Liberation Ecologies: Environment, development, social movements*. New York: Routledge.
Priporas, C. V. and Kamenidou, I. (2003) Can alternative tourism be the way forward for the development of tourism in Northern Greece? *Tourism (Zagreb)*, 51 (1), pp. 53–62.
Puhakka, R., Sarki, S., Cottrell, S. and Silkamaki, P. (2009) Local discourses and international initiatives: sociocultural sustainability of tourism in Oulanka National Park, Finland, *Journal of Sustainable Tourism*, 15 (5), pp. 529–549.
Reid, D. (2003) *Tourism, Globalization and Development: responsible tourism planning*. London: Pluto Press.
Richards, G. and Wilson, J. (2004) *The Global Nomad: Motivations and Behaviour of Independent Travellers Worldwide*. Clevedon: Channel View Publications.
Robbins, P. (2011) *Political Ecology: A Critical Introduction*, 2nd edn. San Francisco, CA: Wiley-Blackwell.
Runte, A. (1997) *National parks: the American experience*. Lincoln, NE: University of Nebraska Press.
Ryan, C. and Saward, J. (2004) The zoo as ecotourism attraction – visitor reactions, perceptions and management implications: the case of Hamilton Zoo, New Zealand, *Journal of Sustainable Tourism*, 12 (3), pp. 245–266.
Scarpaci, C., Parsons, E.C.M. and Lück, M. (2008) Recent advances in whale-watching research: 2006–2007, *Tourism in Marine Environments*, 5 (1), pp. 55–66.
Singer, P. (1975) *Animal Liberation: Towards an End to Man's Inhumanity to Animals*. New York: Harper Collins.
Taylor, P. (2011) *Respect for Nature: a theory of environmental ethics*, 2nd edn. Princeton, NJ: Princeton University Press.
Vojnovic, I. (1995) Intergenerational and intragenerational equity requirements for sustainability, *Environmental Conservation*, 22 (3), pp. 223–228.
Watts, M. and Peet, R. (1996) Towards a theory of liberation ecology, *Liberation ecologies: Environment, development and social movements*, pp. 260–269.
Wearing S. L. and Neil, J. (2009) *Eco-Tourism: impacts, potential and possibilities*, 2nd edn. Oxford: Butterworth-Heinemann.
Wearing, S. L. and Wearing, M. (2006) Re-reading the 'subjugating tourist' in neoliberalism: postcolonial otherness and the tourist experience, *Tourism Analysis*, 11 (2), pp. 145–162.
Wearing, S.L., and Wearing, M. (2015) Local, just and sustainable eco-tourism: decommodifying grass roots struggle against a neoliberal tourism agenda. In: J. Mosedale (ed.), *Neoliberalism and Tourism: Projects, Discourses and Practices*. London: Ashgate.

Weaver, D. (2002) Hard-core ecotourists in Lamington National Park, Australia, *Journal of Eco-tourism*, 1 (1), pp. 19–35.

Weaver, D. and Lawton, L. (2007) Twenty years on: the state of contemporary eco-tourism research, *Tourism Management*, 28, pp. 1168–1179.

West, P. and Carrier, J. (2004) Eco-tourism and authenticity: getting away from it all? *Current Anthropology*, 45 (4), pp. 483–498.

White, L. (1967) The historical roots of our ecologic crisis, *Science*, 155 (3767), pp. 1203–1207.

Wight, P. (1993) Eco-tourism: ethics or eco-sell? *Journal of Travel Research*, 3, pp. 3–9.

Xu, H., Cui, Q., Sofield, T. and Li, F. (2014) Attaining harmony: understanding the relationship between eco-tourism and protected areas in China, *Journal of Sustainable Tourism*, 22 (8), pp. 1131–1150.

Part III
Development and conflict

Introduction to Development and conflict

Mary Mostafanezhad

'Indispensable tools for making sense of rural livelihoods', Tania Li suggests, include 'who owns what, who does what, who gets what, and what do they do with it' (following Marx, Li, 2014, p. 6). Each of these questions is equally relevant for making sense of tourism-based livelihoods and livelihoods affected by tourism development around the world. While the chapters in this section consider a broad geographical scope, which spans Africa, the Arctic, South America and South Asia, they are linked by the shared experiences of tourism development and related conflicts. Local lives, as these chapters suggest, are embedded in broader struggles over the dialectical relationship between economic development and conflict over the control of territory and natural resources. Questions about who wins and who loses in the development game more generally have been at the forefront of political ecology debates (Escobar, 2001; Escobar, 2005; Peluso, 1992; Robbins, 2012). Chapters in this section illustrate how political ecology is enacted in and through tourism development, and how human and nonhuman actors produce nature through politically mediated tourism development initiatives. Building on Robbins's inclusion of 'all producers of nature' – actors that range from Indigenous community members, transnational business executives, tourists on mopeds and local tour guides – this section examines the networks and circuits of ecological production as they are linked with local, national and transnational tourism development practices.

The analytic of conjuncture is developed by Li to explain how myriad elements come together to create a constellation through which we experience the world. She explains how the object of analysis in a conjunctural approach includes 'the set of elements, processes, and relations that shaped people's lives at this time and place, and the political challenges that arise from that location' (Li, 2014, p. 4). This analytic is particularly useful for the study of the political ecology of tourism development in that it allows for a more fine-tuned understanding of the myriad relations of production and consumption that collide in tourism development practices as well as concurrent conflicts over resources. These conjunctures are illustrated in foundational works on the political ecology of tourism, such as Susan Stonich's examination of the relationship between tourism, conservation and development in the Bay Islands (2000) and Stefan Gössling's illuminating illustrations of the relationship between environmental change and Robbins's 'all producers of nature'. This relationship, as Gössling so aptly showed in the context of small island tourism, is critical to the creation of more sustainable development pathways (2003).

Over the past decade, many tourism-focused scholars have been preoccupied with the role of neoliberalism, both as economic policy and as ideology. Fundamental aspects of neoliberalism include the expansion of unregulated market exchange and the privatization of social services. Thus, ongoing processes of enclosure as well as the privatization of common goods such as education, water, air and health care are characteristic of neoliberal practices (see Chapter 13). While the details of neoliberalism continue to be debated (Ferguson, 2010; Harvey, 2005; McCarthy and Prudham, 2004), neoliberal practices are united through the widespread belief that privatized market exchange is preferred over government regulation. Neoliberal ideology extends into individual behavior where it is assumed that individual perseverance and self-development lead to economic success. A result of these practices is the growth of private organizations that fill in where the state has left off, or perhaps where it never was. Through 'pro-poor', responsible, and ecotourism practices, such growth has taken many forms, such as microcredit schemes not originally embraced by the private sector.

Part III contains Josh Fisher's examination of the urban revitalization 'Live Clean' campaign in Nicaragua and its attempts to make Managua a palatable space for touristic consumption. Drawing on urban political ecology, Fisher investigates political strategies of 'aesthetic governmentality' in tourism development and illustrates how tourism development contributes to fixed tourism imaginaries and the creation of new environmental subjectivities. In Chapter 12 Megan Holroyd considers the development of international tourism in Tanzania, now the top foreign exchange earner in the country. Employing more than one out of every ten citizens, tourism in the country is often perceived as a panacea for economic and social challenges. Holroyd thus examines the conjuncture of conservation, tourism and land dispossession at Mount Kilimanjaro. Illustrative of persistent trends in tourism, she examines how local elites benefit from its development at the cost of the majority of the local population's further marginalization and exclusion from access to land and natural resources. In Kaokoland (Kaokoveld), north-west Namibia, Jarkko Saarinen investigates the ways in which a country facing chronic drought has been proactive with initiatives that use community-based natural resource management as an economic development strategy. This strategy aims to link economic incentives with environmental management and wildlife conservation. Tourism, Saarinen surmises, is increasingly regarded as a force that can be used for poverty alleviation through the creation of income and employment. This is, in part, because there are no 'brokers' between the mainstream tourism industry and local community members. Yet, he also illustrates how the legacy of colonial conservation efforts linger and thus complicate the postcolonial tourism development encounter. In Chapter 14 Lill Rastad Bjørst, Carina Ren and Dianne Dredge transport us north to the Arctic, where they address one of the most pressing issues of our time – climate change. Their examination deconstructs the links between tourism development and the popular tourism imaginary of the Arctic. In doing so, they also reveal political, economic, social and environmental conjunctures and relations of power in which tourism development in Greenland is embedded and how discourses of economic sustainability become a proxy for global and climatic concerns. In Chapter 13 Kevin Hannam and Anya Diekmann shift our attention to India, where they ethnographically examine the political ecology of slum tourism, now a fast growing niche tourism market throughout the so-called Global South. In their attention to the meanings slum tourists ascribe to their experiences, Hannam and Diekmann seek to denaturalize what they describe as the toxic ecologies of slum environments, thus pointing to a significant yet rarely examined focus of inquiry: the political ecology of health in tourism. Thus, they illustrate how, rather than engage with the politics of issues such as the privatization and pollution of water, slum tourists contribute to the broader depoliticization of uneven geographies of health and sanitation as well as global economic inequality more broadly.

Collectively, the chapters in this section consider how various forms of development and conservation agendas articulate with broader circuits of ecological production that extend well beyond the 'local'. Additionally, the capacities of tourism development both to empower and marginalize local ecological traditions are theoretically and empirically investigated. The essentialization and exotification of environmental degradation in economically poor communities is denaturalized against the backdrop of broader national and transnational politics of human rights to fundamental resources. Ultimately, the section's shared focus is the complex relationship between development (ranging from grassroots to international) and conflict (spanning local to global) over land, water and the natural resources that are embedded within both.

References

Escobar, A. (2001) Culture sits in places: Reflections on globalism and subaltern strategies of localization. *Political Geography*, 20, pp. 139–174.
Escobar, A. (2005) Imagining a Post-Development Era. In: Edelman, M. and Haugerud, A. (eds) *The Anthropology of Globalization and Development: From Political Economy to Contemporary Neoliberalism*. Oxford: Blackwell Publishing.
Ferguson, J. (2010) The uses of neoliberalism. *Antipode*, 41, pp. 166–184.
Gössling, S. (2003) *Tourism and Development in Tropical Islands: Political Ecology Perspectives*. Cheltenham: Edward Elgar.
Harvey, D. (2005) *A Brief History of Neoliberalism*. New York: Oxford University Press.
Li, T. M. (2014) *Land's End: Capitalist Relations on an Indigenous Frontier*. Durham, NC: Duke University Press.
McCarthy, J. and Prudham, S. (2004) Neoliberal nature and the nature of neoliberalism. *Geoforum*, 35, pp. 275–283.
Peluso, N. L. (1992) The political ecology of extraction and extractive reserves in East Kalimantan, Indonesia. *Development and Change*, 23, pp. 49–74.
Robbins, P. (2012) *Political Ecology: A Critical Introduction*. New York, Blackwell.
Stonich, S. C. (2000) *The Other Side of Paradise: Tourism, Conservation and Development in the Bay Islands*. Putnam Valley, NY: Cognizant Communication Corporation.

10 Political ecologies and economies of tourism development in Kaokoland, north-west Namibia

Jarkko Saarinen

Introduction

Namibia is the driest country in sub-Saharan Africa. Over 90 percent of the country's land is characterized by arid conditions with the climate being generally hot and dry, with sparse and unpredictable rainfall. In spite of the relatively tough environment, ethnic groups that have traditionally lived in the region have adapted to the conditions and harsh landscapes. Along with its vast wilderness environments and charismatic wildlife resources, local cultures and communities form a basis for the attractiveness of tourism in the country. Especially in rural and natural areas the tourism industry is seen as the prominent avenue for development and a tool for economic diversification. As a result, tourism has been increasingly introduced to local communities during the past two decades. This has been based not only on the growth of tourism in the region, and the country more generally, but is also the result of a policy emphasis on community-based tourism (CBT) development in Namibia. The country has been particularly proactive with its aim to integrate communities in tourism development (Lapeyre, 2011a; Novelli and Gebhardt, 2007). The key tool here has been a Community-Based Natural Resource Management (CBNRM) program which aims to serve as an enabling platform that links 'economic incentives with environmental management and wildlife conservation' (Brown and Bird, 2010, p. 2). In most cases such economic incentives are based on tourism development and consumption.

As a result of ongoing environmental change, some communities' tourism has become a potential alternative source of additional income or even a new replacement livelihood. While local communities have adapted to the harsh conditions and natural fluctuations in weather, recent extreme changes such as the 2012–2013 droughts (see van den Bosch, 2013) have challenged the viability of traditional livelihoods and created the need for local residents to seek out benefits from newly emerging economic activities. The impacts of recent droughts have been especially serious in the Kunene region of north-west Namibia, where the Ovahimba people (approximately 25,000) live. Indeed, over the past decades, the Ovahimba communities have faced several environmental and also sociopolitical crises (Henrichsen, 2000). The area is demarcated by a veterinarian fence preventing the movement of livestock, meat and related diseases from north to south of the country. Thus, large-scale livestock herding is not an economic option, and this together with the arid climate and common droughts hinder the viability of traditional livelihoods, making Ovahimba communities vulnerable to change. In the case of Ovahimba, vulnerability refers to the degree to which their communities are susceptible to, or even unable to cope with, opposing effects of change and the effectiveness of certain adaptive mechanisms that act as bases for adaptive capacity (Smit and Wandel, 2006). Thus, vulnerability also influences the ability to adapt to tourism and associated changes. The relationship between vulnerability and adaptation is significantly affected by power relations between the tourism industry and communities.

Traditionally the Ovahimba are considered to be semi-nomadic pastoralists, dependent on cattle and small-stock herding (Bollig and Heinemann, 2002). While representing just 2 percent of the country's population, they are highly visible in Namibian tourism promotional materials (Saarinen and Niskala, 2009). At a local scale tourism can provide additional sources of livelihood and employment possibilities for the people. Yet, even with the aim to create benefits and development, it may also contribute to inequalities and marginalization (Saarinen, 2011, 2012). With the exception of the tourist practices of South African and domestic (white) independent travelers, the tourism industry in the Kunene region is highly organized and non-locally coordinated. Communities who have opened up to tourism have no means to control the industry. Instead, they can aim only to be 'attractive' for visitors and relocate their settlements closer to roads and therefore be more accessible for visitors catered for by the mainstream tourism industry. This makes them more dependent on tourism as well as more vulnerable to the vagaries of the tourism industry.

This chapter examines the relationship between tourism development and Ovahimba communities in the Kunene region's northern areas known as Kaokoland (Kaokoveld), where tourism is increasingly regarded as a strategy for poverty alleviation through the creation of income and employment. My aim here is to scrutinize the critical issues in tourism benefit sharing in such a situation, where no 'brokers' or other coordinating or controlling actors exist between the mainstream tourism industry and extremely poor and potentially vulnerable ethnic groups, such as the Ovahimba. The empirical examples used herein are based on a combination of previous studies on host–guest relations and community benefits from tourism in Kaokoland (see Saarinen, 2011, 2012) utilizing community interviews (individual and focus groups) in four communities near the town of Opuwo, the administrative center of the Kunene region.

While tourism in the Kunene region and Ovahimba communities may not officially be labelled as 'Pro-Poor Tourism' (PPT), it does involve core elements from the principles of PPT (Ashley and Roe, 2002). Tourism in Ovahimba communities focuses on the poor. It involves an organized benefit-sharing model with people who are considered to be 'poor' and live in places where community-based organizations (CBOs) or similar projects do not already exist and thus might hinder the capacity and practices of the mainstream tourism industry to create development in local communities and destinations. Goodwin (2009, p. 91) argues that 'the radicalism of the PPT approach was seeking to use mainstream tourism to achieve the objective of poverty elimination'. Therefore, it is also important to consider how the emerging new responsibility discourse in tourism development in which PPT is often situated can be seen as a product of neoliberal 'self-organising' modes of new governance with certain understandings of environment and its uses and values (Hall, 2014). In this respect, a political ecology framework provides fruitful avenues to think about the nature of tourism-community relations and related benefit-sharing models and outcomes.

As an approach, political ecology aims to open up unequal power relations that guide and control benefits from the utilization of natural and/or cultural resources (see Bryant and Bailey, 1997). It is highly intertwined with questions of political economy. As defined by Blaikie and Brookfield (1987, p. 17) political ecology 'combines the concerns of ecology and a broadly defined political economy'. Thus, this chapter utilizes perspectives of both political ecologies and economies; the changing environment and uses of natural resource by local communities and emerging tourism industries are highly interlinked with both. Previous studies have noted that the expected socio-economic benefits and promised development impacts of emerging tourism practices are not always realized (see Britton, 1991; Lapeyre, 2010; Mosedale, 2011; Saarinen, 2012), but are instead unevenly distributed based on a structural marginalization and exclusion where powerful actors, such as the mainstream tourism industry, have control over the operational environments (Hall, 2007; Telfer and Sharpley, 2007; see Robbins, 2004). This process further marginalizes communities in development, which indicates that certain groups of people may be excluded from such development and related decision-making (Lenao, 2014; Watts, 2000; Zapata et al., 2011).

Knowledge and power relations are critical to the process of inclusion and exclusion from the costs and benefits of tourism development. The processes of tourism development often are managed, supported and controlled by international or other non-local actors and organizations (Hall, 1994), and the potential resulting imbalances and inequalities between the tourism industry actors and communities become determining focal points for a political ecology approach. The chapter first presents an overview of the connections between tourism and development with a specific focus on a relatively new emphasis on poverty alleviation. The tourism development context in Namibia is then discussed, followed by a case study demonstrating the political ecologies and economies of tourism development in north-west Namibia. Finally, I conclude by indicating how tourism-led economic growth is not easily translated into major benefits for local people and their living environment.

Global tourism and local development

Tourism is a global-scale activity in which expansion is based on general socio-economic development including local improvements in technologies of mobilities and accessibilities (see Hall, 2005) but also on increased knowledge of the world and its cultures. In this respect tourism is a product of, and a vehicle for, so-called cultural globalization. Nijman (1999, p. 148) has defined cultural globalization 'as acceleration in the exchange of cultural symbols among people around the world, to such an extent that it leads to changes in local popular cultures and identities'. Ideally, ethnic tourism to Ovahimba communities should be concerned with equitable cultural exchange, in which tourists are motivated to experience and learn about different cultures and local needs, and with the ability to gain economically and in other ways from tourists. However, while the increasingly global tourism industry brings together different people and cultures, it also constructs and changes physical and symbolic landscapes of distant places.

Based on the industry capacity to create positive changes in destination environments, tourism has become an important element for many governments searching for new paths for socio-economic development (Sinclair and Stabler, 1997). Globalization and the corollary transformation of rural economies has contributed to tourism development that is increasingly seen as a promising tool for economic diversification, especially in the peripheries of the Global South where there are few alternatives (see Mowforth and Munt, 1998; Saarinen *et al.*, 2009). Ashley and Maxwell (2001, p. 395) argue that 'Poverty is not only widespread in rural areas, but most poverty is rural'. In addition to being a mechanism for employment and revenue creation in rural and peripheral areas, tourism can stimulate nature conservation and initiate cultural exchange and understanding (Smith, 2003; Smith and Richards, 2013). Tourism development also can assist ethnic groups in showcasing their culture and revitalizing their traditions (Mbaiwa and Sakuze, 2009).

However, if development policies and plans are practiced in unsustainable and irresponsible ways, tourism can often have negative consequences. Sinclair (1988), for example, suggested that the economic aspects of tourism should always be placed in an equation consisting of both the advantages and disadvantages of tourism development. This means that economic benefits should be considered with an appreciation of the known or estimated costs. Indeed, tourism development can cause serious cultural, social and environmental change, both positive and negative. In relation to ethnic groups these changes can include land use and access issues, commodification of natural and cultural resources and marginalization of local communities (Smith, 1989; Yang, 2011). Thus, while the touristic production system has a great potential to contribute positively to the socio-economic basis of destination areas, it can also involve major risks for communities and their environments (Mowforth and Munt, 1998), especially countries in the Global South, where tourism systems often are characterized by dependency, inequalities, enclavization and leakages (Britton, 1991), and many of which are facing the challenges of uneven tourism growth (Scheyvens, 2011). In the African context Harrison (2000, p. 37) has emphasized this challenge by stating that, historically speaking, the tourism industry on the continent has been developed 'by colonialists for colonialists'.

In an effort to ameliorate some of the negative impacts of tourism in the area, many scholars underline the need for the industry to be more responsible, espousing the explicit aim to benefit the local people and the very places tourists visit (Rogerson, 2006; Scheyvens, 2011). This has led to a search for more locally beneficial and sustainable tourism development models (Butler, 1999) such as CBT and PPT initiatives. There is an increasing amount of academic and policy literature, as well as development projects, that focus on CBNRM, CBT and PPT in southern Africa (Adams and Hulme, 2001; Jones and Murphree, 2004; Scheyvens, 2009), including Namibia. In general CBNRM aims to involve local communities in natural resource management by stipulating that local communities should have control over the uses and benefits of natural and cultural resources located in their living environment (Blaikie, 2006). CBT initiatives are based thinking along similar lines that local communities should benefit from the use of their environment and/or their culture and heritage resources in tourism. This requires local participation in, awareness of, and control over tourism development (Holden, 2007; Saarinen, 2012; Stronza, 2007). These kinds of community-involvement needs have become increasingly highlighted and have called into debate issues in both global and local responsible tourism development policy discussions. These issues are briefly introduced in the following section.

Emerging poverty alleviation focus in tourism development

Over the past several decades there has been an increasing interest among policy practitioners and academic researchers in integrating tourism development discussions with the aim of global poverty alleviation (Brickley *et al.*, 2013; DFID, 1999; Hall, 2007; Novelli and Hellwig, 2011; Saarinen *et al.*, 2013; Scheyvens, 2011; Spenceley and Meyer, 2012; UNWTO, 2002). This global-scale connection to poverty alleviation often references the UNMDGs that aimed, in part, to halve the number of people suffering from extreme poverty by 2015 (UNWTO 2006). Whether this goal has been achieved is still up for debate, but one key issue here is related to its potential connection with tourism development, especially when thinking beyond 2015 (Saarinen and Rogerson, 2014). Beyond the reduction of poverty (and hunger), the UNMDGs comprised seven additional aims, including facilitating universal primary education, promoting gender equality and the empowerment of women, and developing paths towards environmental sustainability, globally speaking.

In a tourism policy context, Francesco Frangiolli, then Secretary General of the World Tourism Organization (UNWTO), stated in 2006 that the tourism industry can play a major role in the achievement of the UNMDGs (UNWTO 2006; see also World Bank, 2012). This potential role for tourism in the context of UNMDGs, especially in relation to poverty alleviation, has become popular in recent policy discourses. However, it has also raised critical and important questions about whether and how the tourism industry can meet these expectations. Additionally, the use of tourism as a development strategy brings up a number of important questions, considering how it is ultimately a commercial concern that aims to benefit its owners, who are rarely members of poor rural communities on the peripheries of the Global South (Britton, 1982, 1991). The industry operators often have control over their operational environment, which includes natural and/or cultural resources that are also used by local communities. This control may not only exclude people from their traditional resource access and related decision-making, but also marginalize locals from tourism-related development processes. Indeed, as noted (even) by the World Bank (2012, p. 7), the tourism industry 'comes with its own set of risks and challenges'.

In this context, Scheyvens (2009, p. 193) has asked why we should 'assume that [the mainstream tourism industry's actors] have some ethical commitment to ensuring that their businesses contribute to poverty-alleviation?' Without questioning the overall aim of, and need for, an ethical commitment within tourism development – a legitimate debate in and of itself (see Butcher, 2003) – it may be extremely difficult to provide a justifiable answer to the 'why' question per se; such an answer would go beyond a marginal number of tourism operators having a moral stand and an active aim and commitment towards genuine corporate social responsibility (CSR). In addition, it is often difficult to locate who actually holds responsibilities in (or for) tourism development (see Saarinen, 2014; Sharpley, 2013). In contrast, the 'how' question may be relatively easier to answer – but perhaps only on the level of principle. For example, UN Secretary General Ban Ki-moon has stated that 'insofar as it is developed in a sustainable manner, tourism has the capacity to contribute to the alleviation of poverty . . . in order to achieve the Millennium Development Goals' (UNWTO, 2007). Therefore, the purported 'global responsibility' to alleviate poverty through tourism development is based on the relevance, (intended) applicability and policy success (see Hall, 2014) of what has come to be called, 'sustainable tourism'.

Obviously, the call for sustainable tourism principles and manners leaves open the 'real-world' dimension (e.g. whether the industry can meet these high aims and expectations in practice). While there are serious frustrations concerning the idea and implementation of sustainability in tourism development (Butler, 1999; Saarinen, 2014; Sharpley, 2009), there is also increasing interest among tourism practitioners to create development models that contribute to sustainable development. In the context of poverty alleviation and local communities in the Global South, the previously mentioned – although contested – PPT represents this policy and its implementation aim (Ashley and Roe, 2002; Lapeyre, 2011b; Mitchell and Ashley, 2010; Rogerson, 2006; Zapata et al., 2011). The PPT refers to a way of organizing 'host–guest relations' that would lead to net benefits for the poor (Ashley and Roe, 2002; Meyer, 2008; Scheyvens, 2009, 2011). The emphasis towards direct community collaboration and connections with the mainstream industry separates PPT from other 'alternative approaches', such as CBT, which are viewed as a small-scale, public-oriented development approach without sufficient capacity to contribute to the wider poverty reduction agenda (see Goodwin, 2009). Tourism development and community relations in Namibia serve as a demonstrative case of both dimensions: the community- and the industry-driven approaches in benefit sharing.

Tourism and development in Namibia

In Namibia the tourism industry is increasingly used as a medium for attaining a diverse scale of economic and social goals (Ashley, 2000; Republic of Namibia, 1994). After independence in 1990 the tourism industry, together with the mining sector, agriculture and fisheries, became a cornerstone of the country's moves for socio-economic development. In 2011 the direct contribution to the country's GDP was estimated to be 2.9 percent, with relatively high growth expectations (Christie *et al.*, 2014). In contrast to other key economic sectors, tourism is visibly used for the promotion of nature conservation and community development in Namibia (MET, 2005, 2008; see Novelli and Gebhardt, 2007). The CBNRM program and CBT are particularly highlighted in the country's development policies (Kavita and Saarinen, 2015; MET, 2008).

The Ministry of Environment and Tourism (MET) initiated a national CBNRM program in 1991, with a realization that there would be a further need for a specific CBT policy. Based on this CBT policy, implemented in 1995 (MET, 1995), and with created legislative support, rural communities are currently able to register conservancies which provide them rights to manage and use wildlife and other natural resources (Lapeyre, 2010; Long, 2004). By definition a conservancy is a territorial unit where natural resource management and utilization activities are undertaken by a group of people who have organized themselves based on governmental guidelines and acting under a management body (NACSO, 2008). The resources managed by conservancies are effectively based on the same resources on which the overall attractiveness of tourism in Namibia is based. Thus, the conservancies can serve as a tool to encourage the participation of rural poor communities in tourism development, and create benefit-sharing processes.

There has been extensive study of community-tourism relations and ample evaluations of benefit sharing in Namibian conservancies (Bandyopadhyay *et al.*, 2011; Kavita and Saarinen, 2015; Lapeyre, 2010, 2011a, 2011b; Massyn, 2007; Saarinen, 2010). Previous research has demonstrated both successes and failures, making community-tourism relations in conservancies an interesting case to discuss from a political ecology perspective. Here, however, my specific interest is a context that specifically does not involve a communal conservancy policy environment with legislative support, but rather a situation where poor rural communities are integrated into the tourism system without a broker and external advisors or other such agents that are characteristic of conservancies operating as CBNRM and CBT oriented units. Therefore, the case study that follows represents a pro-poor context where local Ovahimba communities in north-west Namibia are directly operating with mainstream tourism businesses with an agreed benefit-sharing model.

Political ecologies and economies of tourism development in Kaokoland

The Ovahimba

Ovahimba are increasingly both directly and indirectly involved with tourism in Namibia. Although they are geographically, politically and socio-economically isolated (see Bollig and Heinemann, 2002), their images and representations are extensively used in the country's tourism promotion (van Eeden, 2006; Saarinen, 2010; see Echtner and Prasad, 2003). In advertising they are characterized as 'primitive', 'exotic' and 'passive' objects for tourists to visit and gaze upon (Saarinen and Niskala, 2009). Typical Ovahimba images include 'half-dressed' women or women with children (van Eeden, 2006; Kanguma, 2000). In both advertising materials and the safari tourist products themselves, the Ovahimba's role as 'tribal people' is to remain 'good' natives (Saarinen, 2012; see Neumann, 1998). This kind of primitivization of ethnic groups has been a characterizing feature in travel literature since the colonial era (Pratt, 1992; Wels, 2004) and contemporary tourism promotion represents a continuation of that process (see Edwards, 1996).

In tourism promotion and popular literature the Ovahimba are usually described using terms referring to the past, authenticity and nostalgia (Jacobsohn, 1998), and are seen to be unchanged and untouched by modernity (Lange, 2004). In various travelogues and ethnographic volumes this is regarded as an outcome of geographical isolation in a harsh environment but moreover resulting from their active refusal of external influences. Jacobsohn (1998, p. 9), for example, states in her book *Himba: Nomads of Namibia* that the 'Himba strategy to generally reject outside values was made possible by their economic independence'. While this is one potential explanation for their isolation, there may be other, less nostalgic or glorifying reasons (e.g. the current situation may not be solely explained by a social cohesion and resulting strong traditions and values). The Ovahimba continue to be largely isolated in Namibian society (Mupya, 2000; Saarinen, 2012). However, in addition to their remote location and the veterinarian fence preventing the movement of livestock and meat sales, there are also important politics of space taking place. Such political geographies originate from a historical context of ethnic relations and painful processes of relatively recent independence. Before the independence in 1990 the Kunene region's northern stretches (e.g. Kaokoland) was a so-called operational area – effectively a 'war zone' between the South African Defence Force (SADF) and the South West Africa People's Organization (SWAPO). Due to the war operations, access to Kaokoland as a border area between Angola and Namibia was restricted (see Bollig and Gewald, 2000).

The SWAPO-led forces and the members of political independence movements comprised the largest population group of Ovambos (approximately 50 percent of the Namibian population), while many Ovahimba either served in (or were associated with) the SADF or did not involve themselves in the conflict at all (see Jacobsohn, 1998; Mupya, 2000). After independence SWAPO became the dominant political party in the country. While SWAPO is politically against tribalism, history (including the process of independence) has left the Kaokoland partially economically underdeveloped, therefore marginalizing the Ovahimba and (possibly) excluding them from the modernization of Namibian society. Thus, while the Ovahimba have succeeded in maintaining their traditions, values and traditional livelihoods, not all related decisions have been in their own hands.

Tourism and benefit sharing in Kaokoland

Kaokoland represents a 'periphery of a periphery' in the global and Namibian national tourism system (Weaver and Elliott, 1996). The main elements of touristic attraction in Kaokoland are the region's wilderness landscapes and Ovahimba culture. The total number of tourists to the region is comparatively modest but Novelli and Gebhardt (2007, p. 450) note that during the 2000s the region 'experienced considerable tourism growth'. Rothfuss (2000, p. 135) estimated that at the turn of the millennium there were some 9,000 tourists visiting the region annually. However, there are no official statistics available to indicate past or current numbers.

The benefit-sharing model between the tourism industry and local communities is notable for several reasons. Benefits are not based on monetary payments, but rather 'gifts' that are given by the visiting organized tourist groups mainly in the form of foodstuffs. Typically, the foodstuffs consist of three main elements: maize flour, sugar and cooking oil, which are the core of the Ovahimba diet. The amount of expected gifts/foodstuffs is relatively fixed, based on the size of the visiting group, but it is influenced also by the size of the community. In addition, visiting tourists, at times, are said to offer gifts of medicine, tobacco and small sums of money (Saarinen, 2011) but this usually happens without the influence of guides or operators. There is also an increasing craft sale activity in the villages, which provides monetary payments to individuals (Saarinen, 2012). Like occasional payments for the opportunity to take photographs, some of these souvenir sales take place outside the Ovahimba villages, which makes the level of monetary income very difficult to evaluate. However, these monetary payments from souvenir sales, while relatively small, are considered valuable by the Ovahimba since money can be used when visiting nearby towns, including for medical and other consumption needs.

The background of this atypical benefit-sharing model in tourism-community relations originated in the late 1990s. After independence, and Kaokoland's opening up for tourism, visitations to the Ovahimba villages were based on monetary payments to the village chiefs or headmen. This practice did not always lead to benefit sharing among the wider community, however, which created various tensions and problems. Thus, in order to both ensure Ovahimba communities welcomed visitors and develop more equal and transparent modes of benefit sharing, the foodstuff model was created by a collective agreement. Eventually, by the end of the 1990s and early 2000s, it replaced monetary payments.

Culturally any visitation to the Ovahimba villages requires a permit from a chief or headman. However, in an organized tourism context, the permit process has only a symbolic role as the foodstuff arranged by the operators and guides work as an offering in exchange for entry into the village. When entrance to the village is granted, the gifts are placed next to the chief or headman for all to see. After the tourist group leaves the village the foodstuff is distributed by the people equally among every woman-led hut in the community (Figure 10.1). A household led by a male can consist of several huts.

In the Ovahimba villages the visits are highly organized; a majority of groups arrive with a guide or guides provided by the tour operator, and some communities exclusively receive visits from organized groups. These operators are mainly non-local but there are local hotels that also provide guided tours. The

Figure 10.1 Sharing the benefits from tourism: distributing maize flour to households after the tourist group has left the community.

Photo: Jarkko Saarinen.

local establishments, however, are not necessarily 'locally' owned or managed. For example, the Opuwo Country Lodge (Figure 10.2) handles the reservation bookings via the capital city of Windhoek; the lodge's owners reside in the colonial-style coastal town of Swakopmund. The lodge has an African safari-themed interior with a large number of luxury rooms and the lodge provides guided excursions (for approximately US$ 50 per person) to meet 'Namibia's last remaining nomadic tribe, the Ovahimba'. According to their characterization the Ovahimba are:

> dictated by seasonal rainfall and constantly in search of new grazing for their cattle, the Himba remain nomadic and primitive. Their unique lifestyle and adaptability to survive the harsh Kaokoland environment has earned them their ethnic individuality.
> (Opuwo Country Lodge, http://www.opuwolodge.com/index.htm)

They further state that these visits are part of the Lodge's 'responsible tourism community programme'. While there are also few individual guides from the local community, these actors are exceptions in the regional tourism system.

The Ovahimba communities value the tourism-based economic opportunities and benefits, and the growing role of tourism is generally considered positive by local people (Saarinen, 2011). In order to receive increased visitors some of

Tourism development in Kaokoland 223

Figure 10.2 The Opuwo Country Lodge's swimming pool and a view to dry 'Himbaland'.
Photo: Jarkko Saarinen.

the communities have decided to re-settle either on the outskirts of the town of Opuwo or relatively close to the town and/or tourist routes. At the same time, access to water has become more challenging. In some cases the industry has 'provided' Ovahimba with a touristically beneficial location that nevertheless lacks the water resources necessary for the people and their livestock in the harsh environment. This problem is making communities dependent on external water deliveries (e.g. in Puros), while the visiting tourists are never short of water (see Figure 2). Furthermore, the presence of locals' herding pastures can degrade the environment, in a form of erosion, due to higher concentration of livestock closer to tourism resources and routes. Thus, during drier periods, and especially during the droughts, men have to move the cattle and other livestock very far away from the communities. This creates land-use conflicts elsewhere, among other problems. In addition, those people who have settled next to the town become increasingly dependent on begging and donations from international tourists.

Altogether, the lure of emerging tourism, despite negative environmental changes, has turned many communities increasingly towards this industry. In the villages the received foodstuff provides the main benefits from tourism. Estimates of the received shared gifts vary greatly between communities as some of them are more accessible than others from the main hotels and tourist routes. Previous studies indicate that a given annual sum community benefits can range from US$ 60 to US$ 1,000 (when the value of foodstuffs is calculated at local market prices), while the size of communities varies between 11–15 households (Saarinen, 2012). The contribution of tourism to the communities and individual households does not seem to be particularly extensive. This has been noted by some community members who do not mention tourism employment, joint ownership or control over tourism businesses (for example) as actual or potential benefits to the communities of tourism development. Thus, the position where people place themselves in the production system of tourism has become the same as given in the place promotion, for example, as being passive recipients of tourists visiting their villages.

In contrast to the relatively unorganized initial period of tourism development in the 1990s, community members do not perceive there to be major problems caused by tourism and tourism-related changes in their living environment. The degradation of the environment is seen as a separate issue, and the organized nature of tourism, with a fixed benefit-sharing model in a form of foodstuff, is primarily seen as positive, despite the fact that some members would prefer additional monetary payments based on craft sales and donations. As there have not been major evident or perceived problems, community members tend to share a view that increased tourism would be beneficial for the villages. However, the villagers have thus far not independently attracted more tourists to their village. This is, in part, because attracting tourists to the village is widely seen to be the role of the industry and travel agents. While the people do not seem to have a vision of how to actively influence the markets, many of the community members have developed an understanding of why tourists want to visit the villages, that is, to come into contact with Ovahimba culture and observe people's way of life (Saarinen, 2012). Therefore, community members stressed the importance of maintaining their way of living in order to remain attractive to tourists. In practice, this refers primarily to the position of women, as well as how they dress and what they wear. Based on this perception, some community members indicated that they were reluctant to send local children, especially girls, to school. The logic was that the educated children would not necessarily dress traditionally and, thus, they would potentially lose the attractive nomadic and primitive habitus used as a selling point in Namibian tourism promotions (see Saarinen and Niskala, 2009). That would be the end of ethnic tourism, it is widely perceived.

What is important to notice here is that, in order to receive benefits from the mainstream tourism industry and to support their traditional livelihoods, the communities are willing to accept these trade-offs, such as to relocate themselves to an area that has no natural resources or to exclude themselves (or their children) from systems of formal education. In this context a key question is one of socio-economic viability: Are the benefits from tourism sufficient? Or are the costs of vulnerability to environmental change and exclusion from Namibia's modernization processes too high in the longer term? Based on the indicated benefits-sharing levels, the involved communities received an amount of foodstuffs that represented a maximum of three to four weeks of annual consumption needs. The community that received the smallest contribution from visiting tourists, based on the amount of tourist 'gifts', would not survive for more than one week (Saarinen, 2012). In order to satisfy the needs of the mainstream tourism industry, the Ovahimba are 'trapped' but remain nevertheless willing to play and stay as an 'unchanged', 'primitive' and 'exotic'.

Conclusion

The tourism industry in the Global South is increasingly seen to be beneficial for local and regional development. By aiming to involve local communities, the benefits of tourism are expected to more efficiently trickle-down to a local level. However, the current system of ethnic tourism in areas such as Kaokoland does not operate in the ways referred to by international policy makers when they promote tourism as a path to (sustainable) development – nor does it result in the postulated outcomes. As a result, it is has become clear that global- and national-scale policy formulations do not easily turn into local-scale practices that follow higher development aims. The needs of local people and the tourism industry are not inherently in conflict. Still, as Ringer (1998, p. 9) has indicated, tourism is often an industry 'that satisfies the commercial imperatives of an international business, yet rarely addresses local development needs'. While this may seem to be an overly critical view of the industry, it is apparent that tourism-led economic growth is not fully translating into benefits for local people such as the Ovahimba or for their environment.

The tourism system in the Ovahimba villages is based on the direct relations between the poor communities and the mainstream tourism operators. While this resembles the principles of PPT, it does not mean that the wider regional tourism system is pro-poor based. The larger actors in the industry receive many more benefits from ethnic tourism than do the Ovahimba communities (see Saarinen, 2012). However, the local communities seem to be satisfied with their share, and there remain no brokers to mediate the extremely unfair level of benefit 'sharing' with the industry. It would therefore seem that this situation does not encourage the mainstream tourism industry to go beyond the current benefit-sharing levels.

More broadly, the case study also reinforces arguments that the PPT framework represents a continuation of a tourism-centric development approach grounded in contemporary neoliberal agendas (Hall, 2007, 2014). From this perspective tourism spaces are primarily articulated to serve the interests of non-locals (e.g. tourists, foreign investors, organizations). However, the Ovahimba example also highlights that some of the local 'beneficiaries' of tourism may nevertheless accept such a configuration, at least in the short term. This situation suggests that although local and non-local needs are not necessarily contradictory, critical questions need to be addressed concerning possible (and fairly common) conflict situations, unequal power relationships and benefit-sharing models. Such relations may also begin to be challenged and transformed as community groups develop greater knowledge and reflexivity of their situation, and are empowered to ask for – and indeed, effect – real change. Nevertheless, long-term capacities to respond may depend far more on their market power as an attraction as opposed to the ethical power and interests of the tourism industry.

References

Adams, W.M. and Hulme, D. (2001) If community conservation is the answer in Africa, what is the question? *Oryx*, 35 (3), pp. 193–200.
Ashley C. (2000) The Impacts of Tourism to Rural Livelihoods: Namibia's Experience. *ODI Working paper 128*. London: Chameleon Press.
Ashley, C. and Maxwell, S. (2001) Rethinking rural development. *Development Policy Review*, 19 (4), pp. 395–425.
Ashley C. and Roe, D. (2002) Making tourism work for the poor: strategies and challenges in southern Africa. *Development Southern Africa*, 19, pp. 61–82.
Bandyopadhyay, S., Humavindu, M., Shyamsundar, P. and Wang, L. (2011) Benefits to local communities from community conservancies in Namibia: an assessment. *Development Southern Africa*, 26, pp. 733–754.
Blaikie, P. (2006) Is small really beautiful? Community-based natural resource management in Malawi and Botswana. *World Development*, 34, pp. 1942–1957.
Blaikie, P. and Brookfield, H. (1987) *Land Degradation and Society*. London: Methuen.
Bollig, M. and Gewald, J-B. (eds) (2000) *People, Cattle and Land: Transformaton of a Pastoral Society in Southwestern Africa*. Köln: Köppe.
Bollig M. and Heinemann H. (2002) Nomadic savages, ochre people and heroic herders: visual presentation of the Himba of Namibia's Kaokoland. *Visual Anthropology*, 15, pp. 267–312.
van den Bosch, S. (2013) Northern Namibia suffers worst drought in 30 years. Reuters July 18, 2013. Retrieved from http://www.reuters.com/article/2013/07/18/us-namibia-drought-idUSBRE96H0B820130718
Brickley, K., Black, R. and Cottrell, S. (eds) (2013) *Sustainable Tourism and Millennium Development Goals*. Burlington, MA: Jones & Bartlett Learning.
Britton, S. (1982) The political economy of tourism in the Third World. *Annals of Tourism Research*, 9, pp. 331–358.
Britton, S.G. (1991) Tourism, capital, and place: towards a critical geography of tourism. *Environment and Planning D: Society and Space*, 9, pp. 451–478.
Brown, J. and Bird, N. (2010) Namibia's Story: Sustainable natural resource management in Namibia: Successful community-based wildlife conservation. Overseas Development Institute, London. Retrieved from http://www.developmentprogress.org/sites/develop mentprogress.org/files/namibia_environment_widlife_conservation.pdf
Bryant R. and Bailey S. (1997) *Third World Political Ecology*. New York: Routledge.
Butcher, J. (2003) *Moralisation of Tourism: Sun, Sand . . . and Saving the World?* London: Routledge.
Butler R. (1999) Sustainable tourism: a state-of-the-art review. *Tourism Geographies*, 1, pp. 7–25.
Christie, I., Fernandes, E., Messerli, H. and Twining-Ward, L. (2014) *Tourism in Africa: Harnessing Tourism for Growth and Improved Livelihoods*. New York: World Bank Publications.
DFID (1999) *Tourism and Poverty Elimination: A Challenge for the 21st Century*. London: HMSO.
Echtner, C.M. and Prasad, P. (2003) The context of Third World tourism marketing. *Annals of Tourism Research*, 30, pp. 660–682.

Edwards, E. (1996) Postcards: Greetings from Another World. In: Selwyn, T. (ed.), *The Tourist Image: Myths and Myth Making in Tourism*. Chichester: Wiley & Sons, pp. 197–222.

van Eeden, J. (2006) Land Rover and colonial-style adventure. *International Feminist Journal of Politics*, 8, pp. 343–369.

Goodwin, H. (2009) Contemporary policy debates: Reflections on 10 years of pro-poor tourism. *Journal of Policy Research in Tourism, Leisure and Events*, 1, 90–94

Hall, C.M. (1994) *Tourism and Politics: Policy, Power and Place*. Chichester: Wiley & Sons.

Hall, C.M. (2005) *Tourism: Rethinking the Social Science of Mobility*. Harlow: Prentice Hall.

Hall, C.M. (2007) Pro-poor tourism: do 'tourism exchanges benefit primarily the countries of the South'? *Current Issues in Tourism*, 10, pp. 111–118.

Hall, C.M. (2014) You can check out any time you like but you can never leave: Can ethical consumption in tourism ever be sustainable? In: Weeden, C. and Boluk, K. (eds), *Managing Ethical Consumption in Tourism: Compromise and Tension*. Abingdon: Routledge, pp. 32–56.

Harrison, D. (2000) Tourism in Africa: The Social and Cultural Framework. In: Dieke, P. (ed.), *The Political Economy of Tourism Development in Africa*. New York: Cognizant, pp. 135–151.

Henrichsen, D. (2000) Pilgrimages into Kaoko. Herrensafaris, 4x4s and Settler Illusion. In: Miescher, G. and Henrichsen, D. (eds), *New Notes on Kaoko*. Basel: Basler, pp. 159–185.

Holden, A. (2007) *Environment and Tourism*. London: Routledge.

Hulme, D. and Murphree, M. (eds) (2001) *African Wildlife and Livelihoods*. Cape Town: David Phillip.

Jacobsohn M. (1998) *Himba: Nomads of Namibia*. Cape Town: Struik.

Jones, B.T.B. and Murphree, M.W. (2004) Community-Based Natural Resource Management as a Conservation Mechanism: Lessons and Directions. In: Child, B. (ed.), *Parks in Transition*. London: Earthscan, pp. 63–103.

Kanguma, B. (2000) Constructing Himba: The Tourist Gaze. In: Miescher, G. and Henrichsen. D. (eds), *New Notes on Kaoko*. Basel: Basler, pp. 129–132.

Kavita, E. and Saarinen, J. (2015) Tourism and rural community development in Namibia: Policy issues review. *Fennia* (in press).

Lange, K. (2004) Himba: consulting the past, divining the future. *National Geographic*, January 2004, pp. 32–47.

Lapeyre, R. (2010) Community-based tourism as a sustainable solution to maximise impacts locally? The Tsiseb conservancy case, Namibia. *Development Southern Africa*, 27, pp. 757–772.

Lapeyre, R. (2011a) The Grootberg lodge partnership in Namibia: towards poverty alleviation and empowerment for long-term sustainability? *Current Issues in Tourism*, 14, pp. 221–234.

Lapeyre, R, (2011b) Governance structures and the distribution of tourism income in Namibia Communal Lands: a new institutional framework. *Tijdschrift voor economische en sociale geografie*, 102, pp. 302–315.

Lenao, M. (2014) Rural tourism development and economic diversification for local communities in Botswana: The case of Lekhubu Island. *Nordia Geographical Publications*, 43, pp. 1–53.

Long, S. (ed.) (2004) *Livelihoods and CBNRM in Namibia*. Windhoek: Directorate of Environmental Affairs and Ministry of Environment and Tourism.

Massyn, P.J. (2007) Communal land reform and tourism investment in Namibia's Communal areas: a question of unfinished business. *Development Southern Africa*, 24, pp. pp. 381–392.

Mbaiwa, J.E. and Sakuze, L.K. (2009) Cultural tourism and livelihood diversification: The case of Gcwihaba Caves and XaiXai village in the Okavango Delta, Botswana. *Journal of Tourism and Cultural Change*, 7, pp. 61–75.

Meyer, D. (2008) Pro-poor tourism - from leakages to linkages. A conceptual framework for developing linkages between the tourism private sector and 'poor' neighbours. *Current Issues in Tourism*, 10, pp. 558–583.

MET (Ministry of Environment and Tourism) (1995) Promotion of Community-based Tourism. Policy Document. Windhoek: Government of Namibia.

MET (Ministry of Environment and Tourism) (2005) A National Tourism Policy for Namibia (First Draft, Feb 2005). Windhoek: Government of Namibia.

MET (Ministry of Environment and Tourism) (2008) National Tourism Policy for Namibia (Draft, Sept 2007). Windhoek: Government of Namibia.

Mitchell, J. and Ashley, C. (2010) *Tourism and Poverty Reduction: Pathways and Prosperity*. London: Earthscan.

Mosedale, J. (ed.) (2011) *Political Economy of Tourism*. Abingdon: Routledge.

Mowforth, M. and Munt, I. (1998) *Tourism and Sustainability: A New Tourism in the Third World*. London: Routledge.

Mupya, W. (2000) Political Parties and the Elite in Kaoko. In: Miescher, G. and Henrichsen, D. (eds), *New Notes on Kaoko*. Basel: Basler, pp. 207–219.

NACSO (Namibian Association of CBNRM Support Organisations) (2008) Namibia's Communal Conservancies: A Review of Progress and Challenges in 2007. Windhoek: NACSO.

Neumann, R. (1998) *Imposing Wilderness: Struggles over Livelihood and Nature Preservation in Africa*. Berkley, CA: University of California Press.

Nijman, J. (1999) Cultural globalization and the identity of place: the reconstruction of Amsterdam. *Cultural Geographies*, 6 (2), pp. 146–164.

Novelli, M. and Gebhardt, K. (2007) Community based tourism in Namibia: 'Reality Show' or 'Window Dressing'? *Current Issues in Tourism*, 10 (5), 443–479.

Novelli, M. and Hellwig, A. (2011) The UN Millennium Development Goals, tourism and development: the tour operators' perspective. *Current Issues in Tourism*, 14, pp. 205–220.

Pratt, M.L. (1992) *Imperial Eyes: Travel Writing and Transculturation*. London: Routledge.

Republic of Namibia (1994) *White Paper on Tourism*. Windhoek: Cabinet, Republic of Namibia.

Ringer, G. (1998) Introduction. In: Ringer, G. (ed.), *Destinations: Cultural Landscapes of Tourism*. Abingdon: Routledge, p. 9.

Robbins, P. (2004) *Political Ecology: A Critical Introduction*. Malden, MA: Wiley-Blackwell.

Rogerson C. (2006) Pro-poor local economic development in South Africa: the role of pro-poor tourism. *Local Environment*, 11 (1), pp. 37–60.

Rothfuss, E. (2000) Ethnic Tourism in Kaoko: Expectations, Frustrations and Trend in a Post-Colonial Business. In: Miescher, G. and Henrichsen, D. (eds), *New Notes on Kaoko*. Basel: Basler, pp. 133–158.

Saarinen, J. (2010) Local tourism awareness: community views in Katutura and King Nehale conservancy, Namibia. *Development Southern Africa*, 27, pp. 713–724.

Saarinen, J. (2011) Tourism, indigenous people and the challenge of development: the representations of Ovahimbas in tourism promotion and community perceptions towards tourism. *Tourism Analysis*, 16, 31–42.

Saarinen, J. (2012) Tourism development and local communities: the direct benefits of tourism to OvaHimba communities in the Kaokoland, North-West Namibia. *Tourism Review International*, 15 (1–2), pp. 149–157.

Saarinen, J. (2014) Critical sustainability: setting the limits to growth and responsibility in tourism. *Sustainability*, 6, pp. 1–17.

Saarinen, J., Becker, F., Manwa, H. and Wilson, D. (eds) (2009) *Sustainable Tourism in Southern Africa: Local Communities and Natural Resources in Transition*. Bristol: Channel View.

Saarinen, J. and Niskala, M. (2009) Local culture and Regional Development: the Role of OvaHimba in Namibian Tourism. In: Hottola, P. (ed.), *Tourism Strategies and Local Responses in Southern Africa*. Wallingford: CABI Publishing, pp. 61–72.

Saarinen, J. and Rogerson, C.M. (2014) Tourism and the millennium development goals: perspectives beyond 2015. *Tourism Geographies*, 16 (1), pp. 23–30.

Saarinen, J., Rogerson, C.M. and Manwa, H. (eds) (2013) Tourism and Millennium Development Goals: Tourism, Local Communities and Development. London: Routledge.

Scheyvens, R. (2009) Pro-poor tourism: is there value beyond the rhetoric? *Tourism Recreation Research*, 34 (2), pp. 191–196.

Scheyvens, R. (2011) *Tourism and Poverty*. London: Routledge.

Sharpley, R. (2009) *Tourism Development and the Environment: Beyond Sustainability?* London: Earthscan.

Sharpley, R. (2013) Responsible Tourism: Whose Responsibility? In: Holden, A. and Fennell, D. (eds), *The Routledge Handbook of Tourism and Environment*. London: Routledge, pp. 382–391.

Sinclair, T. (1998) Tourism and economic development: a survey. *Journal of Development Studies*, 34 (5), pp. 1–51.

Sinclair, T. and Stabler, M. (1997) *The Economics of Tourism*. London: Routledge.

Smit, B. and Wandel, J. (2006) Adaptation, adaptive capacity and vulnerability. *Global Environmental Change*, 16 (3), pp. 282–292.

Smith, M. (2003) *Issues in Cultural Tourism Studies*. London: Routledge.

Smith, M. and Richards, G. (eds) (2013) *The Routledge Handbook of Cultural Tourism*. London: Routledge.

Smith, V.L. (ed.) (1989) *The Host and Guests: The Anthropology of Tourism*. Philadelphia, PA: University of Pennsylvania Press.

Spenceley, A. and Meyer, D. (2012) Tourism and poverty reduction: theory and practice in less economically developed countries. *Journal of Sustainable Tourism*, 20, pp. 297–317

Stronza, A. (2007) The Economic promise of ecotourism for conservation. *Journal of Ecotourism*, 6 (3), pp. 210–230.

Telfer, D. and Sharpley, R. (2007) *Tourism and Development in the Developing World*. London: Routledge.

UNWTO (2002) *Tourism and Poverty Alleviation*. Madrid: UNWTO.

UNWTO (2006) UNWTO's *Declaration on Tourism and the Millennium Goals: Harnessing Tourism for the Millennium Development Goals*. Madrid: UNWTO.

UNWTO (2007) Tourism and Poverty Alleviation. Retrieved from http://www.step.unwto.org/content/st-ep-initiative-1

Watts, M.J. (2000) Political Ecology. In: Barnes, T. and Sheppard, E. (eds), *A Companion To Economic Geography*. Oxford: Blackwell, pp. 257–275.

Weaver, D. and Elliot, K. (1996) Spatial patterns and problems in contemporary Namibian tourism. *The Geographical Journal*, 162, pp. 205–217.

Wels, H. (2004) About Romance and Reality: Popular European Imagery in Postcolonial Tourism in Southern Africa. In: Hall, C.M. and Tucker, H. (eds), *Tourism and Postcolonialism: Contested Discourses, Identities and Representation*. London: Routledge, pp. 76–94.

World Bank (2012) Transformation through Tourism: Development Dynamics Past, Present and Future (draft). Washington, DC: World Bank.

Yang, L. (2011) Ethnic tourism and cultural representation. *Annals of Tourism Research*, 38 (2), pp. 561–585.

Zapata, M.J., Hall, C.M., Lindo, P. and Vanderschaeghen, M. (2011) Can community-based tourism contribute to development and poverty alleviation? *Current Issues in Tourism*, 14, pp. 725–749.

11 Cleaning up the streets, Sandinista-style

The aesthetics of garbage and the urban political ecology of tourism development in Nicaragua

Josh Fisher

Introduction

In urban Nicaragua, garbage is a growing problem. Many Nicaraguans joke that the country's national flower is the plastic bag, which dot the urban landscape in a rainbow of impossible hues. The commentary comes from a keen awareness of degraded urban environments around the country, particularly in the capital city of Managua, which is home to two-thirds of the country's population, or about 2.4 million people. Of approximately 1,500 tons of garbage produced daily in the city, more than 20 percent goes uncollected. That garbage clutters the streets, collects in parks, runs off into streams, and clogs drainage ditches.

Recently, Managua's garbage situation has been tied to public health issues and has been targeted by a number of organizations in the country, including the Ministry of Health (MINSA) and the international network of non-governmental organizations (NGOs) with whom they work. For them, the concern is that chemicals are leaching into the soil, contaminating drinking water and breeding diseases like dengue (Choza, 2002; MINSA, 2012; Nading, 2014). For the country's politicians and the still-powerful private business elite, garbage is also a big problem, but for very different reasons. They blame lags in tourism and capital investment on the aesthetic costs of environmental degradation. Indeed, due to the limited viability of other industries, tourism has become an important strategy for national development in Nicaragua, having surpassed all other traditional exports in 2001. Tourism generated an average of 20.6 percent of total exports between 2004 and 2008 and, in 2008, accounted for 6.1 percent of GDP (INTUR, 2009). The Nicaraguan Institute for Tourism (INTUR), accordingly, has become increasingly relevant as the country attempts to market its endogenous natural and historic resources, following the ecotourism model provided by their neighbor to the south, Costa Rica. In that respect, Nicaragua has the potential to excel in tourism development. It has the second largest rainforest in the hemisphere, after the Amazon. Lake Nicaragua, likewise, is home to the world's only freshwater sharks and has been nominated for the Seven Natural Wonders of the World competition on multiple occasions. The city of Granada is one of the first colonial cities in the Americas, and its leaders have worked diligently to preserve the architectural heritage. Acahualinca, a UNESCO World Heritage Site, located within the capital city of Managua, memorializes some 2,000 year-old human footprints in volcanic ash and mud. And then there are the rather more mainstream tourist draws: the tropical sun, the sand, the 40 towering volcanoes, and the world-famous Flor de Caña rum.

The garbage problem, however, begins the moment the tourist steps off the plane at Augusto C. Sandino Airport in Managua. 'Managua is a shambles', says the *Lonely Planet*, 'chaotic and broken' (Egerton and Benchwick, 2013). '[It is] the most chaotic of Central American capitals', claims the *Moon Handbook* (Dobrzensky, 2013, p. 10). And despite attempts to dress it up, the place is 'still among the ugliest capital cities in the hemisphere', writes *New York Times* columnist Stephen Kinzer (2002). The travel website VirtualTourist now even lists littering under a section titled 'Local Customs' (VirtualTourist, 2015). Nearly all such guides recommend that visitors skip Managua entirely and head directly to their actual destination, be this the forest, the beach, or Costa Rica. As one professional tour guide opined during a 2012 interview, 'Why don't people come to Nicaragua? Because of Managua. It's a mess'.

This chapter is about the political ecology of tourism development in Managua, Nicaragua and the aesthetic, infrastructural, and ecological dynamics of the project of remaking the face of the city. There have been many attempts to carry this out over the years. I focus on a recent national development campaign launched in 2013 by President Daniel Ortega and the *Frente Sandinista de Liberación Nacional* (FSLN) party, titled *Vivir Limpio, Vivir Sano, Vivir Bonito, Vivir Bien* ('Live Clean, Live Healthy, Live Beautiful, Live Well'). The 'Live Clean, Live Healthy' campaign, as I will call it for the sake of brevity, is a permanent national campaign that connects the Ministries of Health (MINSA), Environment (MARENA), and Education (MINED) with the Nicaraguan Institute for Tourism (INTUR) and, as such, explicitly links environmental stewardship with a range of issues such as health, well-being, community, national identity, and tourism development. Although the campaign is framed by co-investments in clean energy, public health, and education, the implications of Managua's garbage problem for tourism development are undoubtedly a main focal point. Live Clean, Live Healthy aims not only to clean up the streets, by closing down informal dumps and by constructing a modern waste management infrastructure, but also by changing the cultural habits and mentalities of Nicaragua's citizens through an expansive public awareness campaign, promulgated by leaflets, billboards, and brigades of city workers and students. By promoting a certain aesthetics of 'good living' (*el buen vivir*) and 'pretty living' (*vivir bonito*) in the construction and reconstruction of urban space, in other words, Live Clean, Live Healthy endeavors to 'offer the best image of ourselves, of our Nation, to the world' (Murillo, 2013, p. 8). In the 1980s, such a statement might have been read as pure Sandinista-style nationalism. In the 2010s, the task of cleaning up the streets is driven also by 20 years of neoliberal economic policies, including a sustained effort to 'market' Nicaragua and to develop tourism into a leading industry.

Yet, in a highly polarized national political atmosphere, the campaign has also drawn significant opposition. 'The government is trying to create the sensation of well-being, to apply makeup to poverty, which is something fascist governments do', charges politician Eliseo Núñez. In his reckoning, the Live Clean, Live Healthy campaign is the twenty-first century 'green' equivalent of the Family Code, which institutionalized the Sandinista definition of the family in 1981, after the revolution (Potosme and Picón, 2013). The opposition newspaper *La Prensa*, similarly, has accused the Ortega administration of appropriating the educational system as an ideological apparatus to indoctrinate students in Sandinista values and ideas (Silva and Potosme, 2013). In my reading, these claims offer still more evidence that aesthetics, in tourism as in politics, is a highly contested field of discourse (cf. Rancière, 2004).

After a brief review of the literature about an urban political ecology of tourism, I track the development of tourism in Managua and the rise of the Live Clean, Live Healthy campaign as a urban revitalization-cum-tourism development project, beginning with the problematization of the iconic La Chureca, once the largest unregulated, open-air dump in Central America before it was finally closed in February 2013 (one month into the campaign). Based on longitudinal ethnographic research in Managua between 2008 and 2014, including a dozen semi-structured interviews and multiple periods of extended participant observation with recyclers, economic intermediaries, and policy makers, I demonstrate that the campaign employs what Ghertner (2010, 2011) terms 'aesthetic governmentality' as a key political strategy for tourism development. The campaign problematizes the aesthetics of garbage in the urban landscape and seeks to convert Managua into a potential site for tourism consumption by installing new infrastructure, new disciplinary apparatuses, and new techniques of governance that target the habits and mentalities of environmental subjects as 'multipliers of knowledge' (*multiplicadores de conocimiento*), in the words of the Ministry of Health (see MINSA 2012, p. 15).

In that respect, my analysis of Live Clean, Live Healthy, an urban revitalization project with profound motivations and implications for tourism development in Nicaragua, confirms urban political ecology's thesis about the 'social nature' of thickly settled urban spaces like Managua (Desfor and Keil, 2004). It also complicates an already multidimensional theory of power, which in the case of Live Clean, Live Healthy spans some of the 'big questions' posed by political ecology: the many historical forces that produce degradation and marginalization in cities in relation to the larger national and international processes of which they are part; the many conservation and control efforts that seek to maintain built and natural, aesthetic and economic environments; and the 'environmentalization' of urban social policies that direct modes of state and developmentalist power to the project of cultivating environmental subjectivities among denizens in order to marshal the full resources of the city for tourism (Robbins, 2012).

Toward an urban political ecology of tourism development

Political ecology poses a way of thinking about the power-laden relationships between different groups of human beings and the environment. Historically, the framework emerged out of an attempt to bring political economy into the discussion of cultural ecology and thus to explicate the role of regional or global systems in the production of local ecological problems (Wolf, 1972). The traditional subject matter of political ecology, in that respect, has retained a predominant focus on rural landscapes while engaging a wide range of new issues, including the dynamic connections between socio-economic marginalization and environmental degradation (Blaikie, 1985; Grossman, 2014, 1993; Hecht and Cockburn, 1989; Schmink and Wood, 1987, 1992); the unintended consequences of conservation, preservation, and management of natural resources for local populations (Neumann, 1998; Peluso, 1995; St. Martin, 2001; West, 2006); and the conflicts that flare up between different social, economic, and ethnic groups over access to these resources (Martínez-Alier, 2002). More recently, drawing inspiration from urban studies and urban ecology, political ecologists have also challenged the subject's implicit focus on the natural and rural landscapes to consider the dynamics of urban and other built environments. This second approach has opened up an expanded range of issues for consideration under the heading of *ecology* or *environment*; the production and social distribution not only of 'resources' but also of 'hazards' of both natural and technological origin (Auyero, 2009; Kaika, 2005; Klinenberg, 2002; Mustafa, 2005; Robbins, 2007); also the different 'metabolic' processes of cities in which social forces are closely intertwined with biochemical and physical ones (Heynen *et al.*, 2006; Monstadt, 2009), and the entanglement of human and nonhuman actors in urban environments (Nading, 2014; Swyngedouw and Heynen, 2003).

The above varieties of political ecology, as a kind of 'epistemic community' rather than a single discipline, tend to pivot on different definitions of 'the political', and so it is worth examining very briefly these conceptions of power in terms of how they might also shape different perspectives on the political ecology of tourism. The classic approach comes from the historical materialist tradition of Marxist political economy and works through the dialectical relationship between the political and ecological qua the social and material. In this dialectic, one arrow points from the material to the social and describes how the environment provides the 'base' upon which law, politics, and society are founded, while the other points from the social to the material and accounts for how human beings also manipulate those natural systems in accordance with social, political, and economic imperatives. Power arises from the inevitable contradictions between those two systems, producing environmental degradation alongside social marginalization through unequal access to the means of production (Foster, 2000; O'Connor, 1996; Schmink and Wood, 1987, 1992). In the political ecology of tourism, that optic has served to shed light on the repercussions of tourism development for local populations, who lose control over resources, are exposed to new social and environmental risks, and must grapple with the degeneration of environmental health (Cole, 2012; Gössling, 2003; Stonich, 1998; cf. Sharpley, 2009). In that respect, the still inchoate literature parallels scholarship in the political ecology of conservation and preservation. Neumann (1998), for example, demonstrates that when landscapes in Tanzania are converted into nature reserves, which are *de facto* intended for tourist consumption and worth hundreds of millions in tourism receipts each year, there are resounding consequences not only for the ecological systems that are artificially demarcated by the boundaries of the new reserves but also for those who once laid claim to such lands as a matter of subsistence. Now – once locals of that area – such people are identified as 'environmental threats' when they 'trespass'.

Upon closer inspection, however, the conservation and preservation literature also highlights a somewhat more complex play of power that is instructive for the political ecology of tourism. That is, landscapes are converted into conservation or preservation territories not only by virtue of power as a property of the state and other institutions, they are also made 'legible' (Scott, 1998) in symbolic and aesthetic terms, through discourse. They become objects of the tourist gaze and are tied to consumer imaginaries through an aesthetics of 'wildness' which requires the absence of humans (Cronon, 1996; Merchant, 2003; Neumann, 1998). Far from being a simple matter of interplay between separate domains, the material and social, political ecological analysis may examine the 'construction' of complex, dynamic, and self-generating ecological assemblages in terms of their delicate intertwining with discourse, wherein they become also objects of knowledge, imagination, and action (Escobar, 1999, 2008). That theoretical framework has been productively applied to the study of rural landscapes, but recent scholarship has demonstrated also that cities and urban environments undergo such a construction as well. They are political landscapes not only because they form complex 'metabolic' systems (i.e. of food, waste, housing, industry, and pollution) but also because they emerge as aesthetic and imaginative domains of action (Ghertner, 2011; Heynen et al., 2006; Verón, 2006). In the political ecological study of cities, garbage has emerged as an important point of reflection for thinking about the geographies of inequality, since waste tends to be sited in marginalized neighborhoods, along with other aspects of the 'sociomateriality' of everyday items (Gregson and Crang, 2010; Hultman and Corvellac, 2012; Moore, 2011, 2012). While the implications of this political ecology for tourism in rural landscapes, including ecotourism, are clear (e.g. Fletcher, 2014; Robbins and Fraser, 2003; West and Carrier, 2003), neither cities nor garbage have yet to emerge as sites for considering the political ecology of tourism development.

One final point about *the political* dimension of the political ecology of tourism is that, without denying materiality, discursive politics does not stop with the construction of environments and landscapes as meaningful sites, it also shapes the subjectivities of those who inhabit or interact with them. From 'green consumer' campaigns and certification initiatives to conservation projects that enroll local populations alongside distant ecotourists as caretakers of forests, recent literature in political ecology has also demonstrated the crucial importance of people's identities in relation to particular environments, which are linked through their actions, habits, and ideas about them (Agrawal, 2005; Bryant, 2002; Guthman, 2004; Mutersbaugh, 2002; Sundberg, 2004; Wainwright, 2008). Conservation and preservation, for example, employ a kind of governmentality in which the environment is constructed as 'a critical domain of thought and action' (Agrawal, 2005, p. 16) such that environmental subjects internalize the missions of projects, embody the norms and expectations for behavior, and work to create and maintain the desired environments. As tourism development, in other words, constructs landscapes and cultivates their aesthetic properties so too might it generate environmental subjectivities among both the producers and consumers of those environments.

Ghertner (2010, 2011) refers to this political strategy as 'aesthetic governmentality', which, I argue, becomes a key element of the urban political ecology of development projects, particularly tourism development. Governmentality typically operates through the generation of statistics, maps, and other scientifically rational technologies (Escobar, 1995; Mitchell, 1991). Rather than simply reflecting reality, however, such discursive formations work to 'conduct the conduct' of a population (Foucault, 2007, p. 96), and as such is a common technique in 'participatory' development initiatives that target the well-being of a population and seek to secure a particular kind of action within the populous (Li, 2007). As Ghertner points out, those 'rationalities' may, however, also take the form of '[ascribing] an aesthetic sense of what ought to be improved and what ends achieved' (Ghertner, 2011, p. 289). In aesthetic governmentality, governmental legibility is therefore achieved not through the simplification of complex landscapes as statistical representations, governed by 'rules of evidence' (Foucault, 2007, p. 350) and other 'numericized inscriptions' (Rose, 1991, p. 676) but rather through the propagation of discourses of idealized, aesthetic norms that construct images and ideas about what said landscapes, and the populations that inhabit them, should look like.

In what follows, I draw on this framework of aesthetic governmentality to explore the FSLN's Live Clean, Live Healthy campaign as one such tourism development project for the capital city of Managua, one that constructs images and ideas of a 'clean' and 'modern' urban Nicaragua according to the consumption preferences of tourists, while also expressing state power. First, however, I will trace the emergence of the Live Clean, Live Healthy campaign through the early problematization of the site of Central America's largest unregulated open-air dump, La Chureca.

From dumps to national development

The urban environment of Managua has long been a matter of concern for tourism development, although only recently have the aesthetics of garbage in that environment come under scrutiny. In the 1950s and 1960s, Managua was a veritable metropolis, perhaps the most modern in Central America, replete with high-rises, leafy promenades, and a thriving business district. In 1972, however, that landscape was forever disrupted by a powerful earthquake, centered directly under the city, which killed over 10,000 people, left another 500,000 homeless, and leveled nearly 900 city blocks. The rubble from collapsed buildings was carted off to a site along the shoreline of Lake Managua, located next to the famous archaeological site (and *barrio* of the same name) of Acahualinca which, over the next 40 years would serve as the municipal dump, La Chureca (a Nahuatl word meaning 'old rag'). Meanwhile, the reigning Somoza government redirected a substantial portion of international assistance designated for aid and reconstruction to private ends, including paved roads to the dictator's many estates outside the city, such as Montelimar, now a popular beach destination.

Historically speaking, the political and economic aftershocks, and the civil unrest that followed, laid the groundwork for contemporary political ecologies of garbage and tourism development. In 1979, the FSLN carried the banner for a popular revolution that captured the imaginations of journalists, left-leaning activists, writers, and intellectuals around the world and, by the 1980s, the capital city of Managua was no stranger to international tourists. They flocked to the small Central American country to bear witness to the socialist Sandinista party's historic experiments and to protest the US-sponsored contra war (Hollander, 1986). Managua became their hub. They built hostels and collaborated with the Ministry of Tourism to produce some of the city's first travel guides, which brought a uniquely Sandinista ideological bent to the city's historical sites. Meanwhile, those *internacionalistas* or *sandalistas*, as they are called, served to link the country with a growing network of activists around the world. It was in the border city of El Porvenir that Witness for Peace, the activist turned advocacy organization that now runs educational tourism trips around the world, got its start as North American 'human shields' to stave off contra incursions. The political-economic context of embargo, likewise, gave rise to the fair trade coffee company Equal Exchange. As a small New England food co-op seeking to protest the Reagan administration's embargo levied on Nicaragua during the Sandinista decade, they started exporting 'the forbidden coffee', called Café Nica, to a Dutch alternative trading company where it was roasted and then legally imported to the US (Equal Exchange, 2015). The company would be among the first to organize 'solidarity tours' of Nicaraguan coffee cooperatives for consumers in the US and Europe.

The year 1990, however, proved to be something of a historical watershed for the country. With mounting costs from the economic embargo and the US-sponsored contra war, the FSLN was defeated at the polls and a liberal opposition party came to power, which immediately restructured the economy and turned to more traditional forms of tourism as a basis for economic development. The mayor of Managua, Arnoldo Alemán, launched an urban revitalization project titled 'Nueva Managua' that sought to erase the public signs of the country's troubled past. Patching the bullet-ridden facades and painting over the colorful murals of revolutionary artists and activists, the goal was to make Managua into a 'clean' and 'modern' city, worthy of the same class of traveler's-check-toting tourists who visit places such as Costa Rica, as opposed to continuing to draw the scruffy, left-leaning backpackers of the Sandinista years (Babb, 1999). Into the latter half of the decade, during which Alemán became president, that project expanded to include the construction of a well-lit and high-speed system of roadways and roundabouts that would provide the urban elite access to a 'fortified network' of private spaces and would allow tourists to proceed directly to their destinations without having to pass by peddlers, squatters, and the various and sundry informal economies in which they work (Rodgers, 2004 p. 120). That project changed ever so slightly in the 2000s (Babb, 2004). Without fundamentally transforming the tourist development project or its goals, the aesthetic project of constructing a clean, timeless, and modern city shifted once again to resemble what Rancière (2004) might call a 'postmodern post-politics', one which forecloses on the very possibility of reading politics into the urban landscape. The historical and cultural images of the once rejected past, the revolution and the omnipresent icon of Augusto C. Sandino, Nicaragua's very own 'Che', became the platform for a neoliberal project to develop a kind of national 'brand', now stripped of its once potent political ideology.

During the same period of time, the structural vulnerability of Managua following the earthquake gained new expression in the landscape through garbage. Underlying the dramatic changes in the late 1980s and early 1990s, when the US-sponsored embargo came to an end, was the slow-moving disaster of structural adjustment. That transition is usually depicted in political-economic terms as a time in which the country opened up the floodgates to 'free trade' and in which the economy reoriented to grow service sectors like international tourism (Babb, 2001). However, free trade also had dramatic political ecological consequences. The consumer landscape, and thus the waste and recycling landscape, transformed as a deluge of cheap, new products made their way onto store shelves. In the 1980s, a banana leaf that served as a lunch wrapper and was casually tossed aside would have easily biodegraded in less than a year in the tropical climate. In the 1990s, however, the plastic bag that took its place will stick around for a thousand years. Without a working waste management infrastructure, thanks in large part to deep cuts in public sector spending, the urban landscape changed as well. These new consumables and the various plastics to ship and contain them started to accumulate in the gutters, in the drainage ditches, and in a plethora of informal and unregulated dumps. Later in that decade, in 1998, yet another powerful disaster called Hurricane Mitch struck Central America and dropped upwards of six feet of rain in the region over four days. The storm further weakened the country's infrastructure and flooded entire sectors of the city, taking its largest toll along the shoreline of Lake Managua, along the capital city's northern border, and further highlighting the city's inundation in plastics. The displaced either were resettled to a new encampment in Ciudad Sandino to the west (Nading, 2014), pushed off into marginal peri-urban zones of Managua like La Sobrevivencia (Rodgers, 2011), or forced to scavenge for a living in La Chureca (Hartmann, 2012).

The fetid landscapes of La Chureca tell an important story about the political ecology of Managua and the contradictory forces of tourism development. Every day since 1972, thousands of tons of solid waste, from newspapers, plastic bottles, diapers, and spoiled food to tennis shoes, old clothes, outmoded electronics, dead animals, and medical and industrial waste, some inert if malodorous and others noxious to the environment and human health, have been deposited in the site. As archaeologist Bill Rathje might point out, the excavation of those layers of trash, though undoubtedly a messy business, would speak of Nicaragua's recent history of industrialization and free trade (Rathje and Murphy, 2001). The Nicaraguan poet Juan Sabolvarro Muñoz (2007, p. 31) makes a similar point when he reflects on the iconic plastic bag. Once upon a time, it served as 'makeup' (*maquilla*) to hide the country's incomplete and uneven modernization. Now, anathema to projects for national development including tourism, it flutters along the streets or blows from the toxic dump into the adjoining Lake Managua and is a symbol of modernity's deep contradictions.

Then again, another side of the political ecological story of Managua could be told by real people, the thousands of Nicaraguans termed *churequeros*, who actually live or work in that dump. With changing patterns of urbanization, employment, and consumption in the 1990s and 2000s, La Chureca's working population grew to over 2,500, most of whom live in extreme poverty and earn less than a dollar a day sifting through discarded items in search of something wearable, repairable, sellable, or edible (Grigsby Vergara, 2008). These individuals suffer medical maladies due to their exposure to heavy metals, organic pollutants, and other hazardous materials. Despite working in largely hidden transnational recycling economies, however, their activities are not wholly unregulated but rather are governed by a complex 'urban informality' (Roy and AlSayyad, 2004). Although they remain unprotected by environmental and labor laws, their work is organized through informal cooperatives as well as other, more hierarchical, patronage structures, which divide up and systematize labor and shape access to the worksite (Hartmann, 2012). Yet, policy makers and other onlookers continue to marginalize these groups both economically and socially. Churequeros are not only denied the legal and economic benefits of 'formal' work, they are also referred to in public discourse with animalistic terms like *zopilote* (vulture). In 2008, the mayor of Managua, Dionisio Marenco, even accused churequeros of feeding on the carrion birds with whom they share the dump (Equipo Envío, 2008).

In the late 2000s, the public spectacle of La Chureca provided an important impetus for a new set of tourism development policies in Managua. As La Chureca grew, a new generation of 'poverty tourists' (Scheyvens, 2001) and voluntourists, sponsored by churches, civic groups, and international NGOs mostly from the US and Europe, started visiting the site on a regular basis (Zapata Campos, 2010). These groups brought back photographic testimonials of 'authentic poverty' from the dump, which in turn started to circulate through international aid and development networks as aesthetic evidence of Nicaragua's crisis (cf. Mostafanezhad, 2013). It was in large part that growing visibility that set off a chain of events that would lead to the closing of the dump and the inauguration of an environmental turn in Nicaragua's national and tourism development efforts. In 2007, the Spanish magazine *Interviú* got wind of La Chureca and listed it among 'the 20 Horrors of the Modern World', ranked third after whale hunting and the burka but before the destruction of the Amazon rainforest (Cabrera, 2007). Then, in the following year, a rise in global market prices for plastic and metal recyclables prompted a confrontation between churequeros and poorly paid municipal workers, who had taken to preemptively gleaning the most valuable items before dumping the rest. During what was called the *churecazo* (the fiasco in La Chureca) churequeros linked arms and blocked city garbage trucks from entering what they deemed 'our dump'. Garbage started to pile up on the streets of Managua, raising public health alarms and issuing a political challenge to authorities (cf. Moore, 2008, 2009). The spectacle was a public relations disaster that drew attention to precisely the side of Nicaragua that many INTUR and other national development officials would have rather never seen the light of day, for it was anathema to the images of tropical paradise they were so carefully cultivating for foreign tourists and investors.

Although churequeros won certain concessions, their communities, their work, and their environments were suddenly on the map, in this highly contested political terrain, as 'social problems'. In 2008, claiming that these garbage and recycling economies attracted bugs, rodents, and other pathogen carriers, MINSA launched a municipal-wide campaign called the Plan Chatarra to push these life forms from the city center and its many homes and shops. Then, 'like flies to dung', to borrow the words of one recycler I interviewed in 2012, the spectacle of La Chureca started to attract a range of international actors to its cause. Among them was Spain's primary humanitarian aid organization, the Spanish Agency for International Cooperation and Development (AECID). At the behest of Vice President María Teresa Fernández de la Vega, who had been moved by the *Interviú* article and visited the site later that same year, AECID pledged US$ 60 million to close the dump and convert the site into a modern waste-processing, recycling, and methane extraction facility. The Acahualinca Development Programme, as the project came to be called, in turn linked up with a UN-Habitat project that promised to house La Chureca's dwellers, to provide infrastructural planning for the entire *barrio* of Acahualinca (also the home of a UNESCO World Heritage Site), and to develop a waste management strategy that would have immediate implications for the municipality's tourism economy.

Aesthetic governmentality and tourism development in the Live Clean, Live Healthy campaign

With these events, garbage was only just beginning to become a political rallying point in Nicaragua. La Chureca's conversion was completed in February of 2013, and AECID transferred management responsibilities to a private Spanish firm called Biomass Investment Nicaragua SA (BINICSA), which currently runs the facility. However, the 'crisis' had already cemented a new set of policies to revitalize the urban environment that on the surface appear to be about public and environmental health, but have had the powerful, instrumental effect of expanding the state's role in tourism development. Indeed, as an INTUR official told me during a 2014 interview, although Live Clean, Live Healthy is ultimately about urban revitalization, 'making Managua into a nice place to live and to visit', the development of Nicaragua's tourism industry is one of its most important by-products.

In the month before La Chureca's closing, in fact, Nicaragua's first lady and a lead spokesperson for the FSLN government, Rosario Murillo, announced a comprehensive national development campaign. In FSLN political circles, it was widely referred to as 'the second stage of the revolution'. For the public, it would be known as Live Clean, Live Healthy (Equipo Envío, 2013). From the outset, the campaign was politically controversial. First, it had been pushed through by a constitutional amendment that created a new cabinet post for 'Family, Community, and Life', harking back to the FSLN's Family Code of 1981, which signaled a clear intervention of government in social life. Second, it was rather explicit in its goals to green the landscape by intervening in everyday social and family life. Murillo discusses the ethical underpinnings of the campaign using the language of 'love', 'care', and 'beauty', important elements in the FSLN's re-visioned, twenty-first century political philosophy of *sandinismo*:

> When we convert these areas into trash heaps, when we make our immediate surroundings and public areas ugly, that is aggression – it's a lack of love and we are called to mind the beauty, for our sense of self-esteem and respect, to respect ourselves, to love ourselves, and to care for ourselves, which is also about caring for harmony and for the aesthetics of daily life beyond poverty; we can be poor but clean, poor but honorable, with community and nature.
> (quoted in Equipo Envío, 2013)

Those ideas were further elaborated in a document released with the launch of the campaign, entitled 'A Basic Guide for Clean Living, Healthy Living, Pretty Living, and Living Well...!' (*Guía Básica para Vivir Limpio, Vivir Sano, Vivir Bonito, Vivir Bien...!* see Murillo, 2013). The Basic Guide consists of a lengthy introduction and 14 comprehensive points delineating the principles and strategies for fomenting minor and major revolutions in everyday life. Point Four, for instance, argues for 'greening' the landscape by planting trees and gardens in urban spaces and encouraging the cultivation of medicinal gardens. Others are more far reaching, such as Point Nine, which promotes the 'norms of consumption and garbage elimination that address judicious and respectful conduct... so as to promote the Nice and Better Nicaragua that all men and women want' (capitalization in original). Still others, like Point Seven, are even more encompassing and entreat the public to come together 'to develop our own Tradition, Religion, and Culture' and to 'join efforts to prevent and address this plague called Modernity'.

From the outset, then, the national Live Clean, Live Healthy campaign stood apart from mainstream development efforts, such as those that revolve around the construction of apolitical and technical problems and solutions. Certain elements reflect early experiments in 'autonomous development', such as Julius Nyerere's efforts to cultivate a distinctly African socialism (*ujamaa* or 'familyhood') in Tanzania. In Nicaragua, similarly, in keeping with the re-articulated *sandinismo* and the FSLN's own branding as 'Christian, socialist, and solidary' (*cristiana, socialista, solidaria*), the campaign offers a uniquely Sandinista version of 'comprehensive' or 'sustainable' development (*desarrollo integral*), as a FSLN official explained during a June 2014 interview in the Ciudad Sandino headquarters, yet one that notably finds no contradiction with the project of tourism development begun in earnest under neoliberalism in the 1990s. Weaving together spiritual and cultural issues with health, well-being, community, and environment, the campaign establishes a semantic equivalence between concepts of 'clean', 'healthy', and 'pretty' – the cultural, the medical, and the aesthetic – as pathways for a kind of 'cultural evolution' (Silva and Potosme, 2013). As the Basic Guide further explains, the target of the campaign is the conscience or consciousness (*consciencia*) of the individual as a physical, psychological, and spiritual entity and proceeds along three axes of intervention: (1) 'self-esteem' (*la autoestima*) or care for the self as a dignified person; (2) 'respect' (*la estima*) or care for others, including community and nation; and (3) 'the aesthetic' (*la estética*) or care for the environment and the cultivation of a sensibility for the beautiful.

Since 2013, the campaign has had complex political ecological implications for tourism development. Taking up the charge of the Acahualinca Development Programme, the campaign calls for simultaneous investments in Managua's infrastructure, public health, and education. On the surface, the goal is to alter the city's waste management system and thus the political ecological 'afterlives' of garbage (Gidwani and Reddy, 2011) that have so shaped Managua's urban landscape. Delving deeper, however, that landscape is intertwined with multiple modes of public discourse about health, environment, education, and tourism. Indeed, MINSA, MARENA, MINED, and INTUR each receive directives from the office of 'Family, Community, and Life', under whom garbage is a renewed matter of concern. Tourism development, meanwhile, has become entangled in other modes of national development that recognize garbage as a simultaneous economic, ecological, social, and aesthetic blight that reflects the need for sustained coordination between various political offices.

In neighborhoods all over the city, municipal authorities have built a new infrastructure, including neighborhood collection sites and substations that channel garbage flows to proper processing plants like Acahualinca. The campaign has also extended waste collection and recycling services to informal settlements around the city, a sector that represents 40 percent of Managua by population yet has never received such services. In these urban and peri-urban neighborhoods, streets are oftentimes little more than narrow, dirt alleys with low-hanging wires from illegal electricity hookups and deep potholes that turn into muddy ruts during the rainy season, when they are nearly impassable. Of the city's 80 garbage trucks, the 40 or so large modern compaction trucks were donated by the government of Italy in 2008 (Parés Barberena, 2006). In response, the city forged a novel public-private partnership with an NGO called Manos Unidas and an informal waste collection cooperative consisting of 18 men with horse-drawn carts. Prior to the agreement, these men collected waste from rich neighborhoods for a small fee and typically dumped it in unauthorized locations in poorer parts of the city. Now, they receive payment from the municipality to service the least accessible parts of the city, depositing the garbage and recycling in authorized collection sites (Zapata Campos and Zapata, 2013). Manos Unidas, meanwhile, has triumphed the agreement as a socially inclusive and responsible solution, not only to Managua's revitalization but also to global issues such as climate change. As for tourism, the arrangement has the added benefit that waste excesses from both the rich and poor neighborhoods are less commonly found along the roads or in public parks around historic sites.

But Live Clean, Live Healthy goes beyond the infrastructure of the urban environment. It has also made changes through legislation and education with an eye toward securing popular support for the revitalization project. In March 2013, for example, the Managua City Council passed a municipal ordinance that would fine those who damage the urban environment, either through illegal dumping or through visual or noise pollution (Lara, 2013). With a *mano dura* (heavy hand), municipal authorities issue fines between 100 and 100,000 córdobas (US$ 4 to US$ 4,000) to waste-pickers, businesses, and other organizations found violating the law including, in a widely publicized case, the Organization of American States (OAS) headquartered in Washington DC, which was discovered disposing of their garbage in a local ravine. Managua's citizens, moreover, may participate in the national effort by reporting violators with one of the many green telephones (*líneas verdes*) located around the city.

At the same time, Live Clean, Live Healthy has also reached into the educational system, enrolling students and teachers from preschools to universities, public and private alike, as multipliers of knowledge. In the classroom, MINED has taken an active role in promoting the habits, practices, and principles of 'living clean and pretty' by turning the maintenance of the school grounds itself into a kind of prefigurative, pedagogical politics for the campaign (Bermúdez and Álvarez, 2014). Teachers are encouraged to monitor students' behaviors; on a semi-weekly basis, in many of the schools I visited during ethnographic fieldwork between 2012 and 2014, students and teachers alike are charged with cleaning up trash on school grounds or graffiti on walls and desks. In the curriculum, too, teachers have been required to implement elements of the revitalization campaign, from hygiene and health to care for the community and environment. Teachers undergo training with a 58-page guide to the campaign and are expected to teach from a shorter, six-page document issued by MINED called 'The Basic Guide to Living Pretty' (*Guía Básica para Vivir Bonito*), which includes scripted lectures and class activities. Outside the classroom, those lessons continue. Students take class trips and participate in a range of community greening projects organized by MARENA, from planting trees to 'clean-up days' (*jornadas limpiezas*) in parks or unauthorized dump sites. At the same time, harkening back to the literacy brigades led by student supporters of the revolution in the 1980s, FSLN affiliated movements of young people like the Federation of High School Students (FES), the Sandinista Youth (JS), and the 'Guardabarranco' Environmental Movement (named after the turquoise-browed motmot, the national bird of Nicaragua) have organized 'ecological brigades' (*brigadas ecológicas*) to raise public consciousness of environmental issues, particularly with regard to the proper disposal of garbage.

Ostensibly, the goal of these many educational efforts is much the same as recycling campaigns in the US; to instill an environmental consciousness or a set of habits in the minds of young people so that they might pass them along to present and future generations. Taken as a whole, however, the FSLN's Live Clean, Live Healthy campaign comes closer to what Ghertner terms aesthetic governmentality. On the surface, the revitalization effort is about constructing an image of what the urban landscape *should look like* in order to be fit for tourist consumption. Yet, the term also serves to demonstrate how such images are fixed in the minds of citizens as well as how they translate into the habits and practices of Managuans as environmental subjects in order to create and maintain that landscape. The campaign seeks to activate both a 'sanitary citizenship' (Briggs, 2003; Nading, 2014, pp. 64-5), implicating people who will care for the health of the community, and 'environmental subjects' (Agrawal, 2005) who understand the environment as an important domain of action. Ghertner (2010, 2011), for example, demonstrates that developmental projects to demolish slums in Delhi are driven not by a calculation of well-being for the populous but rather by an aesthetic rationality having to do with the unruliness and messiness of slum space. In Nicaragua, similarly, the FSLN has not marshaled to their cause statistical depictions of garbage in the urban environment, but rather its aesthetic discourses in which cleanliness, health, and beauty are synonymous with national development and tightly intertwined with tourism development.

In fact, one of the consequences of the revitalization campaign has been marked growth in the tourism economy, a 6 percent increase in 2014 alone (Prado Reyes, 2014), which will likely increase even more with the completion of Nicaragua's own transoceanic canal (Chávez, 2015; Marina, 2015). In Managua, the campaign has pushed for the installation of public art projects in parks, new murals with revolutionary and religious imagery on public buildings, and renewed opportunities for national, regional, and international tourism with the arrival of new restaurants and businesses selling Nicaraguan products in the city (Cerón Méndez, 2015a; Ortega Ramírez, 2015; Vidaurre Arias, 2015). The campaign has helped to revitalize degraded urban spaces like the Plaza of the Revolution (home to the National Palace and the National Museum) as well as the Lake Managua waterfront (*El Paseo Xolotlán*), which is now lit with dozens of glitzy, 40-foot metal yellow 'Trees of Life'. The same iconic trees also dot many of the city's roundabouts, including one that has recently been dedicated to the late Hugo Chávez. For MINED, in addition to enrolling students in the project of maintaining the beautified Managuan landscape, tourism appears to be a growing area of interest. The ministry has recently worked with INTUR to build several new municipal tourism schools in Managua and surrounding cities, which will train students in tourism services (Cerón Méndez, 2015b).

Conclusion

The Live Clean, Live Healthy campaign demonstrates several important points about the urban political ecology of tourism development. One is that urban political ecologies are complex and hybrid economic, infrastructural, and socionatural systems. In those systems, politics often plays out through the calculation, management, and distribution of hazards and resources within the landscape. The other is that, with the rapid environmentalization of urban social policy and planning in Latin America and beyond – a by-product of tourism development, as countries like Nicaragua, Costa Rica, and Brazil orient themselves to the reality that aesthetic and sanitary expectations of foreign travelers often translate directly into staid economic categories – we are forced to consider a broader range of political strategies, some of which may not work through sheer calculation but rather are expressed through the propagation of aesthetic norms and policies. In other cases, however, such as in Nicaragua's current campaign, power is instead expressed through the propagation of aesthetic norms and policies. State power now also disciplines aberrant behavior, makes landscapes legible in symbolic and aesthetic terms, and enables a particular kind of action among environmental subjects through techniques of governance that target their habits, mentalities, and self-understandings with respect to maintaining the aesthetics of households, communities, and the nation. For the Live Clean, Live Healthy campaign, the instrumental effect of 'cleaning up the streets' is not only to expand state power but also to align the aesthetic qualities of the city with the imperatives of tourism development. As such, I have argued that tourism development in contemporary Nicaragua cannot usefully be separated from other modes of social and economic development; the political apparatuses of which go well beyond straightforward economic or technical interventions and instead target a broader politics of living or even, as is now commonplace in Latin American environmental discourse, 'living well' (*buen vivir*).

At the same time, however, it is widely recognized that strategies of governance such as these rarely accomplish what they intend. That appears to be the case for the Live Clean, Live Healthy campaign, at least for Nicaraguans whose economic activities and interests do not happen to coincide with the city's tourism development project or its hegemonic aesthetics and who are therefore the unintended victims of conservation efforts (in accordance with political ecology's conservation/control thesis). Although some have found new economic niches, thanks to the project and to organizations like Manos Unidas, thousands of poor Managuans continue to work in 'informal' economies of recycling and waste as garbage pickers, recyclers, or scrap merchants, where they are characterized as blights on the landscape but yet are given no other economic opportunities. Although addressing the question of resistance to tourism development is beyond the scope of this chapter, suffice it to say that such groups have not passively abided by their continued marginalization with the Live Clean, Live Healthy campaign. Hence, political ecology's abiding concern with the social nature, as it were, of political action. Urban landscapes, like nature itself, are clearly not only the medium of political action, they are political actants that shape social and economic life (Latour, 2004). So too are those who are often relegated to the margins of those landscapes. Working through national and international advocacy networks for informal economy workers, such as REDNICA, REDLACRE, and WIEGO, for example, Nicaraguan garbage pickers have taken the opportunity of Live Clean, Live Healthy (and its environmentalization of urban social policy) to push for more 'socially inclusive' solutions to recycling and to resignify themselves as responsible, environmental citizens – strong and productive 'ants' (*hormigas*) or 'grassroots recyclers' (*recicladores de base*), rather than vultures picking through the detritus of society. In the long term, this continues to be the strength of political ecology as an analytical framework: to politicize the landscapes in which seemingly apolitical forces like 'tourism' and 'development' work (cf. Ferguson, 1990; Escobar 1995), to map the driving forces of environmental degradation and social marginalization, and to reveal how such landscapes become contested domains of both calculative and aesthetic governance.

References

Agrawal, A. (2005) *Environmentality: Technologies of Government and the Making of Subjects*. Durham, NC: Duke University Press.
Auyero, A.J. (2009) *Flammable: Environmental Suffering in an Argentine Shantytown*. Oxford: Oxford University Press.
Babb, F.E. (1999) Managua is Nicaragua: the making of a neoliberal city, *City & Society*, 11 (1–2), pp. 27–48.
Babb, F.E. (2001) *After Revolution: Mapping Gender and Cultural Politics in Neoliberal Nicaragua*. Austin, TX: University of Texas Press.
Babb, F.E. (2004) Recycled sandalistas: from revolution to resorts in the new Nicaragua, *American Anthropologist*, 106 (3), 541–555.
Bermúdez, J.C. and Álvarez, R. (2014) 'Vivir Bonito' perdió impulso. *La Prensa*, 4 Feb.
Blaikie, P.M. (1985) *The Political Economy of Soil Erosion in Developing Countries*. New York: Routledge.
Briggs, C. (2004) *Stories in the Times of Cholera: Racial Profiling During a Medical Nightmare*. Berkeley, CA: University of California Press.
Bryant, R.L. (2002) Non-governmental organizations and governmentality: 'Consuming' biodiversity and indigenous people in the Philippines, *Political Studies*, 50 (2), pp. 268–292.
Cabrera, K. (2007) Los horrores del mundo. *Intervi*, [online]. Available from: http://www.interviu.es/reportajes/articulos/los-horrores-del-mundo [Accessed 29 Mar 2015]. A Cerón Méndez (2015a) Gobierno instalará escuelas municipales de turismo [online]. *El 19 Digital*. Available from: http://www.el19digital.com/articulos/ver/titulo:27095-gobierno-instalara-escuelas-municipales-de-turismo [Accessed 7 Oct 2015].
Cerón Méndez, T. (2015b) Managua cada vez más atractiva para el turismo local e internacional [online]. *El 19 Digital*. Available from: http://www.el19digital.com/articulos/ver/titulo:26256-managua-cada-vez-mas-atractiva-para-el-turismo-local-e-internacional [Accessed 29 Mar 2015].
Chávez, K. (2015) Turismo de Negocios incrementa en Nicaragua con desarrollo del Gran Canal y Tumarín [online]. *El 19 Digital*. Available from: http://www.el19digital.com/articulos/ver/titulo:27449-turismo-de-negocios-incrementa-en-nicaragua-con-desarrollo-del-gran-canal-y-tumarin- [Accessed 29 Mar 2015].
Choza, A. (2002) Elementos básicos para la protección de las aguas subterráneas aplicados en el acuífero de Managua, Nicaragua. *Revista Geológica de América Central*, 27, pp. 61–74.
Cole, S. (2012) A political ecology of water equity and tourism: a case study from Bali, *Annals of Tourism Research*, 39 (2), pp. 1221–1241.
Cronon, W. (1996) *Uncommon Ground: Rethinking the Human Place in Nature*. New York: Norton.
Desfor, G. and Keil, R. (2004) *Nature and the City: Making Environmental Policy in Toronto and Los Angeles*. Tucson, AZ: University of Arizona Press.

Dobrzensky, A. (2013) *Moon Nicaragua*. Berkeley, CA: Avalon Travel.
Egerton, A. and Benchwick, G. (2013) *Lonely Planet Nicaragua*. Franklin, TN: Lonely Planet.
Equal Exchange (2015) Equal Exchange – Our Co-op [online]. Available from: http://www.equalexchange.coop/our-co-op [Accessed 1 Mar 2015].
Equipo Envío (2008) Nicaragua Breves: Los Pobres de La Chureca. *Envío*, 313.
Equipo Envío (2013) Vivir bonito: ¿una 'revolución cultural'? *Envío*, 372.
Escobar, A. (1995). *Encountering Development: The Making and Unmaking of Development*. Princeton, NJ: Princeton University Press.
Escobar, A. (1999) After nature: steps to an antiessentialist political ecology, *Current Anthropology*, 40 (1), pp. 1–30.
Escobar, A. (2008) *Territories of Difference: Place, Movements, Life*. Durham, NC: Duke University Press.
Ferguson, J. (1990) *The anti-politics machine: 'development,' depoliticization, and bureaucratic power in Lesotho*. CUP Archive.
Fletcher, R. (2014) *Romancing the Wild: Cultural Dimensions of Ecotourism*. Durham, NC: Duke University Press.
Foster, J.B. (2000) *Marx's Ecology: Materialism and Nature*. New York: New York University Press.
Foucault, M. (2007) *Security, Territory, Population: Lectures at the Collège de France 1977–1978*. New York: Palgrave Macmillan.
Ghertner, D.A. (2010) Calculating without numbers: aesthetic governmentality in Delhi's slums, *Economy and Society*, 39 (2), pp. 185–217.
Ghertner, D.A. (2011) Rule by aesthetics: world-class city making in Delhi. In: A. Roy and A. Ong, (eds) *Worlding Cities: Asian Experiments and the Art of Being Global*. Malden, MA: Wiley-Blackwell, pp. 279–306.
Gidwani, V. and Reddy, R.N. (2011) The afterlives of 'waste': notes from India for a minor history of capitalist surplus, *Antipode*, 43 (5), pp. 1625–1658.
Gössling, S. (2003) *Tourism and Development in Tropical Islands: Political Ecology Perspectives*. Northhampton, MA: Edward Elgar.
Gregson, N. and Crang, M. (2010) Materiality and waste: inorganic vitality in a networked world, *Environment and Planning A*, 42, pp. 1026–1032.
Grigsby Vergara, W. (2008) La 'nueva' Chureca: de la basura a la dignidad humana. *Envío*, 321. Available at: http://www.envio.org.ni/articulo/3749 [Accessed 29 Mar 2015].
Grossman, L.S. (2014) *Peasants, Subsistence Ecology, and Development in the Highlands of Papua New Guinea*. Princeton, NJ: Princeton University Press.
Guthman, J. (2004) *Agrarian Dreams: The Paradox of Organic Farming in California*. Berkeley, CA: University of California Press.
Hartmann, C.D. (2012) uneven urban spaces: accessing trash in Managua, Nicaragua, *Journal of Latin American Geography*, 11 (1), pp. 143–163.
Hecht, S.B. and Cockburn, A. (1989) *The Fate of the Forest: Developers, Destroyers, and Defenders of the Amazon*. Chicago, IL: University of Chicago Press.
Heynen, N.C., Kaika, M., and Swyngedouw, E. (2006) *In the nature of cities: urban political ecology and the politics of urban metabolism*. New York: Routledge.
Hollander, P. (1986) Political tourism in Cuba and Nicaragua, *Society*, 23 (4), pp. 28–37.
Hultman, J. and Corvellac, H. (2012) The waste hierarchy model: from the sociomateriality of waste to a politics of consumption, *Environment and Planning A*, 44 (10), pp. 2413–2427.
INTUR (Nicaraguan Tourism Institute) (2009) *Tourism Statistics Report*, 19.

Kaika, M. (2005) *City of Flows: Modernity, Nature, and the City.* New York: Routledge.
Kinzer, S. (2002) A Faded City Brightens in Nicaragua, *The New York Times*, 17 February, pp. 10–12.
Klinenberg, E. (2002) *Heat Wave: A Social Autopsy of Disaster in Chicago.* Chicago, IL: University of Chicago Press.
Lara, R. (2013) Multas altas a quienes contaminen, *El Nuevo Diario*, 23 Mar.
Latour, B. (2004) How to talk about the body? The normative dimension of science studies, *Body & Society*, 10 (2–3), pp. 205–229.
Li, T.M. (2007) *The Will to Improve: Governmentality, Development, and the Practice of Politics.* Durham, NC: Duke University Press.
Marina, M. (2015) Ana Carolina García: 'El Gran Canal de Nicaragua potenciará el turismo y las inversiones en el país', *elEconomista.es*, 3 Mar.
Martínez-Alier, J. (2002) *The Environmentalism of the Poor: A Study of Ecological Conflicts and Valuation.* London: Elgar.
Merchant, C. (2003) *Reinventing Eden: The Fate of Nature in Western Culture.* New York: Routledge.
MINSA (Ministry of Health), (2012) Plan de Acción de Manejo de Basura 2012.
Mitchell, T. (1991) *Colonising Egypt.* Berkeley, CA: University of California Press.
Monstadt, J. (2009) Conceptualizing the political ecology of urban infrastructures: insights from technology and urban studies, *Environment and Planning A*, 41 (8), pp. 1924–1942.
Moore, S.A. (2008) Waste practices and politics: the case of Oaxaca, Mexico. In: D.V. Carruthers, (ed.) *Environmental Justice in Latin America: Problems, Promise, and Practice.* Boston, MA: MIT Press.
Moore, S.A. (2009) The excess of modernity: garbage politics in Oaxaca, Mexico, *The Professional Geographer*, 61, pp. 426–437.
Moore, S.A. (2011) Global garbage: waste, trash trading, and local garbage politics. In: R. Peet, P. Robbins, and M. Watts (eds) *Global Political Ecology.* New York: Routledge.
Moore, S.A. (2012) Garbage matters: concepts in new geographies of waste, *Progress in Human Geography*, 36 (6), pp. 780–799.
Mostafanezhad, M. (2013) The politics of aesthetics in volunteer tourism, *Annals of Tourism Research*, 43, pp. 150–169.
Murillo, R. (2013) Vivir Limpio, Vivir Sano, Vivir Bonito, Vivir Bien! Gobierno de Reconciliación y Unidad Nacional, Nicaragua.
Mustafa, D. (2005) The production of an urban hazardscape in Pakistan: modernity, vulnerability, and the range of choice, *Annals of the Association of American Geographers*, 95 (3), pp. 566–586.
Mutersbaugh, T. (2002) The number is the beast: a political economy of organic-coffee certification and producer unionism, *Environment and Planning A*, 34 (7), pp. 1165–1184.
Nading, A.M. (2014) *Mosquito Trails: Ecology, Health, and the Politics of Entanglement.* Berkeley, CA: University of California Press.
Neumann, R.P. (1998) *Imposing Wilderness: Struggles over Livelihood and Nature Preservation in Africa.* Berkeley, CA: University of California Press.
O'Connor, J. (1996) The second contradiction of capitalism. In: T. Benton, (ed.) *The Greening of Marxism.* New York: Guilford Press, pp. 197–221.
Ortega Ramírez, P. (2015) Managua, recreación, turismo e historia, *El 19 Digital*, 15 Feb.
Parés Barberena, M.I. (2006) Estrategia municipal para la intervención integral de asentamientos humanos espontáneos de Managua, Nicaragua.

Peluso, N.L. (1995) Whose woods are these? Counter-mapping forest territories in Kalimantan, Indonesia, *Antipode*, 27 (4), pp. 383–406.
Pérez Rivera, A. (2009) La ciudad capital estrena botaderos de basura, *La Prensa*, 5 Jan.
Potosme, R.H. and Picón, G. (2013) Ven fascismo en doctrina de Murillo, *La Prensa*, 8 Feb.
Prado Reyes, Y. (2014) Prevén crecimiento del 6% en el sector turismo, *El 19 Digital*, 28 Feb.
Rancière, J. (2004) *The Politics of Aesthetics*. New York: Bloomsbury.
Rathje, W.L. and Murphy, C. (2001) *Rubbish! The Archaeology of Garbage*. Tucson, AZ: University of Arizona Press.
Robbins, P. (2007) *Lawn People: How Grasses, Weeds, and Chemicals Make Us Who We Are*. Philadelphia, PA: Temple University Press.
Robbins, P. (2012) *Political Ecology: A Critical Introduction*, 2nd edn. Malden, MA: John Wiley & Sons.
Robbins, P. and Fraser, A. (2003) A forest of contradictions: producing the landscapes of the Scottish Highlands, *Antipode*, 35 (1), pp. 95–118.
Rodgers, D. (2004) 'Disembedding' the city: crime, insecurity and spatial organization in Managua, Nicaragua, *Environment and Urbanization*, 16 (2), pp. 113–124.
Rodgers, D. (2011) An illness called Managua: 'extraordinary' urbanization and 'maldevelopment' in Nicaragua. In: T. Edensor and M. Jayne, (eds) *Urban Theory Beyond the West: A World of Cities*. New York: Routledge, pp. 121–136.
Rose, N. (1991) Governing by numbers: Figuring out democracy, *Accounting, Organizations and Society*, 16 (7), pp. 673–692.
Roy, A. and AlSayyad, N. (2004) *Urban Informality: Transnational Perspectives from the Middle East, Latin America, and South Asia*. Lanham, MD: Lexington Books.
Scheyvens, R. (2001) Poverty tourism, *Development Bulletin*, 55, pp. 18–21.
Schmink, M. and Wood, C.H. (1987) The ' political ecology' of Amazonia. *Lands at risk in the Third World: Local-level Perspectives*, pp. 38–57.
Schmink, M. and Wood, C.H. (1992) *Contested Frontiers in Amazonia*. New York: Columbia University Press.
Scott, J.C. (1998) *Seeing Like a State: How Certain Schemes to Improve the Human Condition Have Failed*. New Haven, CT: Yale University Press.
Sharpley, R. (2009) *Tourism Development and the Environment: Beyond Sustainability?* New York: Routledge.
Silva, J.A. and Potosme, R.H. (2013) Ordenan al país 'vivir bonito', *La Prensa*, 7 Feb.
Sobalvarro Muñoz, J. (2007) Agenda del desempleado. Managua, Nicaragua: *400 Elefantes*.
St. Martin, K. (2001) Making space for community resource management in fisheries, *Annals of the Association of American Geographers*, 91 (1), pp. 122–142.
Stonich, S.C. (1998) Political ecology of tourism, *Annals of Tourism Research*, 25 (1), pp. 25–54.
Sundberg, J. (2004) Identities in the making: Conservation, gender, and race in the Maya Biosphere Reserve, Guatemala, *Gender, Place and Culture*, 11 (1), pp. 43–64.
Swyngedouw, E. and Heynen, N.C. (2003) Urban political ecology, justice and the politics of scale, *Antipode*, 35 (5), pp. 898–918.
Verón, R. (2006) Remaking urban environments: the political ecology of air pollution in Delhi, *Environment and Planning A*, 38 (11), pp. 2093–2109.
Vidaurre Arias, A. (2015) Proliferan inversiones turísticas en Nicaragua, *El Nuevo Diario*, 21 Mar.

VirtualTourist (2015) Managua Local Customs [online]. VirtualTourist.com. Available from: http://www.virtualtourist.com/travel/Caribbean_and_Central_America/Nicaragua/Departamento_de_Managua/Managua-1697194/Local_Customs-Managua-TG-C-1.html [Accessed 19 Jun 2015].

Wainwright, J. (2008) *Decolonizing Development: Colonial Power and the Maya*. Malden, MA: Wiley-Blackwell.

West, P. (2006) *Conservation Is Our Government Now: The Politics of Ecology in Papua New Guinea*. Durham, NC: Duke University Press.

West, P. and Carrier, J.G. (2004) Ecotourism and authenticity: getting away from it all?, *Current Anthropology*, 45 (4), pp. 483–498.

Wolf, E. (1972) Ownership and political ecology, *Anthropological Quarterly*, 45 (3), pp. 201–205.

Zapata, P. (2013) Waste in translation: global ideas of urban waste management in local practice. In: *Organising Waste in the City: International Perspective on Narratives and Practices*. Chicago, IL: Policy Press, pp. 83–98.

Zapata Campos, M.J. (2010) Branding Poverty: La Chureca, the 'Slum Project Millionaire': How a Project Becomes a Project. Paper presented at the workshop *Exploring Spaces and Linkages between Services, Markets, and Society*, Lund University, August 25–27.

Zapata Campos, M.J.Z. and Zapata, P. (2013) Switching Managua on! Connecting informal settlements to the formal city through household waste collection, *Environment and Urbanization*, 25 (1), pp. 225–242.

12 The political ecology of tourism development on Mount Kilimanjaro

Megan Holroyd

Introduction

The Tanzanian government has promoted international tourism as a means to diversify its economy and to bring increased development to the country. Tourism is often used as a development strategy in part due to its perceived multiplier effects. These effects include increased visitor spending for food and souvenirs as well as an increase in local businesses and investments that are created to meet the demands of the tourists (Hall and Page, 1999). The promotion of the tourism industry in Tanzania has resulted in its increased economic importance to the country. In 2013, tourism contributed US$ 1.88 billion to the economy, making it the top foreign exchange earner (*Daily News*, 2014a). By 2014, tourism was estimated to make up 4.5 percent of Tanzania's GDP and account for 11.2 percent of employment in the country (WTTC, 2014).

The tourism sector in Tanzania relies predominantly on its natural resources as a means of attracting international tourists. These consist of coastal areas and beaches as well as inland game reserves and national park areas (Barron and Prideaux, 1998; Nelson, 2004). Tourism often contributes to the creation of protected areas that serve as tourist attractions and a means of generating revenue through fees and taxes, in turn creating an interest in maintaining these protected areas (Gössling, 2003). Political ecology can serve as a useful tool in understanding the dynamics of natural resource-based tourism. Bryant and Bailey (1997) saw political ecology as a means to inform how capitalist expansion and political interests have impacted peoples and environments in developing countries, which often results in highly unequal power relationships. These relationships are present in the tourism industry in Tanzania as well. Government institutions that facilitate tourism and tourism businesses have been established to operate in or near areas set aside for tourism. These have a dominating impact in the areas in which they are located, as they often determine who can access certain lands, as well as what activities and places will be prioritized for tourism development.

Tourism in Tanzania is heavily concentrated and promoted in the northern part of the country, and the sector is often identified as the Northern Circuit in the Tanzanian tourism industry. Northern Circuit attractions such as the Serengeti, Ngorongoro crater and Mount Kilimanjaro draw the most tourists and revenue to the country (UNCTD, 2008). Kilimanjaro National Park is especially lucrative to the tourism industry as the park brings in the highest revenue of all of Tanzania's national parks (Mariki, 2013; Mitchell *et al.*, 2009). Additionally, Mbonile and Bart (1999, p. 6) noted that the 'Kilimanjaro region is one of the most active regions of Tanzania, as far as agriculture, trade, and tourism are concerned. In such conditions, there are strong links between the problems of natural resource management and socio-economic development'. This chapter builds on the work of other researchers who have studied the relationships between conservation, tourism and dispossession in Tanzania (Benjaminsen and Bryceson, 2012; Brockington, 2002; Brockington and Igoe, 2006; Honey, 1999; Nelson, 2012; Neumann, 1998) by focusing on the case study of tourism on Mount Kilimanjaro. By looking at the role and impact of tourism in this region, and specifically the impacts that the industry has on the surrounding communities, we can better understand how communities become marginalized or excluded from the revenue and other benefits that are generated from Kilimanjaro as a tourist destination. This chapter utilizes a political ecology analysis by historically situating conservation initiatives in Tanzania and offering a place-based case study of tourism and conservation on Mount Kilimanjaro, while providing a multi-scalar analysis of the different actors involved the tourism sector.

A political ecology critique can offer insight into the practice of creating parks and protected areas for environmental conservation. Robbins (2004) links conservation and control, as well as the exclusions that results from power struggles over access to resources, as major themes with which political ecologists are concerned. Political ecologists have recognized that environmental conservation can serve as a form of environmental and resource control used by the more powerful, dominant actors in order to take control of generated revenue (Peet *et al.*, 2010). I argue that the dominant actors in Mount Kilimanjaro tourism have used the park for similar purposes while marginalizing and excluding local communities from tourism revenues. Additionally, the local communities are compelled to shoulder much of the costs related to tourism development to the region due to the loss of access to resources in the protected area.

Methodology

This chapter draws on 16 months spent living, working and conducting research in Tanzania. Political ecological research often utilizes a variety of methods in order to obtain information from different actors. My methods included interviews, participant observation and informal discussions with individuals and groups, as well as quantitative surveys and questionnaires to complement the qualitative research (Bassett and Crummey, 2003). I began my research in Tanzania by spending 12 months working at an international environmental and conservation NGO in Dar es Salaam with monthly week-long trips to a field site in Mweka, located just outside Kilimanjaro National Park. During this time, I conducted preliminary research consisting of observations and informal conversations with residents and tourism workers. Through these conversations and observations, I was able to get a better understanding of the tourism industry in the region, which informed the direction of my research.

I then relocated to Moshi for two months so that I would be able to work full-time on my research, during which time I conducted surveys and interviews of individuals working in the Mount Kilimanjaro service sector. The surveys and interviews were primarily conducted in Moshi, located at the base of Mount Kilimanjaro. My work with the NGO had connected me with students from the College of African Wildlife Management (CAWM). These students were encouraged and required to complete fieldwork and research outside of their formal coursework. I worked with a few CAWM students in order to conduct the survey portion of my research. These students had previously completed their own research and had conducted surveys previously. They were able to administer the surveys in Kiswahili if needed. The data was then entered into an excel database and analyzed.

Overview of park formation in Tanzania

The formation of Tanzania's parks can be traced to colonial policies and practices under German, then British rule in Tanganyika. Under the premise of conservation, the German administrators created forest reserves, game reserves and later national parks, effectively denying Africans occupancy and land rights in these protected areas (Chachage and Mallya, 2005). The creation of these parks meant that Africans would not be allowed to use their traditional land-use systems and further that it would be illegal for them to shoot animals for food or for protection. By doing so, the colonizers were ensuring they would be the only ones with access to the market for such goods as ivory and fur (Honey, 1999). It was during this time that wildlife effectively became reserved for European hunters, while the costs were borne primarily by Africans (Neumann, 1998). As a result, people who had once counted on hunting wildlife for subsistence were required to look elsewhere to meet these needs, unless they were prepared to break the newly imposed colonial laws, which carried heavy punishments.

These laws not only served to benefit the colonizers in terms of the extraction of resources, they also laid the groundwork for the formation of safari tourism to the region, which led to further marginalization of many Africans. The emergence of tourism as an industry to serve Westerners is exemplified in the creation of the protected areas, such as Serengeti National Park and Ngorongoro Conservation Area. Their creation severely limited – and in places outlawed – local people's access to the land that was needed to support pastoral lifestyles (Charnley, 2005). Much of the land alienation that was occurring under colonial rule took place throughout northern Tanganyika where German immigrants settled. Many of these settlements were located in the north on fertile lands used for farming, such as the already-inhabited slopes of Mount Meru and Mount Kilimanjaro. Colonial settlement on Mount Meru and the creation of Arusha National Park resulted in the alienation of Meru lands, and subsequently led to conflict between the community and park authorities after the Meru were denied access to lands and resources to which they traditionally had access (Neumann, 1998). On Mount Kilimanjaro, the Germans had created a game reserve by the early 1900s, while the British, after gaining control following the First World War, added a designated Forest Reserve on the mountain in 1921 (Durrant, 2009). The creation of these areas also resulted in the loss of access to resources for the people living on the mountain.

Often this loss of access was couched in terms of conservation, as national parks not only drew tourists and hunters, but also attracted conservationists from the West, many of whom maintained that Africans were wiping out their wildlife populations and therefore needed someone else to oversee these resources. The conservationists were able to create a powerful lobby to remove the Maasai from the Serengeti by arguing that the soil and water reserves were too fragile to support them (Honey, 1999). A political ecological history of park formation recognizes that local landscape management practices have often protected environments, and that the exclusion of these groups and the often violent seizure of these lands have taken place due to the interests of more powerful actors (Peet et al., 2010, p. 27).

Protected area enclaves

Game reserves, national parks and other tourist attractions created under colonial rule were built on the alienation of natural resources and lands from the local communities who had lived in these areas and used resources that were now denied them (Chachage and Mallya, 2005). However, after independence in 1961, the government of Tanzania created even more protected areas that continued to prevent ownership and use by the surrounding communities, propagating many of the colonial practices. For example, government policies continued to relocate people from parks, as well as to make use of armed patrols and anti-poaching measures. Bureaucratically, authority and revenue remained concentrated at the top while local communities 'were deprived of access to ancestral homelands, grazing land, water, and wildlife, and saw few tangible benefits from either the parks or tourism' (Honey, 1999, p. 222). Brockington (2002) referred to this forceful eviction or exclusion of local people from the lands in order to restore an imagined untouched wilderness as 'fortress conservation'.

This structure remained in place under Tanzanian socialism and through the economic reforms of the 1990s that were meant to decentralize control and provide benefits to local communities. However, these reforms often proved ineffective in providing more control and benefits to the local communities. In fact, Benjaminsen et al. (2013) argue that neoliberal reforms and conservation measures have helped prompt an increase in protected areas within Tanzania as the resources they enclose rise in value on the global market. By protecting these areas through the use of conservationist discourse, the government has been able to create resource-rich tourism pockets or enclaves that exclude the local communities. Because the creation of these protected areas has helped facilitate tourism to the country, the Tanzanian government has increasingly set aside significant amounts of land for parks and other protected areas (see Figure 12.1). By 2012, at least 40 percent of Tanzania's land was conserved or protected in some form or another (Benjaminsen and Bryceson, 2012). About 34 percent of Tanzania's protected areas consist of national parks and wildlife protected areas, while forest

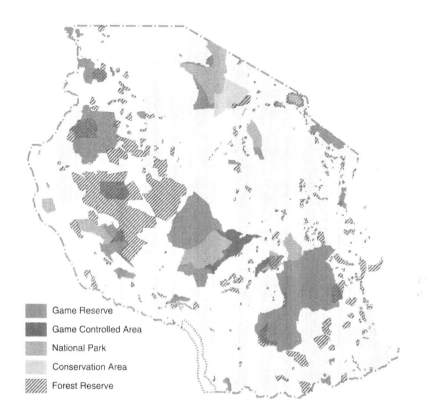

Figure 12.1 Map of Tanzania's protected areas.

Sources: Tanzania GIS User Group (2014) and World Database on Protected Areas (2014) (redrawn by author) and Sémhur (2014b – modified by author).

reserves cover about 15 percent (United Republic of Tanzania, 2014). These protected areas consist of lands and resources that local communities can no longer access. Benjaminsen and Bryceson (2012, p. 336) 'argue that these conservation processes result in dispossession of land and resources from local users, as well as capital accumulation by more powerful actors', creating a form of 'primitive accumulation'. The closing off of park and protected lands from the communities, while at the same time concentrating the benefits of this industry to only the most powerful actors, has resulted in an enclave system of tourism within Tanzania.

Britton (1982) recognized the enclave nature of the tourism industry in many former colonies, due to both a reliance on visitors from developed countries to purchase the tourist product and especially to the spatial organization of the tourist services and infrastructure that segregates the industry from communities in the destination country. His analysis is applicable to Tanzania's tourism industry in several areas. Tanzania relies on international tourists, primarily from Western

Europe and the United States, to fuel its tourism sector, making the industry vulnerable to factors such as economic downturns in these countries or even a change in destination popularity. Additionally, the areas that Euro-American tourists frequent are spatially segregated from local communities through the restricted access to national parks or, for example, locals being priced out of high-end tourist hotels. With enclave tourism, the goods and services provided for the tourists often exclude local communities from their use, as they are unable to afford these services. Additionally, the weak economic linkages to the local economy often result in minimal economic growth or opportunities in the destination country. In this way, the natural resource base on which the industry relies primarily benefits outsiders, resulting in a form of 'internal colonialism' (Mbaiwa, 2003). The result is the direct opposite of one of the supposed benefits of tourism, which proposes increased linkages to, and economic growth in, other sectors of the economy – the so-called multiplier effect.

The dominant actors in Mount Kilimanjaro tourism

The enclave nature of the tourism industry in Tanzania has been constructed and maintained due to the influence of foreign conservation NGOs, the government of Tanzania and international corporations. These institutions have benefitted from the continuation of the enclave nature of tourism and they comprise the dominant actors in the tourism industry in Tanzania. Together, they have created a dominant culture that serves to further the continuation of the enclave tourist system from which they benefit. Cosgrove (1989, p. 128) defined the dominant culture as 'a group with power over others'. However, a dominant culture does not simply mean a governing body; it can also encompass a group that has 'control of the means of life: land, capital, raw materials, and labor power'. On Mount Kilimanjaro, specific institutions or dominant actors that comprise the dominant culture can be seen in the Kilimanjaro National Park Authority (KINAPA) and the many tour companies that take international tourists to Mount Kilimanjaro (Kokel, 2015; Lovelock, 2015).

Kilimanjaro National Park (KINAP)

Tanzania's Ministry of Natural Resources and Tourism (MNRT), and the parastatal organizations that the ministry oversees, are largely responsible for managing the country's natural resource base. Within MNRT, Tanzania National Parks Authority (TANAPA) is the parastatal organization responsible for the management of Tanzania's 16 national parks. Among these parks, KINAPA is responsible for managing the national park on Mount Kilimanjaro. The park was officially created in 1973 and comprised the 75,575 hectares above the tree line. The Kilimanjaro Forest Reserve consists of an additional 107,828 hectares of forest surrounding the park. However, in 2005, at the urging of the World Heritage Committee, the park was expanded to include the Forest Reserve as well (UNESCO, 2015). KINAP is classified as an area that is protected primarily for ecosystem conservation and recreational purposes. As such, people are prohibited from utilizing natural resources located within the park boundaries (Mariki, 2013). Due to the 2005 expansion, this restriction now includes the Forest Reserve as well. However, over 80 villages either border the park or are located within it (Durrant *et al.*, 2003). These villages have traditionally relied on the forest for their livelihood and to meet their daily living needs. With the increase in restricted land area, these communities are increasingly being denied access to resources in return for tourism-generated development to the country, little of which they see themselves.

This promise of development may not seem particularly unreasonable for the region as KINAP is the highest revenue earner of TANAPA's parks. Even though it receives fewer visitors than some of the other parks, it is estimated that KINAP makes up 38 percent to 45 percent of TANAPA's total income (Mariki, 2013; Nelson, 2008). This income is derived from various fees from tourists, tour operators and investors, and includes entry fees, vehicle permits, resident permits, hotel fees and camping fees (Chachage and Mallya, 2005). In the case of KINAP, there are an estimated 35,000 climbers who visit the park annually. These climbers spend an average of six to seven days in the park due to the time it takes to climb and descend the mountain, and for each day spent in the park, the tourist must pay park fees. In 2014, the fees consisted of the following for international visitors: US$ 70 per day conservation/entry fee, $50 per day camping fee and a $20 per trip rescue fee (TANAPA, 2014). For the average six-day trip, KINAPA receives $740 in fees per person. At the rate of 35,000 climbers spending six days in the park, entrance fees paid directly to the park amount to over US$ 25 million annually.

The local people working in the industry, even at the lowest levels, are well aware of the revenue that the government receives from the tourism industry on Mount Kilimanjaro. Of the individuals surveyed, 34 percent listed either the government or TANAPA/KINAPA as a primary beneficiary of tourism in the region. One respondent commented on the high park entry fee to climb Kilimanjaro, which is then compounded by the number of visitors to the park: 'The government benefits the most from tourism. If someone goes to Kili, it is $600. It is owned by the government, and every day, every hour, people are going to Kili' (Interview, 2010). A porter noted that people working for KINAPA benefit the most from tourism in that 'they charge for each activity conducted inside the park, and there are many penalties' (Survey, 2010). These responses suggest that there is recognition of the monopoly that the government has on the tourism income received through park fees. Additionally, the porter touched on the access to rents in the form of park 'penalties' to which KINAPA workers have access. Revenue through, for example, fines, penalties, fees and licenses comprise natural resource rents to which officials and tourism actors have access (Nelson, 2012). Other residents on Mount Kilimanjaro have accused KINAPA 'of practising an economic sanctuarization: a false protected area only designed to be profitable' (Sebastien, 2010, p. 6). However, these profits are only available to those who have access to the park and the tourism revenue that it brings.

In line with the enclave nature of tourism in Tanzania, most of the villages around Kilimanjaro National Park do not receive tourists. There are no hotels or guesthouses on the mountain; these are all located in Moshi. There are also no campsites, as all camping must take place in the park. Nor are small-scale, local businesses such as curio shops or restaurants present around or within the park; they are instead confined to Moshi. Tourists are transported from their hotels in the towns of Moshi or Arusha and taken directly to the park gates. Once they have climbed and descended the mountain, they are then transported back to their hotels having never actually set foot outside of the confines of the hotel or the park. The Mweka route is one of the most common descent routes of the Kilimanjaro climb as it is thought to be the fastest and most direct. As a result, large numbers of vehicles transporting tourists or porters drive through Mweka. However, the promised multiplier effect of tourism – in the form of service provision or small-scale businesses that could cater to the needs of these tourists passing through Mweka – never actually trickles down to the village, since the tour company has already provided for them.

Tourism businesses

The majority of tourists who come to Tanzania arrive via a tour package they have purchased in their home country. In 2009, 78 percent of visitors to Tanzania who came for leisure and holiday arrived with a package tour arrangement (MNRT and BOT, 2011). The tour company in the country of origin then contracts out to companies that are located in Tanzania. However, in many cases, Tanzanians may not own these companies, as there is a large amount of vertical integration and foreign ownership in the tourism sector. Chachage and Mallya (2005) argue that the ties between foreign and supposedly local tour operators, along with vertical integration, result in 75 to 90 percent of tourism revenues either leaking out of the country or remaining in the country of origin. Much of this leakage through foreign ownership can be traced to the structural adjustment and privatization policies implemented in the early 1990s. At that time, many of the large hotels in Tanzania were acquired at 'throw-away' prices during privatization and are now either fully or partially foreign-owned (Chachage and Mallya, 2005). As a result, foreign ownership (and thus, power) within the tourism sector in Tanzania is often very high.

In a 2001 MNRT report, data from All Africa Travel and Tourism Association (AATTA) showed that, in Tanzania, foreign ownership controls 99 percent of air travel, 95 percent of hunting, 80 percent of land and hotels and 50 percent of recreation and leisure (Ranja, 2003; Salazar, 2008). This high level of foreign ownership results in much of the tourism revenue leaking out of the country in various ways. Okech (2010) estimated that out of every tourist dollar spent in Tanzania, 20 cents goes to the government, 40 cents is leaked out of the country on imports and another 40 cents goes to privately owned hotels and other tourism businesses. Additionally, when these businesses are foreign-owned, there is an even greater amount of leakage. For example, a 2005 UNCTAD study of tourism in Tanzania found that locally-owned businesses source 85 to 90 percent of their supplies within Tanzania, while foreign-owned businesses import over 80 percent of these supplies (UNCTAD, 2008). Therefore, when there is a high level of foreign ownership, there is a low level of linkage between the tourism sector and other sectors of the Tanzanian economy, such as agriculture and manufacturing.

In addition to the high level of foreign ownership, foreign-owned businesses often receive much of the tourism business. Chachage and Mallya (2005) found most of the hotels and tour companies in Tanzania were either foreign-owned or had foreign connections, and that a handful of the large foreign-owned operations received 50 percent of the tourism business. In fact, these businesses dominate the tourism industry in Tanzania to the extent that 71 percent of tourism's tax revenue comes from foreign-owned companies (UNCTAD, 2008). Chachage and Mallya (2005, p. 23) argued 'what is involved in this industry, given the dominance of foreign concerns, is not simply a question of leakages, but that of companies making super-profits, without benefiting the people and the country in general'. This argument is reflected in my survey results of workers in the Mount Kilimanjaro tourism sector. For example, 80 percent of the tourism workers surveyed listed tourism business owners, tour companies and/or hotels, as benefiting the most from tourism in the region. Some respondents also commented on the leakage that occurs in tourism. One guide explained that 'more generated income goes back abroad since airways, some of the equipment and even tourists come from abroad' (Survey, 2010).

As mentioned, the funneling of money away from the destination country is often exacerbated with package tourism. After purchasing the tourism package in their home country, tourists may bring only a small amount of money to buy souvenirs in the destination country (Chachage and Mallya, 2005). However, in some cases, the foreign operator provides even these small items as well. For example, one of the large curio shops and tour companies in Moshi receives most of its customers through ties to a tour company in the US. The curio shop started out as a tour company and is still at least partly foreign-owned. The company receives all of its customers via the US company, through which customers have to book their tours. The curio shop was created as a way to further corner the tourist market by meeting their customers' demand for souvenirs. Instead of taking their clients around to various tourist shops, or letting them go off on their own, they are encouraged to patronize the curio shop owned by the tour company as part of their tour package.

The Tanzanian workers who do not have such connections are often negatively affected by the implications of this type of arrangement. One street vendor noted that the owners of the tour companies benefit 'as they are the one who has all influence in tourism. Some sell handicrafts to tourists instead of leaving us to do that'. By cornering the market and providing an all-inclusive tourism package, these large, often foreign-owned companies are pushing Tanzanian tourism workers out of work. Additionally, this inclusiveness of the tour package serves to isolate the tourist experience and further exemplifies the enclave nature of the tourism experience in Tanzania, an occurrence that is especially true of the larger tour companies. In reference to the above tour company and curio shop, one respondent stated: 'The big tour companies are all foreign-owned. Tourists come here and can spend their money just with this company and all the money goes back to the owner's country. No one here gets any business'.

The distribution of tourism-generated wealth and benefits is important when considering tourism's role in development. Certain actors receiving the majority of the benefits and others receiving none may result in a tourism oligopoly (Gössling, 2003). In fact, Nelson (2012) argues that the political-economic order in Tanzania places barriers to local communities, impeding ordinary citizens' ability to enter the tourism market due to the fact that the state owns and controls the resources that draw the tourists. As a result, 'state interests often actively compete with or displace local initiatives to capture tourism revenues' (Nelson, 2012, p. 363). Once this occurs and local, small-scale actors are unable to benefit from the tourism industry, it can become increasingly difficult to argue for the benefits of tourism and its multiplier effects, especially when the multiplier effects do not reach those who have disproportionately borne the costs of conservation for tourism development. In this sense, the (neo)colonial structure of the tourism industry that has been created in the region privileges elite and Western access to the benefits of tourism while denying access to non-Western and non-elite actors, thus perpetuating the inequalities between the two.

Impacts and implications of protected areas for tourism development on Mount Kilimanjaro

The geographic expansion of parks and protected areas into communities that have relied on such lands to maintain their livelihoods has led to conflicts and antagonism between the parks or government and local communities (Mkumbukwa, 2008; Neumann, 1998). This conflictual relationship has resulted from the dispossession and displacement of local communities due to park expansion. Even when people are not forcibly removed from an area, they can still be subject to displacement if they are excluded from areas upon which they rely for their economic activity. By being denied access to and use of protected areas people are excluded from their economic activities and are unable to continue their lives in the same way as before (Brockington and Igoe, 2006). The creation of national parks has taken land out of production in Tanzania in order to generate revenue through tourism. A consequence of this is that local communities are being denied their 'moral right to subsistence' because they can no longer access these lands and resources (Neumann, 1998, p. 43). On Mount Kilimanjaro, much of this conflict is centered on access to the forest reserves and an area known as the half-mile strip.

Local access to resources and revenues

The half-mile strip is a belt of land along the southern and eastern edge of the Forest Reserve (see Figure 12.2). This area of the mountain is more heavily populated than others, and the strip was meant to provide for local use of the forest.

Tourism development on Mount Kilimanjaro 261

The Chagga people who live on the mountain have historically used the strip and the Forest Reserve area for survival and maintenance of their livelihoods (Durrant *et al.*, 2003). The Chagga Council established the strip in 1941 after the colonial government created the Kilimanjaro Forest Reserve and prohibited the local communities from using this land. The Chagga Council managed the area for 20 years, during which time local people contributed to its management by planting and thinning trees and fighting fires in return for the use of the land for forest products. However, in 1962 the management of the strip was turned over to the District Council, which in turn emphasized commercial use of the strip over local needs. Misana (1999, p. 55) argued that due to the fact that 'the local people

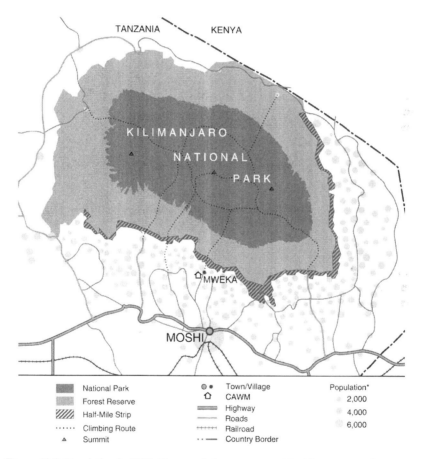

Figure 12.2 Population in 1988. The population represented in this map was taken from the 1988 census and does not reflect current higher population levels in the area. However, it does accurately indicate that the population is concentrated on the southern and eastern slopes of the mountain, which is also where the half-mile strip of the forest has been designated.

Sources: Bart *et al.*, (2006 – redrawn by author); Sémhur 2014a, 2014c – modified by author).

no longer had any right or sense of ownership of the forest strip, they started mismanaging the resources that they had successfully managed for twenty years'. For example, it was claimed that villagers would not participate in government forestry development programs, such as fighting fires, due to their resentment at being seen by the government as the enemy of the forest. This ultimately led to increased degradation of the strip, and people being further marginalized from using the area when the central government took control in 1972 (Misana, 1999).

As part of the decentralization of the late 1980s and early 1990s, management of the strip was again returned to the District Council, which then granted permits to use the strip to collect firewood, fodder and other products (Misana, 1999). In 1994, the Community Conservation Service (CCS), which was initiated by TANAPA and the African Wildlife Foundation, was adopted by KINAPA (Mariki, 2013). A goal of the CCS was to help create *ujirani mwema* or 'good neighborliness' with the communities living next to the park (Chachage and Mallya, 2005; Mariki, 2013). The idea was to involve the local community in conservation efforts and to allow them to benefit by receiving some funds for improvement projects in the community, such as the construction of schools, feeder roads, dispensaries and water pumps (Chachage and Mallya, 2005). However, most of Tanzania's park revenues have continued to be remitted to the park headquarters, with only 7.5 percent of the TANAPA budget set aside for local communities through programs such as CCS. Of this percentage, political influence and leakages through corruption and mismanagement serve to further diminish the amount of money entering local communities (Mariki, 2013). In fact, Benjaminsen *et al.* (2013) argued that these attempts by the state to promote 'community-based conservation' were never intended to actually give up the state's control of these resources, and that these measures have only served to consolidate the wealth and rent-seeking power of the state in conjunction with foreign investment.

Additionally, even when this money is set aside for social and infrastructural investment, Mariki found that often 'local people have no power or influence over its allocation' (Mariki, 2013, p. 11). As a result, the projects that are funded by the CCS may not actually be addressing the needs of the villagers living in the vicinity of the park. Chachage and Mallya (2005, p. 38) have even argued that 'these so-called contributions are mere palliative' in that they do not work to address the root causes of why these services are needed, and instead have been used to justify the continuation of policies that led to the need in the first place. One individual who has been working in Mount Kilimanjaro's tourism sector for several years underscored this sentiment when asked what benefits tourism brings to the people who live on Mount Kilimanjaro:

> Villages on Kilimanjaro do not get anything. KINAPA can say that they will help build a school in a village and they do, but the school fees still need to be paid and this is very difficult for people. Why not make school free? KINAPA depends on porters and guides because tourists cannot enter the park without them, but they get nothing. The children of KINAPA officials go to schools that cost maybe 300,000 tsh [US$ 250] a year, and he just sits in the office. But the porters and guides who freeze in the cold to take tourists

up the mountain, they get nothing. Tourists help only a few people. The local people, we get very little and we do the hard work. The people who sit in the office, they get *all* the benefits.

In his study of natural resource management and local people's support of conservation on Mount Kilimanjaro, Mariki (2013) also found that the communities surrounding KINAP did not find the park to be beneficial. In fact, she argued that 'the failure of the park to allow meaningful local participation and equitable sharing of the park's benefits with affected local people, is leading to hatred, resentment, and illegal harvest of natural resources from the park' (Mariki, 2013, p. 1). Additionally, implementing 'community-based' conservation measures has created a foothold on village lands whereby subsequent dispossessions by the government could take place (Benjaminsen and Bryceson, 2012). For example, after CCS had been adopted by KINAPA in 1994 and implemented for several years, a report in 2005 claimed the greatest threats to Mount Kilimanjaro came primarily from local resource use such as charcoal burning, livestock grazing, farm practices and logging (Mariki, 2013). As a result of the report and at the urging of the World Heritage Committee, the park boundaries were changed to include the Kilimanjaro Forest Reserve. Conservation measures of this sort rely on and perpetuate the narrative that it is local communities that have caused the problems. This narrative then serves to justify the dispossession of land or access to resources (Benjaminsen and Bryceson, 2012). The continued dispossession on Mount Kilimanjaro has come in the form of park expansion to include the Forest Reserve.

However, when the park boundaries were changed to include the Forest Reserve, there was little involvement of local people during this process. As a result, some of the surrounding villages' farmlands ended up within the boundaries of the park. Additionally, there was no clear communication about changes in restrictions, access and use of the forest (Mariki, 2013). Although the CCS is meant to educate individuals about the legal use of park resources and conservation policies, KINAPA staff responsible for this education stated that the 'villagers are expected to know the half-mile strip, Forest Reserve, and national park boundaries themselves' (Durrant *et al.*, 2003). Knowing these boundaries can prove difficult when there are no physical signs or markings of the park boundaries and when the boundaries are often changed without notifying the surrounding residents. In fact, a 2002 study conducted by CAWM and Brigham Young University found that the changing regulations, 'along with poor communication between KINAPA and the villagers is contributing to general confusion and inconsistency concerning legal resource use for residents . . . and the definite boundaries where different types of resource use are allowed' (Durrant *et al.*, 2003, p. 494). This confusion can be especially problematic when coupled with the 'fortress conservation' mentality of park operations and boundary maintenance.

Policing the Park

Maintaining park boundaries is essential in natural resource extractive enclaves. Although these enclaves may be closely linked to multinational corporations and urban centers, they are kept separate from the surrounding society, often through violent means (Ferguson, 2005). This attitude is reflected in the running of the parks, as TANAPA has adopted a military stance in terms of protecting national parks against local communities (Neumann, 1998). Additionally, Malthusian arguments based on population growth or scarcities, such as those used to justify the expansion of the park, are often used by states to justify violence against their citizens (Peluso and Watts, 2001). On Mount Kilimanjaro, the confusion over park boundaries, coupled with the police-like enforcement of forest conservation, has resulted in a zone of conflict and danger for local communities.

In 1999, even before the forest had been placed under KINAPA control, Misana found that the management of the Forest Reserve 'has often been police-like, with the local people not involved at all in its management' (1999, p. 52). By 2005, 300 of the 350 KINAPA employees working in the park and forest were armed guards who constantly patrolled the area for any trespasser who was not a tourist (Sebastien, 2010). One of the individuals with whom I spoke in Mweka explained that the Forestry Department used to manage the land and forests that are in the park. The department would let the villagers come into the park to gather firewood. However, the relationship between the community and the park authorities changed when the government handed the management of the Forest Reserve over to KINAPA and villagers were no longer allowed into the park. Additionally, he said that KINAPA would annex new lands into the park without notifying the residents of Mweka. As a result, people collecting firewood often unknowingly venture onto KINAPA lands. When caught, the villagers, including women and children, would be beaten or taken to jail. Due to these types of occurrences, the relationship between the KINAPA and Mweka residents is tense.

The effects of this tense relationship were also raised during a community group meeting in Mweka in 2009. In this region, women are primarily responsible for household activities and chores, including the gathering of firewood for cooking and heating purposes within the home. Fulfilling these activities often entails the collection of firewood that is located within the park boundaries. As a result, women are often more vulnerable to the repercussions of illegal park entry due to their need to obtain resources located within the park boundaries. One of the most pressing concerns of the community group was the treatment of women who went into the forest to collect firewood. Often, when women go into the forest, TANAPA officials rape them for collecting, or being suspected of collecting, firewood (Research notes, 2009). These government employees were using direct violence against women's bodies as a coercive and punitive tactic to 'safeguard' the Forest Reserve, and they continue to do so. Mariki (2013) spoke with individuals who reported that women were sexually harassed or raped and that men were beaten or arrested for trespassing in the park. The reports of rape became so widespread that, in 2014, MNRT Minister Lazaro Nyalandu called for an investigation into the accusations of KINAPA workers raping women who go into the park to gather firewood or grass (*Daily News*, 2014b). However, this practice may not stop if such forms of violence have indeed become a state-sanctioned strategy used to control a resource (Peluso and Watts, 2001). In fact, these occurrences may support Neumann's (2001, p. 324) argument that 'state violence and the new community-oriented initiatives are integrated forms of social control designed to meet the needs and goals of international conservation organizations and the tourism industry'.

This means of social control through violence has also taken the form of deadly force. In 2009, two men from Mweka went into the forest to collect firewood. They were reportedly caught cutting a tree in the forest instead of collecting dead wood. The KINAPA rangers shot at them, killing one and injuring the other. The rangers responsible for the shooting were immediately taken into custody and placed in jail. This measure may have been done to both reduce community outrage at the shootings and to protect the rangers from any retaliation. However, people in the village were still angered by the shooting and seemed unconvinced that the rangers would be punished for the man's death. Additionally, they did not see the individual rangers as being solely responsible for the incident. They also faulted KINAPA and the exclusion and dispossession that the park's presence has in their daily lives. This was exemplified in the retaliatory action of some individuals tearing down the access bridge in Mweka that allowed KINAPA vehicles and personnel to enter the park. The retaliation was not directed at the individuals responsible for the shooting, as the KINAPA officials had perhaps expected. Instead, the retaliation was directed towards park operations and the prevention, or at least hindrance, of their daily operations. As such, the reaction demonstrates the degree to which the tourism industry, and not just individual park rangers, is perceived as being responsible for the treatment of people in communities surrounding the park.

Discussion and conclusion

The communities adjacent to Kilimanjaro National Park shoulder much of the cost of natural resource protection policies for tourism development through a loss of access to resources and the violent maintenance of park boundaries. When violent tactics are used to coerce groups who challenge or contest the state's authority to control natural resources, 'local resistance to what are perceived as illegitimate state claims and controls over local resources is likely to increase' (Peluso, 1993, p. 216). This resistance to the control and dispossession of resources may come in the form of sabotage, as seen in the tearing down of the KINAP access bridge. The bridge provided access to the protected area enclave of the park from which the surrounding communities have been excluded physically and economically.

This enclave system of tourism on Mount Kilimanjaro has been created and maintained by the dominant actors in tourism. International conservation organizations have helped to establish the idea that conservation must consist of protected, unpopulated areas. The Tanzanian government has in turn been able to benefit from the creation of these government-owned and managed natural resource areas by using them to draw international tourists to the country. This tourism has increased government earnings and provided opportunities for tourism companies. Political ecology stresses the importance of the relationship between environmental protection or conservation, global capital and power.

Peet, Robbins and Watts (2011, p. 27) have argued that conservation initiatives 'can frequently be a mechanism for (or more cynically a "cover" for) powerful players to actually seize control of resources and landscapes, and the flow of value that issue from these'. This flow of value in Tanzania has been controlled through the establishment of an enclave tourism structure that separates the protected area tourist destination from local communities while funneling revenue to the more powerful players, namely the Tanzanian government and the primarily foreign-owned tourism companies. As a result, tourism on Mount Kilimanjaro has not brought increased opportunities and development for those communities living next to the park. Quite the opposite; instead, these communities bear the costs of conservation measures, as they are no longer able to access the forest resources upon which they depend for their livelihood. This has resulted in resistance to park policies through the illegal use of resources such as venturing into the forest for firewood and fodder. This practice has been met with violent methods of park boundary maintenance, which further increases the costs borne by neighboring communities.

This chapter has provided a political ecological analysis of tourism on Mount Kilimanjaro by looking at the historical and political-economic context of tourism development in Tanzania in general, and the mountain region specifically, in relation to conservation and access to land and resources. Such historical and political-economic analyses are central to political ecology research (Neumann, 2014). This case study contributes to the literature on tourism development, communities and conservation by providing an example of how communities that have not necessarily been physically relocated from a protected area can still be directly impacted through their loss of access to resources needed to meet their daily needs. The loss of access to resources can result in significant hardship for individuals, causing them to risk their safety and well-being in order to survive. This risk results from the fact that resources that were once available to and managed by communities has been designated as being in need of protection and conservation by the dominant actors involved in tourism promotion and management in the area. Although these actors may benefit significantly from tourism revenues, there is little evidence that these revenues are in fact trickling down to those areas most impacted by the protected areas on Mount Kilimanjaro. Although this article presents a case study of tourism in Tanzania, the analysis can be applied more broadly to other tourist areas and regions of the world frequented by Westerners. Such an application could be especially informative in former colonies of Western powers or in developing countries where there is a large amount of Western economic and political influence.

References

Barron, P. and Prideaux, B. (1998) Hospitality education in Tanzania: is there a need to develop environmental awareness? *Journal of Sustainable Tourism*, 6 (3), pp. 224–237.

Bart, F., Mbonile, J. M. and Devenne, F. (2006) *Kilimanjaro: Mountain, Memory and Modernity*. Dar es Salaam: Mkuki na Nyota Publishers.

Bassett, T. and Crummey, D. (2003) *African Savannas: Global Narratives and Local Knowledge of Environmental Change in Africa*. Portsmouth, NH: Heinemann.
Benjaminsen, T. A. and Bryceson, I. (2012) Conservation, green/blue grabbing and accumulation by dispossession in Tanzania, *Journal of Peasant Studies*, 39 (2), pp. 335–355.
Benjaminsen, T. A., Goldman, M. J., Minwary, M. Y. and Maganga, F. P. (2013) Wildlife management in Tanzania: state control, rent seeking and community resistance, *Development and Change*, 44 (5), pp. 1087–1109.
Britton, S. G. (1982) The political economy of tourism in the Third World, *Annals of Tourism Research*, 9 (3), pp. 331–358.
Brockington, D. (2002) *Fortress Conservation: The Preservation of the Mkomazi Game Reserve, Tanzania*. Bloomington, IN: Indiana University Press.
Brockington, D. and Igoe, J. (2006) Eviction for conservation: A global overview, *Conservation and Society*, 4 (3), pp. 424–470.
Bryant, R. and Bailey, S. (1997) *Third World Political Ecology*. New York: Routledge.
Chachage, C. S. L. and Mallya, U. (2005) Tourism and development in Tanzania: myths and realities. Paper presented at the Gender Festival (September 6–9, 2005), Tanzania Gender Networking Programme (TGNP) Gender Resource Center, Dar es Salaam, Tanzania.
Charnley, S. (2005) From nature tourism to ecotourism? The case of Ngorongoro Conservation Area, Tanzania, *Human Organization*, 64 (1), pp. 75–88.
Cosgrove, D. (1989) Geography is everywhere: Culture and symbolism in human landscapes. In: D. Gregory, D. and Walford, R. (eds), *Horizons in Human Geography*. London: Palgrave McMillan.
Daily News (2014a) Tourism tops foreign exchange earnings. [Online] 26th August. Available from: http://www.m.dailynews.co.tz [Accessed: 15 January 2015].
Daily News (2014b) Kinapa rangers 'rape, torture women.' [Online] 27th October. Available from: http://www.m.dailynews.co.tz [Accessed: 15 January 2015].
Durrant, J. (2009) A global adventure and conservation icon. In Clack, T. (ed.), *Culture, History, and Identity: Landscapes of Inhabitation in the Mount Kilimanjaro Area, Tanzania: Essays in Honor of Paramount Chief Thomas Lenana Mlanga Marealle II (1915–2007)*. Oxford: Archaeo Press.
Durrant, J. O., Durrant, M. B., Kaswamila, A., Jackson, M. W., Thurgood, L. B. and Udall, S. R. (2003) Mount Kilimanjaro Community Conservation Service. In: *Papers and Proceedings of Applied Geography Conferences* (Vol. 26, pp. 488–495). [np]; 1998.
Ferguson, J. (2005) Seeing like an oil company: space, security, and global capital in neoliberal Africa, *American Anthropologist*, 107 (3), pp. 377–382.
Gössling, S. (2003) *Tourism and Development in Tropical Islands: Political Ecology Perspectives*. Northampton, MA: Edward Elgar.
Hall, C. M. and Page, S. J. (1999) *The Geography of Tourism and Recreation: Environment, Place and Space: Environment, Place and Space*. New York: Routledge.
Honey, M. (1999) *Ecotourism and Sustainable Development: Who Owns Paradise?* Washington, DC: Island Press.
Kokel, M. (2015) Case study 9: The working conditions of *Wagumu* (high altitude porters) on Mt Kilimanjaro. In: Musa, G., Higham, J. and Thompson-Carr, A. (eds.), *Mountaineering Tourism*. Abingdon: Routledge, pp. 285–293.
Lovelock, B. (2015) Climbing Kili: Ethical mountain guides on the roof of Africa. In: Musa, G., Higham, J. and Thompson-Carr, A. (eds), *Mountaineering Tourism*, Contemporary Geographies of Leisure, Tourism and Mobilities Series. Oxford: Routledge, pp. 272–284.

Mariki, S. B. (2013) Conservation with a Human Face? Comparing Local Participation and Benefit Sharing From a National Park and a State Forest Plantation in Tanzania. *SAGE Open*. [Online] October-December. pp. 1–16. Available from: http://sgo.sagepub.com [Accessed: 5 May 2015].

Mbaiwa, J. E. (2003) The socio-economic and environmental impacts of tourism development on the Okavango Delta, north-western Botswana, *Journal of Arid Environments*, 54 (2), pp. 447–467.

Mbonile, D. J. and Bart, F. (1999) Introduction. In: De La Masseliere, C., (ed.), *Mount Kilimanjaro: Land Use and Environmental Management*. Nairobi, Kenya: Institut Français de Recherche en Afrique.

Misana, S. B. (1999) *Deforestation in Tanzania: A Development Crisis? The Experience of Kahama District* (No. 13). Organization for Social Science Research in Eastern and Southern Africa.

Mitchell, J., Keane, J. and Laidlaw, J. (2009) *Making Success Work for the Poor: Package Tourism in Northern Tanzania*. London: Overseas Development Institute. [Online] Available from: http://www.odi.org [Accessed: 5 May 2015].

Mkumbukwa, A. R. (2008) The evolution of wildlife conservation policies in Tanzania during the colonial and post-independence periods, *Development Southern Africa*, 25 (5), pp. 589–600.

MNRT (Ministry of Natural Resources and Tourism) and BOT (Bank of Tanzania) (2011) Tanzania tourism sector survey: The 2009 international visitors' exit survey report. Dar es Salaam. [Online] Available from: https://www.bot-tz.org/Publications/TTSS/TTSS-2009.pdf [Accessed: 5 May 2015].

Nelson, F. (2004) *The Evolution and Impacts of Community-Based Ecotourism in Northern Tanzania*. Issue Paper 131. London: International Institute for Environment and Development. [Online] Available from: http://pubs.iied.org [Accessed: 5 May 2015].

Nelson, F. (2008) Livelihoods, conservation and community-based tourism in Tanzania: Potential and performance. In Spenceley, A., (ed.), *Critical Issues for Conservation and Development*. London: Earthscan.

Nelson, F. (2012) Blessing or curse? The political economy of tourism development in Tanzania, *Journal of Sustainable Tourism*, 20 (3), pp. 359–375.

Neumann, R. P. (1998) *Imposing Wilderness: Struggles over Livelihood and Nature Preservation in Africa*. Berkeley, CA: University of California Press.

Neumann, R. P. (2001) Disciplining peasants in Tanzania: From state violence to self-surveillance in wildlife conservation. In: Peluso, N.L and Watts, M., (eds), *Violent Environments*. Ithaca: Cornell University Press.

Neumann, R. P. (2014) *Making Political Ecology*. New York: Routledge.

Okech, R. N. (2010) Tourism development in Africa: Focus on poverty alleviation *The Journal of Tourism and Peace Research*, 1 (1), pp. 1–8.

Peet, R., Robbins, P. and Watts, M. (eds) (2010) *Global Political Ecology*. New York: Routledge.

Peet, R., Robbins, P., & Watts, M. (2011) Global nature. In: Peet, R., Robbins, P. and Watts, M. (eds), *Global Political Ecology*. New York: Routledge, pp. 1–47.

Peluso, N. L. (1993) Coercing conservation? The politics of state resource control, *Global Environmental Change*, 3 (2). pp. 199–217.

Peluso, N. L. and Watts, M. (eds) (2001) *Violent Environments*. Ithaca, NY: Cornell University Press.

Ranja, T. (2003) *Development of National Entrepreneurship in the East African Tourism Industry. Globalization in East Africa*. Working Paper Series No. 9. Economic and

Social Research Foundation. [Online] Available from: http://global.esrftz.org/output/WPS09_Ranja_Tourism.pdf [Accessed: 5 May 2015].

Robbins, P. (2004) *Political Ecology: A Critical Introduction*. New York: Blackwell.

Salazar, N. B. (2008) A troubled past, a challenging present, and a promising future: Tanzania's tourism development in perspective, *Tourism Review International*, 12 (3–4), pp. 259–273.

Sebastien, L. (2010) The Chagga people and environmental changes on Mount Kilimanjaro: lessons to learn, *Climate and Development*, 1 (2), pp. 364–377.

Sémhur (2014a) Kilimanjaro and Arusha National Parks map. Wikimedia Commons. [Online] Available from: http://commons.wikimedia.org [Accessed 25 November 2014].

Sémhur (2014b) Tanzania location map. Wikimedia Commons. [Online] Available from: http://commons.wikimedia.org [Accessed 25 November 2014].

Sémhur (2014c) Tanzania map. Wikimedia Commons. [Online] Available from: http://commons.wikimedia.org [Accessed 25 November 2014].

TANAPA (2014) Tariffs from 1st July 2013 to 30th June 2015. [Online] Available from http://www.tatotz.org [Accessed: 15 January 2015].

Tanzania GIS User Group (2014) World Database of Protected Areas 2010. [Online] Available from: http://www.tzgisug.org [Accessed 25 November 2014].

UNCTAD (United Nations Conference on Trade and Development) (2008) *FDI and Tourism: The Development Dimension – East and Southern Africa*. [Online] Available from: https://unp.un.org [Accessed: 15 January 2015].

UNESCO (United Nations Educational Scientific and Cultural Organization) (2015) *Kilimanjaro National Park*. [Online] Available from: http://whc.unesco.org [Accessed: 15 January 2015].

United Republic of Tanzania (2014) *Fifth National Report on the Implementation of the Convention on Biological Diversity*. Division of Environment. Dar es Salaam, Tanzania. [Online] Available from: http://www.vpo.go.tz [Accessed: 15 January 2015].

World Database on Protected Areas (2014) [Online] Available from http://www.protectedplanet.net/ [Accessed 25 November 2014].

WTTC (World Travel and Tourism Council) (2014) Travel and tourism economic impact 2014: Tanzania. [Online] Available from: http://www.wttc.org [Accessed: 15 January 2015].

13 'Absolutely not smelly'

The political ecology of disengaged slum tours in Mumbai, India

Kevin Hannam and Anya Diekmann

Introduction

The 2008 Oscar-winning film *Slumdog Millionaire* depicts a teenager from the slums of Mumbai who becomes a contestant on the Indian version of *Who Wants to Be a Millionaire?* He is arrested under suspicion of cheating and, while being interrogated, significant events from his life history, depicting the slums as mafia-controlled sites of entrenched poverty, are shown that explain why he knows the answers to the questions on the quiz show. The 2003 novel, *Shantaram* by Gregory David Roberts, about a Western former convict living in Mumbai's slums, similarly vividly portrays these environments as spaces of death, disease and conflict. The 2014 Urbanist's Guide to the Mumbai Slum of Dharavi (*The Guardian*, 2014) notes that Mumbai's slums have '[t]errible toilets, a cacophony of sound and the ubiquitous big blue drum – welcome to life in this crowded and colourful area'. These dominant images echo many Western representations of one of the largest and poorest urban areas in the world with a slum population estimated at over 6 million people, which includes many lower middle-class people who are unable to source or afford housing (Nijman, 2008). Moreover the slums of Mumbai, and particularly Dharavi, have become one of the most sought after places for Western tourists to visit through guided slum tours.

Developing from these popular Western representations of slums, tours are frequently criticized simply for their voyeuristic and postcolonial characteristics, but such critiques largely ignore the wider role of the neoliberal global political economy and structures of power and inequality from which large-scale urban slums in India have materialized. Thus Bianca Freire-Medeiros (2013, p. 1) argues that such poverty tourism masks how

> ... capitalism has framed the experience of poverty as a product for consumption through tourism. In the megacities of the global South, selected and idealized aspects of poverty which are associated with specific territories are turned into a tourist commodity with a monetary value agreed upon by promoters and consumers in the tourist market ... turning impoverished neighbourhoods into valued attractions for international tourists.

Recent research on slum tours has tended to focus on the visual and mediatized nature of these encounters (cf. Diekmann and Hannam, 2012; Dyson, 2012;

Meschkank, 2012) and has demonstrated the global and multi-scalar nature of slum tours (Frenzel *et al.*, 2015). This chapter, conversely, utilizes a political ecology approach to understand how tourists attempt to make sense of the environments they encounter during slum tours in Mumbai, India.

In terms of tourism, political ecology has been used (cf. Stonich, 1998), as an approach to analyze relationships between tourism development, water, and environmental health in Honduras. Stonich (1998, 2003) identified the various stakeholders involved in the tourism industry, and their relative power with respect to control of water resources, to show that water, land and marine resources in the country's Bay Islands are jeopardized by tourism development. In this regard, Stonich concluded that: 'while significant environmental degradation is attributable to the actions of powerful national and international stakeholders, it is the Islands' impoverished ladino (Mestizo) immigrants and poor Afro-Antillean residents who are the most vulnerable to environmental health risks emanating from those activities' (1998, p. 1). This example is illustrative of how political ecology perspectives seek to 'evaluate the dynamics of material and discursive struggles over natural resources, entitlements and power' (Gössling, 2003, p. 1) in various regional contexts while also highlighting the environmental imaginaries at work (Peet and Watts, 2004; see also introduction to this volume).

A political ecology perspective moves beyond simplistic critiques of the representations and practices of slum tours to shed light on the specific ways in which Western tourists encounter, and are part of, the ecologies of slum environments in Mumbai. Given the above representations of Mumbai's slums it would be expected that tourists would also experience these spaces in terms of being polluted, dirty and unhygienic. Recent research on the 'geographies of shit' has drawn upon ideas of faeces as taboo, following Mary Douglas's (1966) definition of dirt as 'matter out of place' and her conceptualization of pollution and taboo as means by which different cultures socially construct and police social and environmental boundaries (Jewitt, 2011). Thus, in the colonial imagination, binaries separated the sanitary Europeans from their 'dirty' colonial Other and created 'geographies of contamination' linked to malodourous spaces (McFarlane, 2008). In India, moreover, the handling of human waste is taboo for many Hindus and has been traditionally designated as a job for so-called 'untouchable' or Dalit communities that have responsibility, under India's caste system, for disposing of waste (Jewitt, 2011). Nevertheless, as we discuss in our findings, tourists experience the slums of Mumbai not as dirty spaces but as well-organized and relatively clean environments, contradicting the dominant cultural representations and orderings of 'matter out of place'.

Based upon on our research, we argue therefore that tourists, through their relative (dis)engagement with the smell (of the air), the taste (of the water) and the touch of place in their contacts, de-naturalize the everyday ecologies of the slum environments of Mumbai that are not part of a tourism product. As a result, tourists comprehend the inequalities between themselves and slum dwellers by reflecting on life in the slums as 'not so bad' or 'better than on the streets' or 'absolutely not smelly' as one of our respondents put it. We conclude by reflecting on the extent to which these tourist encounters are (dis)connected from or with wider political ecological issues of air pollution, water inequalities and other materialities of place.

Slum tourism

Over the last 20 years, visits to slums have become a 'new' tourism activity performed mainly by Western visitors and situated in developing countries (Steinbrink, 2012; Steinbrink et al., 2012). However, slum tourism is not an entirely new practice, but an experience that was originally created by the European upper classes in the nineteenth century (Heap, 2009) as a recreational pastime that was part of the general trend towards 'flaneurism' – defined as a leisurely pursuit of walking in the city and observing the lower classes in a safe and detached manner (Benjamin, 1997; Tester, 1994). From the late nineteenth century and up until the Second World War, visiting the poor areas of a town or city represented an escape from social constraints, backed by a romantic imaginary, but also a place for well-off, upper-class women to provide charity in order to raise their own self-esteem. Slum tours then became much more organized, with bus tours giving the tourist the 'insights' from a distant and safe perspective (Heap, 2009). In the last decade of the nineteenth century, for instance, Baedeker's *London and Environs* (1887) included, alongside the conventional heritage attractions, excursions to then-slum districts such as Whitechapel and Shoreditch (cited in Koven, 2004). At the same time in the United States, migrant districts became a key attraction for many middle-class white New Yorkers (Heap, 2009). Charles Dickens undertook slum tours of both London and New York in the nineteenth century as part of his research for his novel writing. Indeed, Dickens called this endeavour 'the attraction of repulsion' (Dickens, cited in Stallybrass and White, 1986, p. 14; cf. Seaton, 2012).

In the contemporary period there has been a renewed interest in visiting the poorest areas of cities as a form of everyday cultural tourism. But in contrast to nineteenth-century slum tours, contemporary tourists from the Global North tend to visit slums in the Global South. In the 1990s the first overtly expressed contemporary slum tours began with organized visits to the *favelas* in Brazil. The *favela* tours became a draw for foreign tourists, partly because of a fascination with the drug culture portrayed in various films and books 'together with a charitable, if not voyeuristic desire to 'observe' and 'help' disadvantaged communities' (Williams, 2008, p. 48). Today, the *favela* Rocinha is the third most popular tourist attraction in Rio de Janeiro 'and the growth in bed and breakfast schemes and hostels inside Rio *favelas* attest to the growth in reality tourism in Brazil' (Williams, 2008, p. 485; cf. Freire-Meideros, 2013). Investigations of the political ecology of slum tours in places like Brazil shed light on the structural inequalities that make such commodified sojourns for the mobile tourist.

Slum tourism has been simultaneously criticized and condemned as voyeurism and understood as the only possibility to encounter the 'real' side of a particular country (Hutnyk, 1996; Jaguaribe, 2010). Most of the moral and ethical judgements are journalistic, with many Western-authored articles referring to so-called 'poverty tourism' or even 'poverty porn' in a particularly negative way (Mendes, 2010; Selinger and Outterson, 2009). As Hutnyk (1996, p. 21) pointed out: 'Even among those who acknowledge the realities of economic disparities between travellers and toured, a degree of consumption of poverty is inevitable and can contribute to a maintenance of that poverty as a subject of "observation"'.

Such observations are indeed voyeuristic in terms of the desire to look upon something that is forbidden (Kristeva, 1982; Lisle, 2004). But arguing that voyeurism is simply a bad thing in itself is to miss the perhaps more critical point that increasing numbers of Western tourists do feel compelled to gaze on poverty, catastrophe and even tragedy (Conran, 2011; Freire-Meideros, 2013; Lisle, 2004). Thus, Selinger and Outterson (2009) acknowledge that while some tourists voice concerns about the ways in which slum tours may obscure wider issues about global inequalities, they further point out that such tours also lead some tourists to actually reflect upon and debate issues of global ethics in ways in which they may not have done so beforehand. Indeed, such dilemmas over the social, ethical and moral impacts of tourism have been widely acknowledged in the tourism literature (Mostafanezhad and Hannam, 2014; Smith and Duffy, 2003).

Methodological approach

Political ecology research involves attentiveness to historical and place context as well as the utilization of different scales of analysis. It involves also a consideration of the power relations that underpin ecological relations. Such a perspective informs our own research methods, which used a mobile ethnographic methodological approach to examine the practices of walking tours as experienced by Western tourists in the slums of Mumbai. Mobile ethnographies have recently been distinguished from the conventional localized ethnographies as using multiple methods, on the move and at different scales (Buscher and Urry, 2009; Buscher *et al.*, 2011; Hannam, 2009; Marcus, 1998).

As Leite and Graburn (2009, p. 37) have acknowledged, ethnographic approaches to researching tourism have increasingly attended 'to how actual people understand and conduct their involvement in the interrelated practices of travelling, encountering, guiding, producing, representing, talking, moving, hosting and consuming'. In mobilities research, it is argued that methods of data collection need to become much more 'flexible, informal and context dependent, partly mimicking mobile subjects being studied in their own suppleness' (D'Andrea, 2006, p. 113). Buscher and Urry (2009) further point out that a mobile methodological approach involves observing people's movements through space, and in face-to-face relationships with other people, augmented by the 'scenic intelligibility' of people's meetings, that is, walking with people to understand their worldviews and emotional attachments and engagement with more imaginative tangible technological representations, including the internet and film, as well as intangible representations such as memories. We draw upon these points in our analysis below, having conducted such mobile ethnography in Mumbai's slum spaces using a variety of such sources of material.

Over two months in 2009 and 2010, we conducted participant observation in guided slum tours with a focus on observing the tourists and the guides. Additionally, we interviewed the organizers, both Western and non-Western, and 20 Western tourists travelling in Mumbai. This research was augmented by the use of follow-up online discussions with the interview respondents. Initial interviews with the slum tour organizers in Mumbai were useful in providing a broad context for the research, and necessary for giving informed ethical consent for the rest of the research. The research undertaken was 'mobile' in the sense that we tried to move with the respondents, through their tours and afterwards online in order to capture 'the fleeting, distributed, multiple, non-causal, sensory, emotional and kinaesthetic' (Buscher et al., 2011, p. 1) nature of the tourists' experiences.

Additionally, we conducted a content analysis of the 693 comments left on the website www.tripadvisor.com about slum tours in Mumbai (for more on the increasingly common use of consumer travel websites in tourism research cf. Briggs et al., 2007; Owens, 2012). Research has also highlighted that responses on TripAdvisor are generally credible and influential in terms of making recommendations to other tourists (Ayeh et al., 2013). Although the vast majority of responses on TripAdvisor were cursory or general platitudes such as 'amazing' or 'we had a great time' or 'recommended', other responses offered deeper reflections that we draw upon in this chapter. Comments on TripAdvisor were coded into a number of categories: 'guides'; 'photography'; 'food'; 'media'; 'creativity' and 'environment'. We draw upon comments coded as 'the environment' below.

Mumbai slum tours

Urban planning in cities of the so-called 'Global South' is considerably influenced by the traditional planning systems of cities in developed countries and this is in part due to the fact that they are legacies of a colonial past. The largest slum in the centre of Mumbai, called Dharavi, has around 600,000 households and is a clear example of this (Municipal Corporation of Greater Mumbai, 2010). Dharavi is located on valuable land in the district between Bandra – a northwest suburb with Portuguese architecture, upscale residences and high-brow restaurants – and Sion – a locality which is home to various educational institutions. Highlighting the extreme inequality between the rich and the poor, Dharavi slum dwellers are juxtaposed against the urban elite, such as Bollywood film stars and technology billionaires who live in Bandra (Sharma, 2000). Within the urban landscape of slum dwellings in Bandra, Dharavi is seen as one of the so-called 'better' slums because it benefits by having a water supply for three hours a day as well as electricity – provided mainly in an uncoordinated, incremental and 'para-legal' way (Anand, 2011; Gandy, 2008). The latest development plan, aimed at providing residents with new housing, is currently evolving (Slum Rehabilitation Authority, 2015). Despite its widespread slum dwellings and ongoing sanitation problems, for many residents outside of the slums Mumbai, on the whole, enjoys the representation of being a modern and dynamic city in India (Desai et al., 2015).

Mumbai is frequently portrayed in academic research as a place of 'abject poverty'. For example, O'Hare, Abbott and Barke (1998, p. 270) have argued that '[t]he reality is that over one-half of Mumbai's population lives in conditions of abject poverty, squalor and deprivation'. On the other hand, while the people living in the slums of South Asia, as elsewhere, may not have the greatest access to services, other commentators have argued that they remain people with a great deal on ingenuity and spatial innovation. Sharma (2000) argues that '[w]hat marks Dharavi from other slums is its productivity. It is more like an industrial estate than a slum, except people live and work in the same place'. Indeed, Sharma's point is well noted; Dharavi is perhaps more like a vast innovative industrial estate that transgresses Western socially constructed notions of what is and what is not seen as poverty (Hannam and Diekmann, 2011).

Having experienced slum tourism in the favelas of Brazil, the British co-founder of a local Mumbai company called 'Reality Tours and Travel' attempted to implement in Mumbai the Brazilian model he encountered. Thus, he created – together with an Indian partner – slum tours in Dharavi. In the beginning the two founders encountered strong criticism for developing this form of tourism in India. Very soon, however, they opened a community centre in the slum, providing lessons in English and computer science to local youth, and used the tours as a means to raise funds for various development projects. The philosophy of 'Reality Tours' is to improve and/or change the image of slums and expose the sense of community and organization within the district. The tours are made up of small groups of around six people and the number of groups depends upon demand, but in recent years there has been a significant increase from the original three groups per day. Extensive tourist questionnaire surveys by the company administered in 2010 demonstrated that the vast majority of the approximately 3,500 tourists per year are English-speaking tourists from the UK, the US and Australia, across all age groups. About a third of all tourists join the tour because of recommendations from friends and relatives (often expatriates) whom they are visiting in Mumbai (Hannam and Diekmann, 2011). Others come having learned about the tours from various media, either via the internet or from guidebooks that promote the tours. Most of the tour guides, meanwhile, allegedly come from the slum areas themselves, being able, on one hand, to connect the community and the visitor and, on the other, in order to provide an 'authentic' experience for the tourists. While the website of Reality Tours (http://realitytoursandtravel.com) describes Dharavi as a fairly relaxed and heterogeneous space of play and carefree abandon, the tours themselves follow a set route with a largely scripted commentary and an acknowledged safely controlled experience for the tourist. The power of the tour guide here is vital, and is similar to the role of the director of a film. Indeed, it has often been highlighted that guides have an authoritative position in relation to their audiences (Cohen, 1985; Dahles, 2002) which can be an 'ideologically charged social event' (Brin and Noy, 2010, p. 20). Since 2012 the situation has, however, changed significantly. Reality Tours are no longer the only tour providers and many new local tour operators now offer tours of Dharavi, with less attention to sustainable and responsible behaviour. For instance, other tour companies do not maintain a no-camera policy, and no-go zones – originally instigated by Reality Tours for private housing – are not enforced. As a response to growing competition, Reality Tours purchased a house in the slums that, in the beginning, served as an example of private housing, but this has recently been transformed into an art gallery (Diekmann and Chowdhary, 2015). Most of the tours follow the same path as Reality Tours, leading at some moments to overcrowding in parts of the slums due to too many visitors, and contributing to the denaturalizing of the site. Competition between slum tour operators contributes to reduced moral and physical boundaries for slum tourists eager to experience (and photograph) the 'real thing'. In this sense, the political ecology of tourism more broadly is exposed through an examination of the experience of slum tourism development in Mumbai and the ways in which competition for profit contributes to loosening social, moral and environmental restrictions.

(Dis)embodied political ecologies

Political ecologies are those networks of connections concerned with the practices and paradoxes of pursuing sustainable, universal, urban environmental health and with how one population of informal housing dwellers are affected by the unintended consequences of one aspect to remedy or ameliorate the effects of this degraded setting. In what follows we examine how tourists' engagement with walking tours of Mumbai's slums are enmeshed within a set of tensions regarding such tourism development and locals' lived lives, as the following account conveys:

> We just finished our full day tour (morning slum and afternoon city sightseeing tour in a private car). Having read *Shantaram* (like probably all westerners who travel to India!), I was inspired to see the slum but also conscious of being a voyeuristic foreigner. So going through Reality Tours who, via their sister company, reinvest 80% of their profits back into the slum definitely felt like the right option and we really weren't disappointed.
> (TripAdvisor comment, December 2014)

Western tourists have a fairly ostentatious romanticized representation of the slums filled with compassion, as demonstrated above. In her analysis of volunteer tourism, Conran (2011, p. 1454), highlights how 'intimacy' can in fact overshadow the structural inequalities on which the tourist encounter is based, and 'reframes the question of structural inequality as a question of individual morality and perpetuates an apolitical cultural politics'. Many of our respondents, based upon their previous media experiences, expressed ambivalent emotions about doing the slum tours:

> I expected to be depressed after the tour, I also expected confirmation that Dharavi was seething with criminal activity, organized crime, poverty and hopelessness. I based some of these expectations on descriptions of Dharavi I got from novels and movies.
> (Interview, Female, 69, USA)

The above comment echoed many of the Western representations of slums given in the media, while below, the respondent added the need to make the tour more exciting and more film-like:

> Obviously, Reality Tour tries its hardest to make the experience pleasant and they take you to places within the slum which are interesting and yet safe. Nevertheless, I felt a bit like I was given half the story. Therefore, if I had to change something I would extend the tour to less pleasant areas. Perhaps this is understandably difficult for security reasons therefore don't take it as a criticism.
> (Tripadvisor comment, December 2014)

As highlighted in the quotes above, the slum imaginary comes from novels, media and films but it is significant that the emotional and embodied expectations of the tours are not met. The tourists expect crime, poverty and hopelessness and are surprised when this is not given, and wanted it to be less sanitized and less pleasant. From a political ecology perspective, the pressing question is: what is the effect of Reality Tours and other operators' 'improving' one part of the slum, on those areas and residents who are excluded from participating? Comments from tourists that they were exposed to only 'half the story' are interesting in that they indicate a consciousness about the broader context and experiences of slum life they didn't get to see.

Indeed, many respondents noted that the slums were cleaner than expected, as these respondents testified, in different ways:

> We've all seen *Slumdog Millionaire* meanwhile and of course those pictures spring to mind first. To be flatly honest, I expected a much dirtier, smellier place. The whole trip was the opposite. Tidy, clean, friendly and absolutely not smelly. My expectations were minimal, I just wanted to see a slum, how it operates, the type of life lived there and the activities related to the infrastructure in Mumbai.
> (Interview, Female, 55, Ireland)

This links to a wider literature on the geographies of 'shit' and waste that show how excrement can become metaphorically sanitized, in this case for the benefit of tourists (cf. Jewitt, 2011).

> It is all too easy to think of a slum as a dirty, disease and crime-ridden area, but the tour, led by the charming and enthusiastic Nilesh, dispelled this image. An area of great resourcefulness, determination, humility and human warmth. When the world is too often besieged by bad news, Dharavi is a simple yet powerful symbol of hope and humanity.
> (TripAdvisor comment, December 2014)

Other respondents focused on the issue of community to further sanitize their experiences:

> We were expecting real squalor and depression, however we were surprised to see such organisation and structure to the various locations within the slum. There seemed a genuine community feel about the place considering the conditions in some locations which was pretty grim, however people were getting on with their lives and making the best of what they had.
>
> (TripAdvisor comment, December 2014)

The sense of surprise at the lack of dirt, squalor and disease highlights how the tours overshadow and erase the everyday ecologies of those parts of the slum that remain unvisited. Nevertheless, walking through the slums provided tourists with a feeling of embodiment reinforced by the surrounding conditions and the heat. Being in a small group and accompanied by an English-speaking local guide dressed in a Reality Tours shirt makes the shift from the one place to the other a relatively comfortable experience. From the start, the tourists 'enjoy' a 'sanitized' version of the slums which does not seek to challenge the dominant representations from the films they have watched. The residents and workers are prepared for, and familiar with, the tourists, and there is little begging or hassling. Instead, expectations seem to have been disrupted by the happy children who asked for the tourists' names.

> I was surprised about the fact that they did not pay attention to us. Nobody hassled or followed us. Only a few children wanted to know our names, nothing else.
>
> (Interview, Female, 38, Spain)

During the tour, visitors are taken on a bridge and onto a rooftop where they can view the slums. This experience interrupts the embodied experience of being in the slum where tourists are able to step back and reflect on the slum experience. Moreover, while the tours were arguably 'sensitive' they also highlighted the complexities of contemporary tourism mobilities as the following quotes from respondents demonstrate.

> I have a feeling many are ultimately itinerant. On their way to elsewhere when the opportunity arises – I find myself hoping so, as it seems an oppressive sort of life for the long-haul. The people we saw were friendly throughout, though we had limited opportunity to observe/interact- I wonder how much of their good cheer was 'put on' for our benefit, perhaps in return for monetary reward?
>
> (Interview, Female, 52, Australia)

Respondents also reflected upon their own emotional states and the dangers they expected but then rationalized what they saw as 'not so bad' after all, thus naturalizing the slums rather than Othering these spaces:

> Before: I was worried about being too emotionally affected by the poverty. I also worried about mosquitoes, insects. During: I felt comfortable walking around with the guide. I realized it was not dangerous at all. After: I felt pity for the people there having to work and live in such bad conditions. On the other hand I was happy for them for at least they had work and a roof. I see every day many more people living in the streets of Mumbai in far worse conditions and jobless.
>
> (Interview, Female, 38, Spain)

Embodying the heat through sweat is the only challenge to disembodied engagement of the tourists as they try not to touch too much for fear of 'catching something', recreating a socially constructed contagion, and reproducing the colonial binaries of Othering (Douglas, 1966; Jewitt, 2011). Although they have consciously chosen to experience poverty, the preoccupation of many visitors is their own personal health and safety. Indeed, many respondents expressed their own fears of contagious diseases:

> I felt apprehensive for my own safety, particularly on health grounds. Worried about contracting some disease. During the visit I was stung a number of times by mosquitoes and was glad I had taken my anti-malarial tablets. I was also moved by the apparent good humour of the people we met, humbled by the effort they made to survive in adverse circumstances and impressed by their sense of community and capacity for joy.
>
> (Interview, Female, 52, UK)

While this emphasis on capacity for joy echoes media representations of Mumbai as a *City of Joy*, the attention given by the guides during the visit structures it and comforts the tourists, giving an impression of ethically correct conduct. For example, one tourist commented that:

> The preparation for the tour is also very well thought through, 'slum etiquette' is explained (not to look shocked, not to stare, not to look like you have a bad smell under your nose, be open to the dwellers who want to greet or speak to you, not to stare into their homes). The types of activities and income generated and the number of dwellers is all explained!
>
> (Interview, Female, 55, Ireland)

Preoccupations with one's own health and safety are reflective of the broader political ecology of health in tourism. Tourism is indeed a critically underexamined site from which to examine how, for example, health and safety standards and experiences are structurally conditioned and distributed. The fact that slum tourists in Mumbai had the possibility of exiting the tour if they felt they had come in contact with a contagious Other is critical to the broader relationship between the so-called 'host' and 'guest' in this experience in particular, and tourism experiences more broadly.

Conclusion

Slum tours involve spatial and social partitioning; first, separating the slum from its surroundings and, second, separating those parts within the slum that are visited from those that are not. Tourism development, through slum tours, is sometimes intended to bring into the visited part of the slum money to be used for increased provision of social resources and thus represents the frequent entanglement of neoliberal forms of tourism and social development. Also, tourists report coming to believe slum life is not as bad as they had expected. Instead, the poverty that they witness becomes a naturalized aspect of the 'post' colonial Other. A political ecology approach to this process of the naturalization of poverty is to attempt identify the structural conditions through which slum tour development has come to exist. The influence of the historical and colonial experience in India cannot be overstated. Additionally, the relationship between widespread migration from rural India to Mumbai, as well as Mumbai's position within the broader global political economy is implicated in facilitating slum tourism development in the city. Thus, the ethical and moral questioning about the practice of slum tourism cannot stop at the individual, embodied experience, but requires a historically situated, place-based and multi-scalar analysis. Additionally, the embodied experiences of slum tourists work to perpetuate a larger political ecology discourse around the relationship between the Global South and the Global North. They also shed new light on the political ecology of health (and safety) in tourism – a perspective that would be exceptionally well suited for the analysis of myriad tourism practices.

Moreover, visits are, indeed, voyeuristic, in the same way that tourists gaze upon dark or apocalyptic tourism sites. However, in contrast to the simple generalization of voyeuristic visits with the aim of looking at the poor, the responses we gathered from tourists to the sanitized part of Dharavi emphasized their postcolonial ambivalences and sensitivities to the slum tour experience. A deeper understanding of India and a search for 'authentic' social encounters were both provided by the tourists as explanations for their visits to the slums. Yet, clearly, these Western tourists are also visiting as part of their own journeys into their selves and there is also clearly a romanticized edge to their comments – what Lisle (2004, p. 8) has called 'mediated reverence' – which, in these tours, is largely anaesthetized and disembodied.

As highlighted above the well-organized and controlled contact still is very limited, not to say superficial, allowing for keeping one's distance within the slum environments, which are then seen as less smelly and less toxic than expected. Ultimately, the tours draw upon subtle power relationships through neoliberal self-help narratives, which lead indirectly to the visitors evaluating their own Western positionalities. The managed 'shock' the tourists experience is of the everyday normalities of the visited slum spaces. Without encountering the non-visited part of the slum, and its personal and health dangers, the tourists do not, and are unable to, experience a wider span of political ecologies of air, water and sanitation problems of the environments of Mumbai. This chapter thus exemplifies how political ecology scholars need to address the subtleties involved in tourism encounters in terms of power, place and historical legacy and how these are reconfigured in contemporary tourists' slum tour experiences.

References

Anand, N. (2011) The politechnics of water supply in Mumbai, *Cultural Anthropology*, 26 (4), pp. 542–564.

Ayeh, J., Au, N. and Law, R. (2013) Do we believe in TripAdvisor? Examining credibility perceptions and online travelers' attitude toward using user-generated content, *Journal of Travel Research*, 52 (4), pp. 437–452.

Benjamin, W. (1997) *Charles Baudelaire: A Lyric Poet in the Era of High Capitalism*. London: Verso.

Briggs, S., Sutherland, J. and Drummond, S. (2007) Are hotels serving quality? An exploratory study of service quality in the Scottish hotel sector, *Tourism Management*, 28 (4), pp. 1006–1019.

Brin, E. and Noy, C. (2010) The said and the unsaid: performative guiding in a Jerusalem neighbourhood, *Tourist Studies*, 10 (1), pp. 19–33.

Buscher, M. and Urry, J. (2009) Mobile methods and the empirical, *European Journal of Social Theory*, 12 (1), pp. 99–116.

Buscher, M., Urry, J. and Witchger, K. (eds) (2011) *Mobile Methods*. London: Routledge.

Cohen, E. (1985) The tourist guide: the origins, structure and dynamics of a role, *Annals of Tourism Research*, 12 (1), pp. 5–29.

Conran, M. (2011) THEY REALLY LOVE ME! Intimacy in volunteer tourism, *Annals of Tourism Research*, 38 (4), pp. 1454–1473.

Dahles, H. (2002) The politics of tour guiding, *Annals of Tourism Research*, 29 (3), pp. 783–800.

D'Andrea, A. (2006) Neo-nomadism: a theory of post-identitarian mobility in the global age, *Mobilities*, 1 (1), pp. 95–120.

Desai, R., McFarlane, C. and Graham, S. (2015) The politics of open defecation: informality, body, and infrastructure in Mumbai, *Antipode*, 47 (1), pp. 98–120.

Diekmann, A. and Hannam, K. (2012) Touristic mobilities in India's slum spaces, *Annals of Tourism Research*, 39 (3), pp. 1316–1336.

Diekmann, A. and Chowdhary, N. (2015) Slum dwellers' perceptions of tourism in Dharavi, Mumbai. In: A. Diekmann and M. Smith (eds), *Ethnic and Minority Cultures as Tourist Attractions*. London: Channel View, pp. 113–126.

Douglas, M. (1966) *Purity and Danger*. London: Routledge.

Dyson, P. (2012) Slum tourism: representing and interpreting 'reality' in Dharavi, Mumbai, *Tourism Geographies*, 14 (2), pp. 254–274.

Freire-Meideros, B. (2013) *Poverty Tourism*. London: Routledge.

Frenzel, F., Koens, K., Steinbrink, M. and Rogerson, C. (2015) Slum tourism: state of the art, *Tourism Review International*, 18 (4), pp. 237–252.

Gandy, M. (2008) Landscapes of disaster: water, modernity and urban fragmentation in Mumbai, *Environment and Planning A*, 40, pp. 108–130.

Gössling, S. (ed.) (2003) *Tourism and Development in Tropical Islands: Political Ecology Perspectives*. London: Edward Elgar.

Hannam, K. (2009) The end of tourism: nomadology and the mobilities paradigm. In: J. Tribe (ed.), *Philosophical Issues in Tourism*. Bristol: Channel View.

Hannam, K. and Diekmann, A. (2011) *Tourism and India*. London: Routledge.

Heap, C. (2009) *Slumming – Sexual and Racial Encounters in American Nightlife, 1885–1940*. Chicago, IL: University of Chicago Press.

Hutnyk, J. (1996) *The Rumour of Calcutta: Tourism, Charity and the Poverty of Representation*. London: Zed Books.

Jaguaribe, B. (2010) Favelas and the aesthetics of realism: Representations in film and literature. Available online: http://www.pos.eco.ufrj.br/docentes/publicacoes/bjaguaribe1.pdf [Accessed 31st May 2015].

Jewitt, S. (2011) Geographies of shit: spatial and temporal variations in attitudes towards human waste, *Progress in Human Geography*, 35 (5), pp. 608–626.

Koven, S. (2004) *Slumming: Sexual and Social Politics in Victorian London*. Princeton, NJ: Princeton University Press.

Kristeva, J. (1982) *Powers of Horror: An Essay on Abjection*. New York: Columbia University Press.

Leite, N. and Graburn, N. (2009) Anthropological interventions in tourism studies. In: T. Jamal and M. Robinson (eds), *The Sage Handbook of Tourism Studies*. London: Sage.

Lisle, D. (2004) Gazing at ground zero: tourism, voyeurism and spectacle, *Journal for Cultural Research*, 8 (1), 3–21.

Marcus, G. (1998) *Anthropology through Thick and Thin*. Princeton, NJ: Princeton University Press.

McFarlane, C. (2008) Governing the contaminated city: infrastructure and sanitation in colonial and post-colonial Bombay, *International Journal of Urban and Regional Research*, 32 (2), pp. 415–435.

Mendes, A. C. (2010) Showcasing India unshining, *Third Text*, 34 (4), pp. 471–479.

Meschkank, J. (2012) Negotiating poverty: the interplay between Dharavi's production and consumption as a tourist destination. In: F. Frenzel, K., Koens and M. Steinbrink (eds), *Slum Tourism: Poverty, Power and Ethics*. London: Routledge, pp. 144–158.

Mostafanezhad, M. and Hannam, K. (eds) (2014) *Moral Encounters in Tourism*. London: Ashgate.

Municipal Corporation of Greater Mumbai (2010) *Mumbai Human Development Report*. New Delhi: Oxford University Press.

Nijman, J. (2008) Against the odds: slum rehabilitation in neoliberal Mumbai, *Cities*, 25 (2), pp. 73–85.

O'Hare, G., Abbott, D. and Barke M. (1998) A review of slum housing policies in Mumbai, *Cities*, 15 (4), pp. 269–283.

Owens, T. (2012) TripAdvisor rates Einstein: using the social web to unpack the public meanings of a cultural heritage site, *International Journal of Web Based Communities*, 8 (1), pp. 40–56.

Peet, R. and Watts, M. (eds) (2004) *Liberation Ecologies*. London: Routledge.

Seaton, T. (2012) Wanting to live with the common people . . . ? The literary evolution of slumming. In: F. Frenzel, K. Koens and M.Steinbrink (eds), *Slum Tourism: Poverty, Power and Ethics*. London: Routledge, pp. 21–48.

Selinger, E. and Outterson, K. (2009) *The Ethics of Poverty Tourism*. Boston University School of Law Working Paper No. 09–29.

Sharma, K. (2000) *Rediscovering Dharavi: Stories from Asia's Largest Slum*. New Delhi: Penguin.

Slum Rehabilitation Authority (2015) All slum dwellers residing on the plot prior to 1/1/1995 and are in use of the structure are eligible for rehabilitation. Available online: http://sra.gov.in/pgeDescriptionStage.aspx [Accessed 19th January 2015]

Smith, M. and Duffy, R. (2003) *The Ethics of Tourism Development*. London: Routledge.

Stallybrass, P. and White, A. (1986) *The Politics and Poetics of Transgression*. Ithaca, NY: Cornell University Press.

Steinbrink, M. (2012) 'We did the Slum!' – Urban Poverty Tourism in Historical Perspective, *Tourism Geographies*, 14 (2), pp. 213–234.

Steinbrink, M., Frenzel, F. and Koens, K. (2012) Development and globalization of a new trend in tourism. In: F. Frenzel, K., Koens and M. Steinbrink (eds), *Slum Tourism: Poverty, power and ethics*. London: Routledge, pp. 1–18.

Stonich, S. (1998) Political ecology of tourism, *Annals of Tourism Research*, 25 (1), pp. 25–54.

Stonich, S. (2003) The political ecology of Marine Protected Areas: The case of the Bay Islands. In: S. Gössling (ed.), *Tourism and Development in Tropical Islands: Political Ecology Perspectives*. London: Edward Elgar.

Tester, K. (ed.) (1994) *The Flaneur*. London: Routledge.

The Guardian (2014) An urbanist's guide to the Mumbai slum of Dharavi. Available online: http://www.theguardian.com/cities/2014/apr/01/urbanist-guide-to-dharavi-mumbai [Accessed 31st May 2015]

Williams, C. (2008) Ghettotourism and voyeurism, or challenging stereotypes and raising consciousness? Literary and non-literary forays into the Favelas of Rio de Janeiro, *Bulletin of Latin American Research*, 27 (4), pp. 483–500.

14 Composing Greenlandic tourism futures

An integrated political ecology and actor-network theory approach

Carina Ren, Lill Rastad Bjørst and Dianne Dredge

Introduction

A remote and deserted landscape covered with ice and snow in the winter, Greenland – the world's largest island – becomes a lush, green territory in the summer. Southern Greenland, where long stretches of coast and fjords mark the landscape, is also where we find Ipiutaq Guest Farm, which featured on the *New York Times*' list of '52 Places To Go In 2015'. On a farm owned by French-Greenlandic Agathe Devisme and her husband Kalista Poulsen, agritourism and sheepherding exist side by side. Guests fish in some of the country's best waters for catching char nearby. For dinner, the char is often added to Agathe's famous Greenlandic version of bouillabaisse soup, and the evening coffee can be enjoyed beside cakes baked with berries collected from the surrounding pastures. But all here is not idyllic. Close to the guest house, enormous deposits of rare earth minerals (including uranium, fluoride and thorium) have been discovered by miners. If mined, these extremely lucrative deposits hold enough financial potential to help secure Greenland's independence from Denmark and create a new and prosperous future for its 57,000 residents. Not surprisingly, the uranium find has catapulted Greenland from the global margins to center stage in the international quest for energy resources, and in the process, tourism and nature have come to occupy precarious positions alongside one another. An article entitled 'Pantry or Uranium?' in the Danish newspaper *Weekendavisen* on the Ipiutaq Guest Farm problematizes the establishment of the mine in the vicinity of Ipiutaq. The journalist writes on the prospects of catering for tourists if the mining activities proceed; she contends that 'it is uncertain whether the prestigious paper (the *New York Times*) will still recommend southern Greenland as a tourist destination in a few years, because, as the author rhetorically asks, "how many tourists would like to fish with a uranium mine in the back garden?"'! (Andersen, 2015).

The above story illustrates the complex and controversial relationship between the development of tourism and mining in Greenland. In this chapter, we examine this relationship to explore the intractable connections between tourism, mining and the environment, a relationship that is fraught with tension in other parts of the world as well (Miller *et al.*, 2012). Currently, Greenlandic tourism is still in its formative stages and, while it may not be a small island, it shares many of the challenges of small island developing states in terms of its development, for example, distance from markets, limited economic diversity, lack of expertise, difficulty of attracting investment, high seasonality and highly sensitive environments (Schyvens and Momsen, 2008). Simultaneously, tourism is identified as one of the three pillars, along with fishing and mining, on which the future economic stability of the country should rest.

This chapter unpacks controversial issues that relate to future tourism in Greenland. We explore how Greenlandic tourism futures are composed (or silenced), and how the tensions, conflicts, inconsistencies and contradictions between tourism, mining and nature are enacted. In doing so, we consider the 'relationships between tourism, communities, political processes and the environment' that form the focus of this book (see Introduction, this volume), to further emphasize that these relationships are mediated within a wider and more complex set of social–political–economic–environment relations. While tourism can be a strong and forceful actor, as shown in many of the contributions to this collection, it may also hold a weaker hand – in cases such as Greenland, where the local tourism landscape is not explicitly identified or problematized within broader development discourses. As a result, tourism is enlisted or co-opted into a future not of its own making. How tourism mobilizes and materializes (or not) in local contexts, then, is as much about its presence as it is about its absence in land use, natural resource management and development debates. In current debates about Greenland's future, big dreams are in the making. In this chapter we explore the place of tourism in these debates through a relational approach. This approach allows us to examine the simultaneously emerging industries of tourism and mining, as well as allowing us to consider the agency of human and non-human actors.

To this end, we introduce the antiessentialist political ecology of Actor-Network Theory (ANT), according to which the aforementioned divides are perceived not as ontologically different entities, but as effects of relational and ongoing achievements. In our focus on Greenlandic tourism development, we explore how actors, activities and devices are activated to enable or disable future versions of tourism, and how this is done through hybrid attachments of nature and culture. The concrete composition of Greenlandic tourism futures is explored using an illustrative example of the above mentioned mining prospect in southern Greenland. Specifically, we examine how a promotional video for the Australian mining company Greenland Minerals and Energy (GME) is deployed as a 'futuring device' to compose the landscape of the area. We also explore how its scenario-building prompts a very particular landscape in which tourism, as well as human and animal activity, is only partially present. By doing so, we consider how a seemingly depoliticized enactment of Greenland's future is performed to promote the mining project. The video seamlessly joins mining, nature, science and the social, in a smooth and 'empty' location in south Greenland. By describing and problematizing these connections, we address how discourses of the global, natural science and economic sustainability are enrolled, and display how local tourism and other activities are dislocated from their current position.

The chapter proceeds as follows. First, we introduce the reader to ANT and discuss how the delegation of agency and power alters using such an approach.

Following Holifield (2009), we frame ANT as an alternative approach to Marxism in our engagement with environmental justice. Second, we present an illustrative example of how tourism and mining interfere and contradict each other using the south Greenlandic mining case and the Greenlandic uranium debate, to which it is linked. We describe and analyze how GME composes future mining as smooth, accessible and apolitical in their promotional video, and how in that process other kinds of human activity such as tourism and farming are othered. In the final section, we propose that while large-scale mining may be considered a viable maneuver to boost growth and move toward Greenlandic economic independency, the future scenario claims a disembodied universalizing view of development.

Political ecology and actor-network theory

Tourism is very often characterized as a 'global' phenomenon. As a consequence of its claimed globality, it is perceived as something that concerns us all – but perhaps also as something that none of us can change. As one of the cogs in the grinding machinery of the Anthropocene, tourism is said to influence us all, in one way or the other. In relation to tourism, we may not sense the impacts of the Anthropocene directly and presently, but we are becoming increasingly conscious that its future is fast approaching. This is why discussions framed in terms of political ecology are relevant to push within the tourism academy and hence to foreground that tourism is not only, and never was, only about tourism (Jóhannesson *et al.*, 2015). We contribute here to the work of expanding what it means for something to be about tourism by entering into conversation with a larger field of political ecology. We propose ANT as a relevant analytical tool to engage with the heterogeneous actors who play their part in developing tourism.

In order to appreciate the critical capacities which ANT lends to the field of political ecology, it may be illustrative to show what ANT is not. Holifield (2009) discusses the critical capacities of ANT vis-à-vis Marxist Urban Political Ecology (UPE) demonstrating how the Marxist UPE seeks to explain environmental injustice by identifying inherent unequal power distribution and capitalist structures. While the political project is commendable, Marxist UPE risks 'stepping in to resolve controversies with powerful explanations that appeal to hidden social causes' (Holifield, 2009, p. 653). Hence, the social explanations depoliticize the controversies every bit as much as neoliberalism (which it criticizes) by presenting another set of indisputable social facts. As Holifield notes,

> instead of explaining inequalities by contextualizing and situating them, actor network approaches turn our attention to the forms and standards that make it possible to circulate new associations of entities, to generalize social orders, and to situate actors within a social context – that is, to socialize them in particular ways.
>
> (Holifield, 2009, p. 639)

It is such forms and standards that we investigate in the Greenlandic case.

The ANT approach to political ecology deviates from a Marxist understanding also in terms of agency and power, and where these concepts are researched and allocated. According to Holifield (2009):

> an ANT analysis seeks to register the range of competing accounts of agency in the production of domination: from accounts in which non-human agency takes a deterministic form, to those in which nonhumans are simply passive objects of social mobilization, to those which lie somewhere in between.
> (Holifield, 2009, p. 646)

The role of nature is, however, unsettled, moving from highly deterministic (in relation to climatic change, uranium radiation) to passive (a pristine setting for tourism or as a resource). If at the first pole we find various ways of naturalizing inequality and domination, at the latter pole we find assertions that inequality and domination are completely unnatural: that is, entirely the result of 'social' processes. In between the two poles, we find many other accounts of how non-human agencies participate in the production of inequality and domination (Holifield, 2009, p. 646). In the following, we will also see how the landscape is appointed a certain role in relation to mining and more broadly to social activity.

Surprisingly, the political ecology of ANT claims to have nothing to do with nature (Latour, 2004a, p. 5). Rather, it looks at how nature, science and politics mutually co-constitute, support, challenge and reinforce one another through the principle of so-called general symmetry (Law, 1994, p. 9). The symmetrical approach to human and non-human actors analytically brackets 'common-sense categorization of the entities under investigation' (Jensen, 2003, p. 226) and sees entities that are usually delegated to separate spheres and practices of scrutiny as co-produced. Nature and culture are not a priori different in an ontological sense but are rather made up, often through hybrid assemblages: climate change models in computer programs, uranium samples in labs, business plans in Excel sheets, political debate in the parliament or on the streets. In the same vein the mining landscape, which we present later, is made up as a hybrid assemblage of nature, science and the social.

By tending to how nature, science and politics form part of enacting and resolving controversies and uncertainties in the Anthropocene, ANT offers accounts of 'stabilizing mechanisms so that the premature transformation of matters of concern into matters of fact is counteracted' (Latour, 2005, p. 261). As environmental controversies proliferate in tourism and elsewhere, we can no longer turn to nature (or science) to settle our disagreements. In the case of tourism development in Greenland, science – whether natural or economic – cannot deliver irrefutable answers and solutions, cannot offer hard matters of fact:

> Not many years ago, when we were contemplating the sky above our heads, we thought of nothing but matter and nature. Nowadays, when we look above our heads, we watch a sociopolitical imbroglio, because, for instance, the depletion of the ozone layer brings together a scientific controversy, a political dispute between North and South, and gigantic strategic moves inside industry.
> (Latour, 1994, p. 796)

For that same reason, nature no longer provides us with a certainty of which future to choose. While the scenario of uranium mining in the case of Greenland is being proposed by its advocates as safe, simple and sustainable, a concerned public is not yet willing to let it become institutionalized (Bjørst, 2015c). Hence, what we are left with are highly disputable matters of concern (Latour, 2004b). Our aim in this chapter is therefore to slow down reasoning, creating 'an opportunity to arouse a slightly different awareness of the problems and situations mobilizing us' (Stengers, 2005, p. 994) and to reduce the hectic pace by which mining and tourism (see Bjørst and Ren, 2015) are enacted as large-scale achievements, and to insist on mining risks and impacts as being a complex and relevant public concern, which cannot be turned down by 'mere fact'.

Agency in ANT is distributed and uncertain. We can never know beforehand who – or what – is accorded the power to act or to create effects. Also, these effects are not innate or predictable, but the result of relational work. By tending to this work of assembling the social, ANT abstains from building on assumptions or from outlining social laws to govern, structure and determine the social. By refusing to start with deep structure or power as human actors, we are compelled to collect bits and pieces, looking at how policies, reports, public outcries, scenarios or development models become instituted as matter of fact, and compelled to ask how they come to prevail? Answers to these sorts of questions require new methodological tools.

The methodology of ANT

Opposed to what the name might suggest, ANT is not a theory (Latour, 1999). ANT does not seek to provide clear-cut, static and bounded representations of the social, but instead to promote rich and practice-oriented accounts of constant, relational and complex world building. Also, as noted from the above, it does not seek for universal structures or explanations for the social (for tourism-specific literature, see van der Duim *et al.*, 2012, 2013). The heterogeneous world building of ANT is partial and modest (Law, 1994). As proposed by John Law (2000), researchers must insist on the impossibility of the whole and the necessary commitment to sets of partialities and partial connections. Through this, we may be able to re-appropriate objectivity as local and situated. This objectivist insistence opposes perspectivism, e.g. multiple perspectives to multiple objects/realities 'dialectically jumping between the ideas that reside in the minds of subjects and some objective reality out there' (Mol, 2002, p. 31). So while we must tend to the precarious practices made up of people, discourses and things in specific relations and contexts, we should also insist on the actual existence, workings and consequences of an enacted, partial and mutable reality.

In order to situate ourselves in the middle of things (knowing full well that there is no actual or stable middle), our case study focuses on tourism–mining–nature relations. We explore what is now, what is expected to come, and what can no longer be for tourism in Greenland. While we link our exploration to ongoing mining controversies, other illustrations could also have been used, such as the melting glaciers of Northern Greenland, which paradoxically serve both as climate change icons and tourism marketing symbols (Bjørst and Ren, 2015). In the present case as well as in the glacier example, practices and discourses related to tourism and larger social development coexist, associate or conflict in a number of often unpredictable ways.

The case presented in this chapter substantiates this claim by showing how local and global political discourses entangle, displaying the mining case as a performative but contested representation of the future, which clashes with, contradicts, silences and also, at times, supports tourism. Narsaq, as our case description will disclose, is not a place for easy solutions: it is not necessarily a site for tourism or mining development. However, in whatever form, the activities and controversies around the mining project are part of enacting a particular place and landscape as visitable – or not (Dicks, 2004). As a consequence, future scenarios and discourses entangle with nature and communities in shaping and preparing places to be visited and consumed in certain ways rather than others (Ren, 2015), and will then surely create path dependency for years to come.

Narsaq – a place of tourism or a place of mining?

As the impacts of climate change unfold, global attention has turned to the melting of the Arctic ice sheets and the dramatic changes to Greenland's World Heritage listed Ilulissat icefjord. The focus on climate change has underlined the 'fact' that climate change makes Greenland greener (literally) and 'open for business' (Nuttall, 2013, p. 373). Getting ready for business in the light of the Anthropocene has led to an intense process of scenario-building for the future. This has resulted in the creation of two (at least) configurations of Greenland as a tourist destination. The first was envisioned as serene, majestic, and 'cool' (although melting). This 'cool' position proved difficult to manage when a heavy focus on climate change renders the destination somewhat 'forbidden' but also offering the exclusivity of a 'last chance to see' destination (Bjørst and Ren, 2015).

As a result, Greenland has come to international attention as an emerging tourism destination. Its wild, rocky landscapes, moody weather and inaccessibility make it a 'mystical' and 'forbidden land' (TripAdvisor, 2015, various feedback). In 2011 Greenland received almost 214,000 visitor nights, with most guests arriving from Denmark, Germany and the US. Tourism to Greenland is seasonal, with a summer season (June to August) and a winter season (March and April) being the peaks for visitation. As discussed below, airline seat capacities and limited investment in accommodation accentuate these limits to the industry's development. Cruise tourism, which has experienced a decline between 2010 and 2013 due primarily to external market conditions beyond Greenland's control, has been identified as a potential strategy to grow tourism and circumvent the accommodation supply and airline capacity issues (Visit Greenland, 2015).

But tourism potential, particularly within the Narsaq context, must also be situated within its past. In the 1970s the local Greenlandic economy was primarily based on fisheries (and subsidies from the Danish state). To increase business

Figure 14.1 Narsaq seen from the hillside.

opportunities in the south of Greenland new strategies were initiated such as the formation of a local tourist association and the opening of tourist information centers to accommodate the growing number of tourists (Madsen, 1975, p. 162).

Since the 1980s, Narsaq (see Figure 14.1) has been one of the 'must see' destinations when visiting the south of Greenland. Visit Greenland describes the touristic experiences in Narsaq as ' . . . a combination of agriculture, Norse history and the landscape around Narsaq, which gives a special south Greenlandic flavour' (Visit Greenland, 2015). Likewise Narsaq is promoted as a desirable place for tourists to study mineral deposits and find small pieces of granite, sandstone, lava and tugtupit (a tenebrescent quartz-like crystal) and in promotional material it is explained how 'local people appreciate closeness with nature and consider it as one of the town's core values' (Visit Greenland, 2015). Narsaq has several attractions such as museums, historical monuments and prehistoric Inuit and Norse ruins (see Figure 14.2) and it is possible to visit the inland ice cap and smaller farms on daily boat trips. Additionally, the landscape around Narsaq is excellent for hiking and camping.

In the 1980s it was possible to fly direct from Copenhagen and Reykjavik (Iceland) to the south of Greenland. The once-a-week, direct air route, which operated year-round, was discontinued in 2010 to be replaced with only a limited-access service for a short summer season (June, July and August). As a result, tourism development in the south of Greenland and growth in the local tourism sector has stagnated. In the North of Greenland, on the other hand, an increase in

Figure 14.2 Children in the Narsaq museum.

tourists and overnight guests has taken place (Greenland Statistic, 2012). In other words, the south of Greenland now possesses latent potential for tourism development but local outfitters and entrepreneurs have not pursued this opportunity, in part due to insufficient infrastructure. When one of the authors (LRB) visited Narsaq in 2013 the local tourist information was closed and the former employee had been hired by GME as operations manager. The municipality, together with the local museum, was at that time responsible for developing Narsaq as a place of tourism, but the resources for that particular task were very limited, despite the enthusiasm of the employee. Adding to the challenge, the population in Narsaq has, in the last few years, been in decline. In the 1980s when Narsaq was an independent municipality with its own fish factory it had a growing population. Today the population is only around 1,500 people (Greenland Statistic, 2014). Unsurprisingly, mining has arrived on the scene as tool for economic and social development.

On October 24, 2013, the Greenlandic Parliament, or Inatsisartut, lifted a decade-long moratorium on mining radioactive elements, thereby rescinding a zero-tolerance policy toward uranium extraction. This paved the way for both Greenland and Denmark to become the newest Western (and Arctic) suppliers of uranium (Vestergaard, 2015, p. 153). Until 2009, mining in Greenland was under the institutional control of Denmark. A repatriation of the political and economic responsibility for mineral resources followed the introduction of Greenland self-government in 2009. Greenland has a long history of mining, including the extraction of coal, cryolite, gold, copper and other minerals. Most mining activities took place in a period with limited or no environmental attention on the delicate Arctic ecosystems (Sejersen, 2014b). Mining legacies in the form of social and ecological impacts are known but play a small part in the current political debate in Greenland – a debate framed by questions of agency, respect and Greenland's right to self-determine its development trajectory (Bjørst, 2015a).

With the introduction of the Self-Rule Act from 2009, the people of Greenland were required to seek out new sources of income. The downturn in the economy in 2012, 2013 and 2014 suggests that Greenland might face substantial economic problems in the years to come (Christensen and Jensen, 2014). Scientists, politicians and the Greenlandic business community have more or less accepted this 'inconvenient truth' and are looking for alternative ways to create economic growth and attract investors. Mining and tourism are often mentioned as lucrative possibilities (Fremtidssenarier for Grønland, 2013; Rambøll Rapport, 2014).

The mining company GME began operations in Greenland in 2007. Its activities there are concentrated in a licensed area on Kvanefjeld by the town of Narsaq in the south of Greenland. Kvanefjeld contains 260 million tonnes of uranium and 10.33 million tonnes total of rare earth oxide, according to industry sources (Proactive Investors, 2015). The company is a subsidiary of GME Ltd., which is listed in Australia and has its headquarters in Perth (GME, 2015). GME seems confident that in the years to come it can finalize a cooperation agreement on the regulation of uranium production and commence export operations from Greenland. Yet it still needs 'the social license to operate', which will not be achieved without some level of resistance. Studies from 2013 show that many local actors in Greenland are still in the process of deciding about the mining of the country's uranium (Bjørst, 2015a).

Global NGOs such as WWF, Friends of the Earth, Greenpeace and NOAH have raised concerns about the mining project (Avataq, 2013). While the government of Greenland is preparing for what it characterizes as 'sustainable mineral resource development' in its newest oil and mineral strategy (Government of Greenland, 2014, p. 90), resistance in the urban centers is on the rise, especially in the capital of Nuuk. Conflicting claims about 'what is sustainable for Greenland' is part of this debate (Bjørst, 2015c).

Studies of mining in other parts of the Arctic indicate that the social and cultural cost of mining operations cannot be ignored and need to properly addressed (Tester and Blangy, 2013), especially in terms of the impact of mining on other economic activities such as tourism. Impact and Benefit Agreements (IBAs) have also been mooted as a tool for empowering local people and stakeholders, but science shows that social and economic benefits are highly debatable (Hansen, 2014). Defining what could be of benefit for the local community and Greenland in the future entangles with contested representations of the future, and Narsaq could be a primary 'place of tourism' or a 'place of mining'.

The quest for jobs and development is taking center stage at present in Greenlandic politics and the dominant positive discourse about mining as a 'job machine' (Sejersen, 2014a) is closely linked to the political project of Greenland as a state in formation. As is true of Greenlandic politics in general, this dispute can be read as a negotiation of how that national project is to be configured (Gad, 2009, p. 9). Recent local studies show that mining, and especially the Kvanefjeld project, is considered a solution to Greenland's economic problems. Therefore, the storyline of 'saving the community' (promoted by GME, among others) and doing something for the benefit of Greenland is a common focus of attention among Greenlandic politicians (Bjørst, 2015b). These coexisting political ecologies produce winners and losers and confront former (postcolonial) ideas about the structure of winning and losing in Greenland.

Composing a landscape

In order to enroll the public, politicians and investors in the mining project in Narsaq, GME makes use of a number of highly stratified strategies, such as lobbying, publishing reports and sponsoring the local football team. The company has also created a number of promotional videos to support these activities. While some display the current work at different test sites, others are proposed as scenarios for future development. The video that is the subject of this analysis is a 4-minutes and 42-seconds digital rendering uploaded to Vimeo in January 2015 under the title of Greenland Minerals and Energy Kvanefjeld (https://vimeo.com/111488160). Kvanefjeld is the Danish name for Kuannersuit, the area in southern Greenland where the uranium project is proposed, which is also situated adjacent to the Ipiutaq Guest Farm. In this section, we will take a closer look at this video and its role as a futuring device. We see it as an illuminating example of how the future of a particular place is imagined by the mining company, and by corollary, how other activities such as tourism are (un)imagined and interfered with in this process of futuring.

The video first displays the company name and logo and the Earth appears, rotating in space. An insistent, computerized melody accompanies the video. Zooming in on the Northern hemisphere, the ice-covered island of Greenland emerges. Small dots are plotted in, showing the location of global (Northern) centers: London, Berlin, Seoul, Beijing and New York (Fig. 14.3). In the middle of these, we find Kvanefjeld. We zoom in even further as a small airplane flies over the south Greenlandic coast and lands at the airport of Narsarsuaq, or 'international entry point' as the text informs us. It is summer and the sky is blue. The next frame shows once again 'South Greenland Geography' from above. The location of the town of Narsaq is indicated in relation to the 'Ilimaussaq complex', the prospecting site, its fixed geographical boundaries drawn in yellow. No evidence is visible of the Guest Farm, local sheep farms or other physical structures located in the area. The subtitle reads: 'One of the world's most unique geological environments, host to vast resources of rare earth metals and uranium. Ideally located with direct shipping access, year round'.

Now we are back on the landing strip. Two men stand in from of a helicopter. They shake hands, nod. The helicopter takes off. Sun is still shining. As the helicopter flies into Narsaq Valley, we enter the 'project area', which we read as we

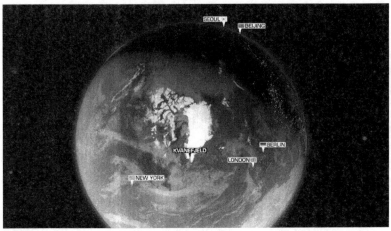

Figure 14.3 Screenshot of *Greenland Minerals and Energy Kvanefjeld*.
GME, 2015: https://vimeo.com/111488160.

approach the bottom of the fjord. The Ilimaussaq complex, covered in yellow, is neatly cut out and is turning around as a 3D model. Different colors penetrate the underground and geological information on rock layers and the discoveries of the outcrop are presented at the bottom of the screen. Three evaluated deposits are presented. We are told that they contain '575 Mlb's of U_3O_8, 10.3 Mt rare earth oxide'. The next frame illustrates by quick progressive cuttings into the fell how Kvanefjeld is 'conducive to simple open-pit mining'. The text informs us that it is 'capable of sustaining large-scale mining for decades'. At the end, the open pit sits back as we again zoom out to get an overview of the Kvanefjeld project geography. On the map, we see the town of Narsaq and Kvanefjeld. A processing plant, a port and an accommodation village are plotted in.

Once again, we hover over Narsaq, with its small wooden houses, all in different colors, and approach the bottom of the fjord, where there is now a large tanker and a port. We do not see the accommodation village, a very controversial issue in the so-called 'large-scale debate': how to house thousands of workers next to very small Greenlandic communities when developing mining projects. Current farms in the area are absent, deleted from the face of the project area. In the same way, no animals – which typically abound in the area in summer – are visible. Norse ruins, which are also present in the same area and which are currently under consideration to enter the UNESCO World Heritage nomination list, cannot be discerned either. Very closely now, we fly over the harbor, power station and processing facilities higher up on the mountainside. The current gravel road running through the area has been replaced by an asphalted road. The terrain is dry and snowless, a condition very unusual for the area.

Composing Greenlandic tourism futures 295

As the helicopter flies over the project area from the top of the fell, right over the open pit, past the port and into the sunset toward the coast, graphical text on the screen informs us: 'Kvanefjeld – a world-class project, ideally located in southern Greenland'. Below, a tanker is sailing off as the text reads: 'Direct shipping access to the world'. We zoom out again to the Northern hemisphere of the globe, and as large cities are plotted in again to encircle Kvanefjeld and Greenland, we read: 'Strategic metals for global industry'. The film ends with the display of the company logo and name.

In his portrayal of the plains of Eastern Africa, Robbins (2012) describes how its landscape is mistakenly perceived as natural and apolitical, as void of power and populated merely by non-human wildlife. Employing the framework of political ecology, Robbins goes on to illustrate how the landscape is connected to humans and nonhumans, which both enable and alter it. While landscape might be portrayed as pristine and disinterested in wildlife documentaries, we should not let ourselves be fooled: landscape is made of politics and power. In our analysis, we follow this argument but also expand it by showing how the landscape is not only connected to or impacted by human activity and global power structures, but rather also as an active participant as proposed by ANT in producing certain forms of landscape, power and agency. Additionally, we shift our attention from present landscapes to the composition of future landscapes, as exemplified in the video.

Following an ANT approach, and the idea of distributed agency (Callon, 2008), the future does not just happen. Nor is it a passive noun. Rather, it is a verb, an ongoing process of futuring and a socio-material outcome of a number of practices. The future can be enacted and tinkered with through the meticulous arrangements of companies, organizations and political interest groups, through various discourses, procedures, technologies and devices, such as the GME video. Below, we first interrogate the video as a futuring device to display the role of nonhumans in delegating power and knowledge. Second, we ask what kind of landscape is being enacted in the video and, more specifically, what is present, what is absent and what is made impossible. Lastly, we sum up our chapter by looking at alternative future landscapes, posing the cardinal question: Where does this leave tourism?

The video as futuring device

In this chapter the future is introduced as a performative achievement. By looking at future management as performative, we ask: 'How and through which practices and procedures do some versions of the future come to prevail over others?' We identified the Kvanefjeld video as a relevant futuring device, seeing it as a conglomerate of GME resources and arguments, such as public and scientific reports, media communications and PR material. Also, it works as a condensed and fairly approachable unit of analysis. According to Jensen (2005) futuring (or future-generating) devices are capable of flexibly gathering people and work practices around them. They 'have very little structure, and few material and discursive limits are (as yet) inscribed in them; to the extent that they do generate more lasting futures, such structure will have to evolve' (p. 247). Their highly diffuse and ephemeral content as well as their vagueness 'enables their flexible distribution and adoption in widely variable material and discursive circumstances' (ibid.).

In that same vein, the video and the proposed future is ephemeral and vague. It does not make clear promises or offer detailed accounts of environmental impact or local job creation, or account for changes in the perception of the place or on perceived quality of life. As such, it is a dream of a 'world-class project', sustaining (open-pit) mining for decades. What it displays are scientific 'facts' and (unwarranted) guarantees of long-term economic security. It manages impressions, discourses and actions toward what is (made) to come.

Future landscape(s)

What is presented in the video is a very particular landscape. First, it is a docile landscape, seemingly welcoming and mild as we cruise over it, detached, in our helicopter or from somewhere outside and above the project area. It is, as stated in the video, 'accessible year-round', yet this is a summer landscape without snow or signs of melting waters. The wind seems not to blow and the sky is blue. This sharply contrasts with the usual local weather and climate, where the weather changes without warning, flights are delayed and roads erode, making any inland movement difficult.

Second, it is a landscape of production, but a very particular kind of production. Although clearly created to suggest ongoing activity and development, the mining scenario is strangely void of people: only the shaking hands of two men at Narsarsuaq airport convey any human (male, Caucasian) presence or interaction. Also animals, traces of culture and history (such as that of the Norse) and any other past or present presences are othered. The activities which (as opposed to mining) currently take place around Narsaq – most importantly tourism and farming, but also fishing, hunting and foraging – are erased from the scene. A large farm situated at the bottom of the fjord is literally deleted from the video and the ever-present sheep and cows in the fjord cannot be seen. Just like the East African plains, this seems to be a landscape populated only by nonhumans: self-operating power plants, machinery and an open pit, which bit by bit digs itself into the underground. Finally, there are no suggestions of tailings, radiation or potential impacts on water, fauna or the nearby town.

Third – and drawing on the two former – it is represented as a safe landscape. The mining future portrayed in the video claims a harmless and welcoming environment and one which is risk-free for investors, for the local environment and the nearby community. The lack of risk not only concerns the deletion of tailings, waste rock dumps, specialized earthworks to contain environmental pollution and radiation, but also the predictability and stability suggested by the decade-long (economically) sustainable mining. The landscape yields a promise of a docile, productive and risk-free landscape: comfortably nested in between global mining hubs for 'strategic mining', yet somehow detached from the vagaries of weather, mineral prices and market forces, and the tiresome burden of co-presence with actors and activities. How could one not hope for and work toward such a future?

The answer to this question lies perhaps in what the video does not convey. By taking a closer look at the hybridity of landscape, we notice how its ephemeral, yet effective, composition conflates science, nature, economy and politics into a seamless joining of 'facts': of mineral composites, of sustained decades, of global market access tying neatly together. While the scenario expresses a true modern landscape, a smooth backdrop for human activity and development, it neglects the complex relationships that make it possible and the concerns such a complex assemblage must necessarily entail. An example of this is the absence in the video of the thousands of workers required to establish the eventual mine, itself a highly controversial issue.

The composition of the landscape displays not only a multiplicity of actors unaccounted for, such as the migrant workers, farmers, local hunters and fishermen and women, but also the controversial character of the landscape. Or rather, the controversy of what does not fit into the mining future at Kvanefjeld. By performing a hegemonic mining landscape, the two additional pillars of the Greenlandic economy – fishing and (more relevant in this context) tourism – are completely othered. While the landscape is hospitable to mining purposes, it would not be so for tourism, where denied access and the perception of radiation risks pose but two of the many challenges. For users seeking to consume the landscape or engage in other relations of production with it, the soil, water, fauna and animals would now enact other risks. In these relations, the landscape is enacted as unwelcoming, unproductive and unsafe – as hostile.

Alternative landscape and staying concerned, or why is this about tourism? Concluding remarks

On Tripadvisor.com past guests have commented upon their stay at the Ipiutaq Guest Farm. Headlines such as Wonders of nature; Scenic beauty beyond compare; Unforgettable adventure encompass the impression of the tourist when visiting the landscape. While one reviewer thinks back on ' ... the sheep at the horizon, the blanket of blueberries under our feet', another sums up her impressions in this way:

> The view is sublime over the fjord and mountains. Shades pass over the pure waters of the torrent. Lady nature spoils us just like our hosts with their kindness and their exquisite dishes. We had an enchanted stay and will be back.
> (Tripadvisor.com, our translation)

These depictions will surely seem out of place in a few years, should the landscape be overtaken by mining.

As part of composing the future, the scenarios proposed by GME in the video (and elsewhere) contribute to a specific shaping of Narsaq to be visited and consumed in a certain way – or some may argue, not to be visited or consumed at all. The proposed development of mining is clearly controversial. Significantly, it does not constitute the only endeavors to enacting a resilient landscape. While the future mining landscape is enacted as docile, productive and risk-free, other alternative enactments of the place exist. On TripAdvisor, tourists talk of a landscape of wonder, of inspiration, of magnificence, of pioneering. These accounts dovetail with the journalist's comments presented in the beginning of this chapter as they display the landscape as unpolluted, wild and consumptive. It is a landscape for tourism (primarily adventure or ecotourism), which, as mentioned earlier, can be retrieved in how Visit Greenland describes the touristic experience in Narsaq.

Unfortunately, the tourism future cannot compete with the scale offered by the mining future: a large scale. Tourism only offers modest and uncertain revenue and does not lend itself well to decade-long promises of steady income. Successful tourism development requires skills, innovation, political will, determination and infrastructure. Unlike the mining prospect, all or most of this must come from national or local entrepreneurs and investors. While literature suggests that tourism may benefit from mining, for instance through mining (heritage) tourism or business tourism in the shape of technical visits (Conlin and Jolliffe, 2011), we suggest that this may not be the case in Narsaq.

As stated by Robbins (2012), political ecology is not merely a disinterested endeavor, but one that seeks to point to how things 'are tractable to challenge and reform' (2012, p. 13). In a similar vein, this agrees well with what Anne-Marie Mol has termed ontological politics (2002). According to Mol, our research practices are important building blocks in creating some realities rather than others. While such endeavors are laborious, precarious and uncertain, they may enable us 'to participate in a modest way in the re-composition of the common world' (Holifield, 2009, p. 655). It is a world that is common not only to other humans but also to nonhumans, who together contribute to the assembling of community and relations of inequality and power.

We have shown in this chapter that envisioning the future is far from an innocent or apolitical endeavor. The future offered in the video not only represents but also composes a place in a highly strategic way. While we do not offer advice with regard to the social, economic or environmental value of mining or tourism, we seek to offer tools by which to engage in slowing down the instituting of matters of fact by unpacking the controversial way by which the future is composed. ANT scholars remind us that in the quest for developing sustainable tourism, non-human actors have, if not a say, then at least a presence and often very large effects on human activity. We should recognize 'that nonhumans are already involved in democracy, through the competing accounts that mobilize them – often in the determinist form of nature – and through the disputable matters of concern at the center of so many controversies' (Holifield, 2009, p. 652).

White (2006, p. 69) argues that the question remains: 'If socio-natures are enacted through situated knowledge practices, what nature to save, and on whose behalf, become vital questions of environmental politics' (White, 2006, cited in Blok, 2011, p. 76). While we do not claim to hold the answer to this, we do advocate the move to slowing down the instituting of matters of fact. This, we argue,

References

Andersen, M. (2015) Uran eller spisekammer? [Uranium or pantry?]. *Weekendavisen* #23, 4 June 2015.
Avataq (2013) Appeal to the Greenlandic and Danish governments not to abolish the uranium zero tolerance policy in the Danish realm. Accessed 9 January 2014: https://www.nirs.org/international/westerne/Statement%20on%20uranium%20mining%20in%20Greenland%2026%20April.pdf
Bjørst, L. R. (2015a, in press) Arctic Resource Dilemmas: Tolerance Talk and the Mining of Greenland's Uranium. In: Körber, L.-A., MacKenzie, S. and Stenport, A. W. (eds), Critical Arctic Studies: Cultures, Environments, Politics, and Practices. Montreal: McGill-Queen's University Press.
Bjørst, L. R. (2015b accepted/In press). Arctic Resource Dilemmas: Tolerance Talk and the Mining of Greenland's Uranium. In: R. C., Thomsen & L. R. Bjørst (eds.), Heritage and Change in the Arctic. Aalborg: Aalborg Universitetsforlag
Bjørst, L. R. (2015c, in press) Uranium – the Road to 'Economic Self-Sustainability for Greenland'? Changing Uranium-Positions in Greenlandic politics. In: Fondahl, G. (ed.), *Northern Sustainabilities: Vulnerability, Resilience, and Prosperity in the Circumpolar World*. New York: Springer Publishers.
Bjørst, L. R. and Ren, C. (2015) Steaming up or staying cool? Tourism development and Greenlandic futures in the light of climate change. *Arctic Anthropology*, 52 (1), pp. 91–101.
Blok, A. (2011) War of the whales: post-sovereign science and agonistic cosmopolitics in Japanese-global whaling assemblages. *Science, Technology & Human Values*, 36 (1), pp. 55–81.
Callon, M. (2008) Economic Markets and the Rise of Interactive Agencements: From Prosthetic Agencies to Habilitated Agencies. *Living in a Material World: Economic Sociology Meets Science and Technology Studies*. Cambridge: MIT Press, pp. 29–56.
Christensen, A. M. and Jensen, C. M. (2014) 'Aktuelle tendenser i den grønlandske økonomi.' Danmarks Nationalbank, Kvatalsoversigt 2. kvartal 2014 53. årgang nr. 2, pp. 71–76.
Conlin, M. V. and Jolliffe, L. (2011) *Mining Heritage and Tourism: A Global Synthesis*. Abingdon and New York: Routledge.
Dicks, B. (2004) *Culture on Display. The Production of Contemporary Visitability*. London: McGraw-Hill International.
Fremtidsscenarier for Grønland (2013) Instituttet for Fremtidsforsknings scenariebeskrivelser for Grønland [Future Scenarios for Greenland (2013). CIFS future scenario descriptions for Greenland]. September 2013. Accessed 9 January 2014: http://www.ga.gl/LinkClick.aspx?fileticket=0iDes11d%2Ft0%3D&tabid=2033&language=da-DK.
Gad, U. P. (2009) Post-colonial identity in Greenland? When the empire dichotomizes back – bring politics back in. *Journal of Language and Politics*, 8 (1), pp. 136–158.
GME (2015) About Greenland Minerals and Energy. Accessed 1 June 2015: http://gme.gl/en/about-greenland-minerals-and-energy

Government of Greenland (2014) Greenland's oil and mineral strategy 2014–2018. FM 2014/133, IASN-2013-093824. Accessed 28 March 2014: https://www.govmin.gl/images/stories/about_bmp/publications/Greenland_oil_and_mineral_strategy_2014-2018_ENG.pdf

Greenland Statistic (2012) Turismestatistik i perioden 1. okt. 2007-30. sep. 2012. Accessed 10 January 2015: http://www.stat.gl/publ/da/TU/201301/pdf/Turismestatistik%20i%20perioden%201.okt.2007-30.sep2012.pdf

Greenland Statistic (2014) Befolkningen i lokaliteterne efter tid, lokalitet og fødested. Accessed 10 January 2015: http://www.stat.gl/dialog/main.asp?lang=da&version=201401&sc=BE&colcode=z.

Hansen, A. M. (2014) *Community Impacts: Public Participation, Culture and Democracy*, Unpublished working paper, Open Access, University of Copenhagen. Accessed 10 January 2015: http://vbn.aau.dk/ws/files/186256309/Community_Impacts.pdf

Holifield, R. (2009) Actor-network theory as a critical approach to environmental justice: a case against synthesis with urban political ecology. *Antipode*, 41, pp. 637–658.

Jensen, C. (2003) Latour and Pickering: Post-Human Perspectives on Science, Normativity, and Becoming. In: Ihde, D. and Selinger, E. (eds), *Chasing Technoscience: Matrix for Materiality*. Bloomington, IN: Indiana University Press, p. 225.

Jensen, C. (2005) An experiment in performative history: electronic patient records as a future-generating device. *Social Studies of Science*, 35, pp. 241–267.

Jóhannesson, G., Ren, C. and van der Duim, R. (eds) (2015) *Tourism Encounters and Controversies: Ontological Politics in Tourism Development*. London: Ashgate.

Latour, B. (1994) On technical mediation. *Common Knowledge*, *3* (2).

Latour, B. (1999) On recalling ANT. *The Sociological Review*, 47, pp. 15–25.

Latour, B. (2004a) *Politics of Nature. How to Bring the Sciences into Democracy.* Cambridge, MA: Harvard University Press.

Latour, B. (2004b) Why has critique run out of steam? From matters of fact to matters of concern. *Critical inquiry*, 30 (2), pp. 225–248.

Latour, B. (2005) *Reassembling the Social. An Introduction to Actor-Network-Theory.* Oxford: Oxford University Press.

Law, J. (1994) *Organizing Modernity*. Oxford: Blackwell.

Law, J. (2000) On the subject of the object: narrative, technology, and interpellation. *Configurations*, 8 (1), pp. 1–29

Madsen, Vagn Eyvind (1975) *K'AK'ORTOK Julianehåb 1775–1975*. Julianehåb Museumsforening 1975. Det Grønlandske Forlag.

Miller, E., Van Megen, K. and Buys, L. (2012) Diversification for sustainable development in rural and regional Australia: how local community leaders conceptualize the impacts and opportunities from agriculture, tourism and mining, *Rural Society*, 22 (1), pp. 2–16.

Mol, A. (2002) *The Body Multiple: Ontology in Medical Practice*. Durham, NC: Duke University Press.

Nuttall, M. (2013) Zero-tolerance, uranium and Greenland's mining future. *The Polar Journal*, 3 (2), pp. 368–383.

ProactiveInvestors (2015) Greenland Minerals and Energy MD John Mair talks with Proactive Investors. Accessed 20 February 2014: http://www.proactiveinvestors.com.au/companies/news/60143/greenland-minerals-and-energy-md-john-mair-talks-with-proactive-investors-60143.html.

Rambøll Rapport (2014) Hvor skal udviklingen komme fra? Potentialer og faldgrupper i den grønlandske erhvervssektor frem mod 2015 [Where should development come from? Potentials and pitfalls in the Greenlandic sector until 2015].

Ren, C. (2015) Possible Greenland. On 'Futuring' in Nation Branding. In: Jóhannesson, G., Ren, D. and van der Duim, R. (eds), *Tourism Encounters and Controversy. Ontological Politics of Tourism Development*. London: Ashgate, pp. 221–238.

Robbins, P. (2012) *Political Ecology: A Critical Introduction*. Second Edition. Oxford: John Wiley & Sons.

Scheyvens, R. and Momsen, J. (2008) Tourism in small island states: from vulnerability to strengths, *Journal of Sustainable Tourism*, 16 (5), pp. 491–510.

Sejersen, F. (2014a) A Job Machine Powered by Water. In: Hastrup, K. and Rubow. C. (eds), *Living with Environmental Change: Waterworlds*. Abingdon: Routledge Falmer, pp. 102–105.

Sejersen, F. (2014b) Efterforskning og udnyttelse af råstoffer i Grønland i historisk perspektiv [Exploration and exploitation of resources in Greenland in a historical perspective]. Working paper, Open Access København. University of Copenhagen. Accessed 9 January 2014: http://nyheder.ku.dk/groenlands-naturressourcer/rapportog baggrundspapir/Efterforskning_og_udnyttelse_af_r_stoffer_i_Gr_nland_i_historisk_ perspektiv.pdf

Stengers, I. (2005) The Cosmopolitical Proposal. In: Latour, B. and Weibel (eds), *Making Things Public: Atmospheres of Democracy*. Cambridge, MA: MIT Press, pp. 994–1003.

Tester, F. J. and Blangy, S. (2013) Introduction: industrial development and mining impacts. *Études/Inuit/Studies*, 37 (2), pp. 11–14.

TripAdvisor (2015) Accessed 13 July 2015: http://www.tripadvisor.com/Hotel_Review-g1136822-d2707776-Reviews-Ipiutaq_Guest_Farm-Narsaq_Kujalleq_Municipality. html (authors' translation).

van der Duim, R., Ren, C. and Jóhannesson, G. (eds) (2012) *Actor-Network Theory and Tourism: Ordering, Materiality and Multiplicity*. Abingdon: Routledge.

van der Duim, R., Ren, C. and Jóhannesson. G. (2013) Ordering, materiality and multiplicity: enacting ANT in tourism. *Tourist Studies*, *13*, pp. 3–20.

Vestergaard, C. (2015) Greenland, Denmark and the pathway to uranium supplier status. *The Extractive Industries and Society*, *2* (1), pp. 153–161.

Visit Greenland (2015) Musse McCartney and Narsaq's Cowboys, website of Visit Greenland. Accessed 27 July 2015: http://www.greenland.com/en/destinations/south-greenland/narsaq

White, D. F. 2006. A political sociology of socionatures: revisionist manoeuvres in environmental sociology. *Environmental Politics*, 15, pp. 59–77.

Conclusion
Towards future intersections of tourism studies and political ecology

Roger Norum, Mary Mostafanezhad, Eric J. Shelton and Anna Thompson-Carr

> Nature is perhaps the most complex word in the English language.
> (Raymond Williams, 1983, p. 219)

Calls for a political ecology of tourism were made first by Susan Stonich and Gerald Sørensen, whose talk at a Society for Applied Anthropology meeting in the early 1990s further solidified the coinage of a rather appropriate name for a field deeply linked to the need for political action and ecological impact in tourism contexts.[1] Calls for political engagements with the tourism industry and with tourists themselves continued through the end of the millennium, when discussions about the various connections between globalization, tourism, and the environment began to reach their peak. By the dawn of 2000, tourism had emerged as the 'world's biggest industry', the environment had become the world's biggest 'cause', and the world was in a state of serious ecological crisis.

Since then, however, political ecology and tourism have been only obliquely or peripherally linked by academics. Hollinshead (1999) observed that scholars were making early headway by exploring the '"governmentality" of tourism'. Hall's (1996) work on tourism's political apparatus, for example, explored 'the norms and ideologies which conceivably underpin the policy platforms by which dominant groups act and regulate the social, economic and environmental world about them' (Hollinshead, 1999, pp. 9–10). Those practitioners deeply embedded within the tourism industry know about these links all too well. As noted in the introduction to this book, Fiji acknowledged nearly a decade ago the integral relationship of tourism and ecology by merging its ministries of the environment and tourism. As Banuve Kaumaitotoya, the first permanent secretary of the new Ministry put it, 'Tourism is *the* vehicle for poverty alleviation in Fiji . . . Without it, our economy would collapse. So we have to plan to mitigate and adapt to climate change' (Rosenthal, 2007; emphasis in original). Still, Gössling and Hall (2006) noted the painful irony that, despite the multitude of writing over the past three decades on sustainable tourism and the patent links of cause and effect between tourism and the environment, there seems to have been more damage done to the planet's ecologies during this time than ever before. The claims made in all this scholarly work look to have fallen on deaf ears.

If the 'end of nature' (McKibben, 2006) – that is, as a force independent of humankind – is already upon us, tourism, for its part, shows zero signs of slowing down; pay a visit to any major world city these days and the exponential proliferation of hovering selfie sticks and idling red city tour buses is as much confirmation as anyone needs. While it is problematic for many that tourism has become the world's largest industry, it is just as reassuring for many nations and communities that it has become their most central economic development strategy (Stonich, 2000), infusing local people with the hope that luring to their shores those of means will make possible positive economic – and consequently social – change. The fact that countries struggling against the vulnerability and precariousness of unsustainable agriculture industries, natural resource extraction, unfavorable market prices, fluctuating world trade agreements, and devastating natural disasters has led to tourism becoming heralded as a crucial development alternative.

Yes, there may be hope yet, and the honed lenses of political ecology can offer the possibility of wading through these complex, muddled waters towards marked clarity, action, and even change. In the way that tourism encourages struggling communities to envision hopeful futures, so does political ecology provide us with new means of reflecting upon and critiquing relations of power and difference, interactions between human groups and their biophysical environments, and political, economic, and ecological concerns. And political ecology is a framework for enquiry available to scholars of many persuasions. As we suggested in the introduction, via Robbins, it isn't necessary to self-define as a card-carrying political ecologist to benefit from the insights that its frameworks offer. This book, for example, showcases the work of social and environmental anthropologists, political and cultural geographers, management experts, ethnobotanists, as well as humanities and tourism scholars. Over the past decade alone, the fields both of political ecology and of tourism studies have experienced unprecedented growth, with scores of dissertations being written and monographs published in each across numerous diverse disciplines. Indeed, many works, while they may not have 'political ecology' in their title, do indeed draw on political ecology analyses (cf. Takeda, 2015). This growth is not surprising, and, undoubtedly, this surfeit of recent work is due to the pre-eminent relevance of both in our world today. What is perhaps surprising is the fact that there has not been more scholarship that binds political ecology and tourism studies as a cohesive field of study.

If tourism is a useful investment for economically marginal communities across the Global South and the Global North, it is useful also for us as an epistemic community of scholars. Across all the chapters in this book, each author has sought to address different aspects of tourism, through the lenses offered by, or underpinning themes relevant to, political ecology. Globalization has made such research more nuanced; in fact, it has made such research necessary to be so. If the ethnographic studies in this book are local in their particular interests, they most certainly are global in scope. And yet, they acknowledge also the immense, fundamental importance of grassroots agency and the role that 'perceptions, motivations, and values that inform this agency' play in human environmental relations (Hollinshead, 1999, p. 26). In between the local and the global, of course, and in addition to the many middling actors, stands the nation state itself. In looking at local practices and institutions, these chapters also have turned a critical eye towards the global influences on, and effects of, these local contexts. Scholarship that brings political ecology critiques to tourism studies must explicitly consider not just the global ramifications of various forms of tourism – including, say, the co-efficiencies of distance and carbon emissions – but also, for example, the more nuanced cultural, social, and psychological involvement in the tourism encounter. Only by doing so can the work of scholars lead to suggestions for ways of encouraging and improving local forms of travel and tourism, and ways of rethinking the globalized forms through which they are conceived of and provisioned (Gössling and Hall, 2006, p. 315).

Ecology, much like tourism itself, is a study area of an ever-shifting academic landscape. In our globalizing and interconnected world, we push the Earth, with our terraforming and fracking, just as, through its quakes, weather events, and rising waters, the Earth pushes us (Huggan and Norum, 2015). Indeed, if recent scholarly and public debates around anthropogenic climate change are any measure, such movements mobilize us into political and environmental action (Ardoin *et al.*, 2015; Peters *et al.*, 2015). Now that we are knee-deep in the Anthropocene – even the Pope has given the epoch his blessing(!), though the coinage *Anthropopescene* remains up for grabs – an age in which human influence affects nearly everything 'natural' on the planet. It is well time we rethink not just how we protect contested ideas of nature, but how such protection can allow us also to continue to explore, experience, enjoy, and enhance the planet's natures for millennia to come.

Political ecology may be harnessed to help us develop a better understanding of the crisis of sustainability. Generally speaking, sustainability refers to the ability to reproduce a given arrangement (Hartman, 1998) and the so-called unsustainable predicament in which we currently find ourselves is an impediment of the reproduction of ecological and social relations. Understanding environmental degradation requires understanding social relations (such as class, gender, and culture) to see the multiple roles they play. Doing so allows for changes that are not just technical in nature, but social as well.

Yet, the 'crisis' of 'sustainability' does not just refer to the pressing need to figure out how to make our natural world sustainable; the crisis is also that we often fail to consider what we even mean by the term *sustainable* in the first place. This buzzword is now lobbed around by everyone from environmental activists to resource-grabbing corporates, and such 'greenwashing' has led it to lose much of the meaning it may have once had (Font *et al.*, 2012). Beyond sustainability politics, a more radical environmental politics may be in order if we are to address ongoing critical and indeed life-threatening environmental challenges.

The environmental movement of the twentieth century contributed to the currency of political ecology. Frankfurt School thinkers were notoriously miffed about the idea that human progress could be in any way positive, rather drawing attention to modernity's dehumanizing tendencies (Merchant, 1994). Modern and industrialized societies, they argued, are inherently unsustainable because of the relations of productions in which they are embedded. The materials necessitated by goods and service provisioning are degrading habitats and depopulating the species that inhabit them. Human control and transformation of the environment has side effects of environmental domination that include, but are not limited to, global climate change and species eradication.

If our ecological relationships are unsustainable, so are the social relations that produce them (Hartman, 1998; O'Connor, 1988).² Capitalism is socially unsustainable because it necessitates social relations that are based on inequity and domination across local and global scales. For such domination to occur, 'an ethical premise is needed to permit or sanction the "just" subordination of that which is subordinate' (Warren, 1990, p. 27). To remedy this lack of a suitable ethics we must seek out sustainability in relationships, not just between all humans, nonhuman species, and our material setting, but also relations among various humans based on respect and tolerance, freedom of thought, equality, and equity. Such sustainability is socially, culturally, and ecologically influenced; 'a product of the class, gender and cultural relations I am part of; and is influenced by the ecological relations I perceive', as Hartman (1998, p. 340) has phrased it. These relations and processes must be examined closely and we need to spend time thinking about ways to better understand and, if necessary, transform them.

Tourism is itself useful as an epistemological and heuristic device in part because it helps so patently to illuminate some of the contradictions of capitalism. Like capitalism more broadly, global tourism also is complicit in perpetuating deep divisions between rich and poor, under-employment, lack of resource access, and unequal concentrations and distributions of wealth. To remedy this we must first be aware of the intricate interconnections between politics, culture, economy, and the physical environment. Only then will we, scholars, practitioners, and citizens, be on a properly shared planet and be prepared to address and rectify the structures, institutions, beliefs, and practices that actively reproduce our currently unsustainable social relations. This remedy means treating the causes, not just the symptoms; rewiring structural faults, not just patching them up. Development work, all too often, just manages the latter. Once we figure out what we mean by sustainability can we then think more about tourism as an encompassing practice by tourists and also practices by industry stakeholders.

Tourism undoubtedly is a key player in human-induced environmental change, though its effects may be indirect, for example in the ecological footprint of international air travel, particularly long-haul routes, and importation or export of goods and supplies for the industry. Our carbon footprints contribute to sea level rise, land submergence, beach erosion, and reduced fresh water availability, among many other dangerous consequences (Gössling, 2003; Gössling and Hall, 2006). Tourism contributes to climate change yet climate change is opening up new paths and new destinations for tourism (Huggan and Norum, 2015; Kristoffersen, et al., 2015). Corporate Social Responsibility (CSR) and the greening of tourism businesses and destinations (through, for instance, environmental certification schemes and practical measures) are attempts to improve the physical and cultural impacts of the tourism industry – but, in numerous destinations, many of these may be too late.

Yet, while tourism is regularly blamed for environmental degradation, it might also be harnessed as a force for positive environmental change – and indeed for longer-term environmental sustainability (Higgins-Desbiolles, 2005). In fact, one glimmering advantage of well-managed, contemporary tourism is that it is at once immensely global and local at the same time. Tourism is valuable as much for its consideration of itself as a practice in its own right as it is a lens for looking at connections – not just for us as scholars, but also for the people who practice it. Furthermore tourism effects change in modes of economic and social relationships, most notably in turning subsistence economies into service industries. This shift can have long-lasting (and complex and destructive) consequences, not just for the physical environment and humans relations with it (Britton, 1991; Gössling, 2003) but also for humans' relations with each other more generally. If tourism displaces agricultural, fishing, and other traditional economic activities, it effects also vulnerable changes in the 'ecological identity' (Gössling, 2003) of people and the places they inhabit, for tourism too is itself a mode of production. In being so, tourism moves humans within and across many different scales of power relations: local villagers, middle-class brokers, tour operators, well-to-do globetrotters, multinational tourism stakeholders, national governments, international tourism and heritage bodies, to say nothing of the tourisms themselves. The stakeholders or actors in tourism's power relations bear increasing responsibility to contribute to ethically managed businesses and destinations (Jamal and Camargo, 2014; Moskwa, *et al.*, 2015).

Some questions the chapters in this book also have considered, albeit at times implicitly, are: How can tourism be harnessed to effect positive change on naturally occurring environmental phenomena? How can globalizing touristic forces be applied to economic development, social solidarity, and the opening up of a space for asserting indigenous rights? And, how can touristic experiences be looked to for examples of the potential for tourism to enact positive change in local and global communities?

Tourism as an academic discipline remains still heavily under-theorized (Tribe, 2004; Franklin, 2004; Hollinshead, 2012, 2014), and if tourism studies is at an impasse, as Ateljevic, Pritchard and Morgan (2012) have suggested, it has at its disposal a number of possible egresses. Deep considerations of political ecology, as the papers here have deftly demonstrated, are effective modes out of the discipline's theoretically impoverished position. Wielded wisely, political ecology can further disrupt tourism's hegemonic ontologies and the tourism industry's equally hegemonic practices. Our hope is not just that this book will advance political ecology and help us, as scholars, planners or practitioners, to transform our study and practice of tourism. Inspired by calls for new modes of conceptualizing planetary and personal imaginaries and practices in the Anthropocene, we hope this collection will spark dialogue between scholars and students of tourism, politics and the environment.

Notes

1 It merits noting that these latter two terms – action and impact – have in recent years taken on four-letter-word connotations among many academicians, given their place in

increasingly neoliberal institutions within their own higher educational establishments. 'Politics', for its part, has long been a bête noire for many ecologists and environmentalists, as has 'the environment' among politicians. The term 'political ecology' itself was first coined by Frank Thorne in 1935.

2 There is subterfuge here, too. As Robbins has pointed out, the 'characteristics that make the market economy vibrant are those that contain, within them, contradictorily, the seeds of ongoing social environmental crisis' (2012, pp. 54–55). This of course is one of the tensions in global tourism today.

References

Ardoin, N.M., Wheaton, M., Bowers, A.W., Hunt, C.A., and Durham, W.H. (2015) Nature-based tourism's impact on environmental knowledge, attitudes, and behavior: A review and analysis of the literature and potential future research. *Journal of Sustainable Tourism*, 23 (6), pp. 838–858.

Ateljevic, I., Pritchard, A., and Morgan, N. (2012) *The Critical Turn in Tourism Studies: Promoting an Academy of Hope*. Oxford: Elsevier.

Britton, S. (1991) Tourism, capital, and place: towards a critical geography of tourism. *Environment and Planning D: Society and Space*, 9 (4), pp. 451–478.

Font, X., Walmsley, A., Cogotti, S., McCombes, L., and Häusler, N. (2012) Corporate social responsibility: The disclosure–performance gap. *Tourism Management*, 33 (6), pp. 1544–1553.

Franklin, A. (2004) Tourism as an ordering: towards a new ontology of tourism. *Tourist Studies*, 4 (3), pp. 277–301.

Gössling, S. (2003) *Tourism and Development in Tropical Islands: Political Ecology Perspectives*. Cheltenham: Edward Elgar Publishing.

Gössling, S. and Hall, C.M. (2006) Conclusion: Wake Up . . . This Is Serious. In: Gössling, S. and Hall, C.M. (eds) *Tourism and Global Environmental Change: Ecological, Social, Economic and Political Interrelationships*. New York: Routledge.

Hall, C. M. (1996) *Tourism and Politics: Policy, Power and Place*. Chichester: Wiley.

Hartman, F. (1998) Towards a Social Ecological Politics of Sustainability. In: Keil, R. Bell, D.V.J., Penz, P., and Fawcett, L. (eds) *Political Ecology: Global and Local*. London: Routledge.

Higgins-Desbiolles, F. (2005) More than an 'industry': The forgotten power of tourism as a social force. *Tourism Management*, 27 (6), pp. 1192–1208.

Hollinshead, K. (1999) Surveillance of the worlds of tourism: Foucault and the eye-of-power, *Tourism Management*, 20(1), pp. 7–23.

Hollinshead, K. (2012) The Under-Conceptualisations of Tourism Studies: The Case for Postdisciplinary Knowing. In: Ateljevic, I., Pritchard, A., and Morgan, N. (eds) *The Critical Turn in Tourism Studies: Promoting an Academy of Hope*. Oxford: Elsevier, pp. 55–72.

Hollinshead, K. (2014) Time for Conceptual Unsettlement: A Call for Improved Tourism Studies Conversations with the Transitionalising World. In: Chien, P.M (ed.) *CAUTHE 2014: Tourism and Hospitality in the Contemporary World: Trends, Changes and Complexity*. Brisbane: School of Tourism, The University of Queensland, pp. 893–896.

Huggan, G. and Norum, R. (2015) Introduction: The Postcolonial Arctic. *Moving Worlds: A Journal of Transcultural Writings*, 14 (2).

Jamal, T. and Camargo, B. A. (2014) Sustainable tourism, justice and an ethic of care: Toward the just destination. *Journal of Sustainable Tourism*, 22 (1), pp. 11–30.

Kristoffersen, B., Norum, R., and Kramvig, B. (2015) An Ecotourism Ethics in the Anthropocene Arctic. In: Huijbens, E. and Gren, M. (eds) *Tourism & the Anthropocene*. London: Routledge.

McKibben, B. (2006) *The End of Nature*. New York: Random House.

Merchant, C. (1994) *Ecology: Key Concepts in Critical Theory*. Atlantic Highlands, NJ: Humanities Press.

Moskwa, E., Higgins-Desbiolles, F., and Gifford, S. (2015) Sustainability through food and conversation: The role of an entrepreneurial restaurateur in fostering engagement with sustainable development issues. *Journal of Sustainable Tourism*, 23 (1), pp. 126–145.

O'Connor, J. (1988). Capitalism, nature, socialism: A theoretical introduction. *Capitalism, Nature, Socialism*, 1 (1), pp. 11–38.

Peters, M.A., Hamilton, D., and Eames, C. (2015) Action on the ground: A review of community environmental groups' restoration objectives, activities and partnerships in New Zealand. *New Zealand Journal of Ecology*, 39 (2), pp. 179–188.

Robbins, P. (2012) *Political Ecology: A Critical Introduction*. Chichester: Wiley-Blackwell.

Rosenthal, E. (2007) Changing climate haunting tourism. *The New York Times*, 30 October. Available at: www.nytimes.com/2007/10/30/business/worldbusiness/30iht-tourism.4.8118330.html (Accessed 24 September 2015.)

Stonich, S. (2000) *The Other Side of Paradise: Tourism, Conservation and Development in the Bay Islands*. New York: Cognizant Communication Corporation.

Takeda, L. (2015) *Islands' Spirit Rising: Reclaiming the Forests of Haida Gwaii*. University of Vancouver: British Columbia Press.

Tribe, J. (2004) 3 Knowing about tourism. *Qualitative Research in Tourism: Ontologies, Epistemologies and Methodologies*, 46.

Warren, K. (1990) The power and promise of ecological feminism, *Env. Ethics*, 17 (2), pp. 125–146.

Williams, R. (1983) *Culture and Society, 1780–1950*. New York: Columbia University Press.

Afterword

James Igoe

Getting in touch with uneven development's cultural side

I would like to finish this volume as Rosaleen Duffy began, with the assertion that this is a very important book. It is important for tourism studies no doubt, and a host of other disciplines besides. More importantly, in my opinion, it calls attention to the crucial importance of tourism as a key realm of expansion, and thus potential contestation, in contemporary culture(s) of capitalism. My goal in this afterword is to highlight that importance.

I would like to begin by revisiting a point with which the editors of this volume open their introduction: a rapidly expanding and intensifying tourist industry is a major mediator of capitalist contradictions to which it is also a major contributor. This, we might say, is a contradiction of contradictions. Recall the stunning example of the Maldives, where revenues from an emerging luxury tourist industry are being earmarked for relocations of Maldivian climate refugees, as their island homes become submerged by projected rising sea levels in the coming century. Tourist travel is of course a contributing cause of the oceanic shifts that are threatening island nations such as the Maldives.

Consider this example in relation to Henri Lefebvre's (1976, p. 21) crucial insight that capitalism succeeds in achieving growth not by resolving its myriad contradictions, but by attenuating them through the production of space. Once the Maldives disappear – according to this logic – governments, tour operators, and tourists can set their sights on new island destinations. Tropical islands will continue to become scarce, but perhaps some new ones will appear as pieces of coastline become separated from main lands. For that matter, islands in more temperate climes may well become more tropical. The continuous movement of capital to take advantage of these transformations, while circumventing their ill effects, figures centrally in Neil Smith's (1984) elaboration of Lefebvre's insights in formulations of *Uneven Development*.

In my work *Spectacle of Nature, Spirit of Capitalism,* I argue for the importance of attending to uneven development's 'cultural side'. If, to paraphrase Clifford Geertz (1973, p. 448), culture consists of stories we tell ourselves about ourselves, then uneven development's cultural side relates to stories that we tell ourselves about ourselves in relation to impending crises of capitalism. But some qualifications need to be made. For instance who is the 'we' that is telling these stories? When Nobel laureate Paul Crutzen proclaims, 'it is *we* who decide what nature is and what it will be', is he really speaking of the whole of humanity?[1] Perhaps on some level, but almost certainly the 'we' who make these decisions will be 'we' experts. 'We' consumers are invited to play a much more limited role by purchasing certified products and making charitable contributions. And while 'we' consumers are certainly telling ourselves stories about ourselves, those stories eddy and swirl in a maelstrom of mass-produced and disseminated stories. These stories hail 'us' as a particular 'we', defining those who are 'not we' into the bargain.

For many readers these arguments will be familiar enough, but they are still only part of the story. These (hi)stories are not only told in the abstract, but appear in the form of fantastic images that seem to mimic the realities they portray. Not only this, but such images are intertwined with circuits of space, intentionally designed – to varying degrees – to eschew aspects of reality that are confusing or frightening, while simultaneously highlighting – in fact often embellishing – aspects that appear certain, comforting, and generally pleasant.

These spatial circuits are attenuated – in the sense of being stretched out and thin – connecting and excluding in insidious ways. They form connections at global scales, while passing through, over, under, and around, people and places that do not fit with their vision or suit their purposes. These are of course the very networks of communication, transportation, and recreation on which global capitalism depends. They are the means of attenuating capitalist contradictions – in the sense of rendering them less immediately viral – while also smoothing them over with pleasantly distracting experiences.

As a growing global industry, dedicated to managing people's experiences of space and place, tourism obviously has a major part to play in these ubiquitous arrangements. I will now elaborate these points through a series of theoretical sketches. These are necessarily brief, and I offer them as guiding propositions rather than definitive statements. These are derived from longer versions in *Spectacle of Nature, Spirit of Capitalism*, but I offer them here to underline the important contributions of this volume, and to sketch some possibilities for future research.

Tourism and government

In an essay written at the height of the First World War, W.E.B. Dubois (1915, p. 712) opined:

> A white man is privileged to go to any land where advantage beckons and behave as he pleases; the black or colored man is being more and more confined to those parts of the world where life . . . is most difficult to live and most easily dominated by Europe.

While clearly glossing over important nuance and complexity, Dubois's statement nevertheless appears remarkably prescient in the present day. Both Europe and

North America are currently in the midst of so-called 'migration crises'. Dark-skinned people perish in the deserts of northern Mexico and the waters of the Mediterranean. Tourists from the United States jet over the contested and dangerous border zone to the resorts of Acapulco and Cancun, and tourists from Europe take in Mediterranean cruises, often en route to North African destinations. Tourists and migrants traverse the same spaces, but still inhabiting essentially different worlds. And while the color of tourists has certainly changed over time, it is still the case that tourism continues to mediate relationships between privileged people and less privileged ones, and those mediations are consistently racialized. Tourism, in other words, is a significant element in the production of social classes and their putative boundaries.

Of course realities are always far more complex than this, but part of what mass tourism does is to cut through complexity with simplified narratives and standardized experiences.

Histories of mass tourism, as intertwined with the histories of conservation and development, turn on remarkable refinements of techniques and technologies for the production of standardized experiences. The ultimate expression of such arrangements is almost certainly Disney Parks, with themed spaces carefully orchestrated to tell people particular stories about themselves (Bryman, 2004; Wilson, 1991). Evidence of productive labor, conflict, waste, and other unpleasantries are carefully relegated to a labyrinth of hidden tunnels and passageways. Service workers who are out of character for whatever reason are also required to withdraw to these tunnels. Encounters between consumers and workers are carefully managed according to standard fantasies.

I would venture that Disney techniques for the production of human experiences are as significant to twenty-first-century capitalism as Fordist production lines were to the previous era of twentieth-century capitalism. In Fordist production, workers stood still and added pieces to commodities on a moving assembly line. In Disneyist production, consumers are conveyed through space to be transformed into commodity-subjects. In the process they are all encouraged – sometimes forced – to see and feel the world in particular ways. I imagine circuits of global tourism as an attenuated version of Disney Parks. In Disney Parks, featured elements of reality are presented in relatively large expanses of themed space, while concealed elements are relegated to narrow corridors. In global tourism, featured elements are often relegated to narrow corridors of transport infrastructure, small nodes of hotels, resorts, and shopping centers, and occasionally large nodes, like parks – from which wider realities can be excluded with relative ease.

My use of a conveyor belt metaphor here is intended – among other things – to invoke the idea of a 'culture industry', popularized by the Frankfurt School and the Situationists in the decades following the Second World War. Contemporary critics of the 'culture industry' idea note that it was premised on scenarios of mass conformity and standardized modes of consumption that never came to pass. Instead, these critics opine, consumers became entrepreneurial producers of themselves and their own consumptive experiences (Nealon, 2008, pp. 66–67; Vrasti, 2013, pp. 28–29; also cf. Büscher and Igoe, 2013). While I am mostly in agreement with these arguments, I also maintain that it remains important to attend to the potentially unifying effects of 'culture industry' in certain contexts. In *Spectacle of Nature, Spirit of Capitalism*, I argue that Guy Debord's arguments about the mutually constitutive relationships of spectacular images and lived reality have special relevance for attenuated circuits of globalizing space.

Though it is beyond the scope of this short chapter, I would like to at least suggest that this has important implications for Foucauldian understandings of government. Here I am thinking of Foucault's (1982, p. 789) most basic definition of government as 'the conduct of conduct', achieved by structuring 'the possible field of action of others'. Based on this definition, Disney 'techniques' are a singular kind of government, designed as nearly as possible to offer singular 'fields of action'. It is notable that the kinds of spatial circuits I am describing here are designed to be similarly singular. There are only so many ways you can move through an airport, a resort, a freeway, or a national park. Thus, while government in many contexts is becoming multiple, lighter, and more virtual (Nealon, op. cit.) in certain key contexts it remains singular in actual space. While racialized and gendered touristic subjects may still 'go to any land where advantage beckons' their perceptions of that land, and what is possible and plausible to do there, may well be heavily mediated by touristic circuits. And there will always be divisions between tourists, whoever they are, and the workers who facilitate their touristic experiences – not to mention those who are left out of the picture. This applies not only to regular tourists, but also to 'rural development tourists'.

Tourism and development

Tourism is now widely regarded as an excellent engine of economic development, particularly sustainable economic development. But it is important to remember that development itself has long been a touristic enterprise. Consider, for instance, the potential productive synergies between my discussions of Disney above and Raymond Apthorpe's (2013) formulations of Aidland. With an explicit nod to Lewis Carroll's *Alice in Wonderland,* Apthorpe characterizes Aidland as a fantasy world, where things are always getting 'curiouser and curiouser'. Of course Disneyland is also a fantasy world, indeed one that incorporates wonderland themes.

But things go much further than this broad similarity, particularly when we consider the ways in which the movements and experiences of development workers are organized in space. In thinking about Aidland as a spatial – theme park-like – phenomenon, I imagine it as roughly divided into two realms. The first is what Robert Chambers (1983, pp. 7–8) calls 'the urban trap': the everyday urban spaces where development workers spend the bulk of their time. The other can be called 'the field' for lack of a better term, the various (usually) rural spaces where development is imagined to get done. This bipartite arrangement is the basis of Chamber's (1983, pp. 10–11) seminal claim that development work is organized through 'rural development tourism' – 'brief rural visits by "core" personnel [that] can scarcely fail to play a part in forming their impressions and beliefs . . . '

Anyone who has worked in development knows that life in the 'urban' sector of Aidland is organized around particular spaces: office compounds, malls, restaurants, resorts, international schools, swimming pools, and of course expatriate neighborhoods. 'Expatria' is a fitting name for this part of Aidland. Almost every so-called developing country has its own Expatria. As the work of Heather Hindman (2013) aptly demonstrates, Expatria is a translocal phenomenon. Borrowing heavily from tourism, Expatria consists of circuited networks of more or less standardized spaces. This makes it possible for a development worker to relocate from Sierra Leone to Nepal or Honduras without being hugely burdened with uncomfortable processes of cultural readjustment.

Even when they leave these spaces for 'the field', however, the experiences and perceptions of development workers will remain profoundly shaped by infrastructure, built environments, and orchestrated encounters. A key work in this area is Lisa Smirl's (2015) *Spaces of Aid,* which engages the ways that NGO compounds, and even vehicles, shape the experiences and practice of development work.[2] Perhaps Chambers's (op. cit., p. 16–17) most enduring contribution to this literature is his analysis of 'showpiece projects' – 'islands of atypical activity, which attract repeated and mutually reinforcing attention'. Showpiece projects, in other words, are famous for being successful and successful for being famous. Ever-increasing numbers of visitors are managed in the fashion of tourists – a fluent guide follows a standard route and a standard routine. The same people are met, the same books signed, the same polite praise inscribed in visitor books [or these days tweeted into cyberspace with selfies of the visitors posing with project staff and beneficiaries].

These insights – important in and of themselves – are especially relevant to contexts in which regular tourism and rural development tourism have become mutually constituting. Such contexts are of course increasingly common. As a celebrated engine of rural development, tourism is both part of the development story and the means by which development encounters are organized. Visitors to showpiece projects – from student volunteers to visiting dignitaries – combine business with pleasure. They learn about the importance of wildlife to the local economy by going on a game drive in a park or conservatory. They learn about tourism and women's livelihoods and participate in cultural performances, such as dancing with Maasai women. Distinctions between the experience, and the putative benefits of the experience to local people, are thereby blurred.

The key point here, I think, is that showpiece projects are crucial sites of representation and knowledge production. Their incorporation of touristic mediations means that their representations can be (a) relatively standardized, and (b) unprecedentedly elaborate. Adding social media and volunteers to this mix means that these representations are also (a) seemingly personalized, and (b) able to circulate well beyond the sites of their original production. As such they become available for a wide variety of diverse projects, from marketing soft drinks and computers to the conservation of nature.

Tourism and nature conservation

Western constructions of nature began with tours and pictures. In his groundbreaking work, *The Country and the City,* Raymond Williams (1975, p. 122) outlines how eighteenth-century British aristocrats embarked on 'Grand Tours' to continental Europe. In the course of these tours they collected landscape paintings by celebrated artists, which became 'prospects' for their country estates. British elites transformed the material landscapes in an effort to make them conform to the (admittedly very expensive) tourist souvenir.

In reading Williams description of how these transformations were achieved, and particularly artful concealments of infrastructure, the similarities between the creation of 'nature' landscapes and the techniques of Disney theme parks are stunning. And indeed the 'nature' of these country estates became an abstract ideal, enshrined in nineteenth-century American national parks and twentieth-century Disney nature films.

In the years immediately following the Second World War, an Austrian conservationist named Bernhard Grzimek took up residency in Serengeti National Park. Once in the park, he and his son Michael produced an extravagant nature film called *Serengeti Shall Not Die!* The film became an international blockbuster and was followed in the 1960s by a German television series called *A Place for Wild Animals.* Grzimek used revenues from these media enterprises to establish modern conservation in East Africa. He also used commoditized images to entice western tourists to visit East Africa to see the wonders of nature with their own eyes. Rising European demand for African Nature tours – initially exaggerated by Grzimek – convinced leaders of newly independent African countries to keep the parks and conservation areas established during European colonialism (see Lekan, 2011).

The Grzimeks, to put it succinctly, were pioneers in promoting images and tourism in the service of conservation, and sometimes nature conservation in the service of images and tourism. Today this combination has been refined in remarkable ways, such that entertainment and government, conservation, and development have become indistinguishable enterprises. In *Spectacle of Nature, Spirit of Capitalism*, I describe how images of people on tour in Tanzania's Northern Tourist Circuit are used to promote not only Tanzania's economy and nature conservation, but NGO fundraising, US foreign policy and security interests, corporate social responsibility, entrepreneurial philanthropy, celebrity brands, and luxury watches, and the list could still be expanded.

Let us return now to the point with which I opened this chapter, which is tourism as a major mediator of capitalist contradictions. In the conceptual framework with which I am operating here, the main role of conservation and development workers is to respond to local expressions of capitalist contradictions as they occur. A central assumption of these endeavors is that it will be possible to 'scale up' local successes to solve socio-environmental problems at national and global scales. Touristic spaces – much like parks and showpiece projects – make it possible to imagine that contradictions have been resolved (rather than attenuated) and that scaling up is working.

Of course there is mounting evidence to the contrary, as indicated by the rise of what Büscher and Davidov (2013) call the Ecotourism-Extraction Nexus. As capitalism's environmental contradictions become increasingly difficult to attenuate in space, they are being incorporated into touristic mediations, in which images and simulations of disappearing nature still seem to offset environmental harm. A key question here, of course, is how far can this go? When no new tropical island destinations are available, and existing ones are overrun, will spectacular simulations of island vacations suffice? How long till environmental contradictions become too plug ugly and downright scary to spruce up with touristic mediations? What kinds of limits may exist for touristically mediated attenuations of capitalist contradictions and what alternatives may be possible?

Tourism, memory, and imagination

I realize that this short closing chapter cannot do justice to the breadth and nuance of ideas and insights presented in this volume. However, I do hope to have highlighted some key areas of importance and future enquiry. I would like to close with one final point, and this has to do with memory and the imagination. George Orwell has famously written, 'He who controls the past controls the future, he who controls the future controls the past'. To restate this just a little differently, human memory and imagination are necessarily intertwined. What we remember of the past interacts reciprocally with what we imagine will be possible in the future, but always in a continuously unfolding present.

In my future work, I hope to engage with the political ecology of memory and imagination, which of course overlaps significantly with the political ecology of tourism. Understanding memory and imagination as ecological processes means that they cannot be reduced to objects residing in the heads of individuals. Memory and imagination are collective, emergent properties of relational interactions between human beings. They are also environmental, emergent properties of relational interactions between human beings and other-than-human entities. Memory and imagination reside between and beyond us.

Tourism is an industry dedicated to producing particular kinds of memories and imaginaries, by controlling and shaping interactions between human beings and between human beings and space. While it is always risky to generalize, it is fair to say that there are consistent patterns to these touristic mediations. First of all they divide the human world into consumers and performers. The former generally produce memories and imaginaries for the sake of the latter. There all also others, who are systematically hidden from view, either because their work produces the memories and imaginaries, but also violates them, or simply because there is no place for them in the spaces of tourism. Even in the case of community-based tourism, I dare say there is always pressure to produce memories and imaginaries that match consumer desires. I also dare say there are always people excluded.

Of course none of this is to say that tourism is monolithic, uncontested, or without transformative possibility. After all, tourism is all about the human imagination, and human imagination is an essential ingredient to the realization of alternative futures. It stands to reason, therefore, that techniques and technologies of tourism may have significant potential for imagining and making alternative worlds. In helping to move us past ideologically driven understandings of tourism, as an unassailably good thing or an irredeemably bad thing, this collection is a welcome step in that direction.

Notes

1 Emphasis mine. Atmospheric chemist Paul Crutzen is most famous for introducing and popularizing the idea of a new geological epoch called the *Anthropocene* http://e360.yale.edu/feature/living_in_the_anthropocene_toward_a_new_global_ethos/2363, accessed August 30, 2015.
2 In 2013, Lisa Smirl's life was cut short by cancer at the age of 37. The website designated to her life and intellectual legacy is well worth exploring. Though left in its formative stages, there is much in this archive on which future scholars may build. I especially recommend the work on 'drive-by development' and the role of SUVs. https://spacesofaid.wordpress.com/category/objects/suv/, accessed August 31, 2015.

References

Apthorpe, Raymond (2013) Coda: With Alice in Aidland: a Seriously Satirical Allegory. In D. Mosse (ed.) *Adventures in Aidland: The Anthropology of Professionals in International Development,* London: Berghahn Books.
Chambers, Robert (1983) *Rural Development:Putting the Last First.* Essex: Longman.
Bryman, Alan (2004) *The Disneyization of Society.* London: Sage Publications.
Büscher, Bram and Veronica Davidov (2013) *The Eco-Tourism-Extraction Nexus: Political Economies and Rural Realities of (Un)Comfortable Bedfellows.* New York: Routledge.
Büscher, Bram and James Igoe (2013) Prosuming Conservation: Web 2.0, Nature, and the Intensification of Value Producing Labour in Late Capitalism. *Journal of Consumer Culture* 13 (3), pp. 283–305.
Dubois, W.E.B. (1915) The African Roots of War, *Atlantic Monthly,* May.
Foucault, Michel (1982) The Subject and Power. *Critical Enquiry* 8 (4), pp. 777–795.
Geertz, Clifford (1973) *The Interpretation of Culture.* New York: Basic Books.
Hindman, Heather (2013) *Mediating the Global: Expatria's Causes and Consequences.* Stanford, CA: Stanford University Press.
Igoe, James (forthcoming) *Spectacle of Nature, Spirit of Capitalism.* [In press.]
Lefebvre, Henri (1976) *The Survival of Capitalism: Reproduction of the Relations of Production.* New York: Allison Busby.
Lekan, Thomas (2011) Serengeti Shall Not Die: Bernhard Grzimek, Wildlife Film, and the Making of a Tourist Landscape in East Africa, *German History*, 29 (2), pp. 224–264.
Nealon, Jeffry (2008) *Foucault Beyond Foucault: Power and its Intensifications Since 1984.* Stanford, CA: Stanford University Press.
Smirl, Lisa (2015) *Spaces of Aid: How Compounds, Hotels, and Cars Shape Humanitarianism.* London: Zeb Books.
Smith, Neil (1984) *Uneven Development: Nature, Capital, and the Production of Space.* Athens, GA: University of Georgia Press.
Vrasti, Wanda (2013) *Volunteer Tourism in the Global South: Giving Back in Neoliberal Times.* London: Routledge.
Williams, Raymond (1975) *The Country and the City.* London: Oxford University Press.
Wilson, Alexander (1991) *The Culture of Nature: North American Landscapes from Disney to the Exxon-Valdez.* Toronto: Between the Lines.

Index

Abbey, Edward 190
Abbott, D. 275
Aboriginal Australians 29, 50, 53, 55–66, 195–196
Acahualinca 232, 240, 242
activism 5, 42–43, 51, 100, 105; *see also* environmentalism
Actor-Network Theory (ANT) 285, 286–289, 295, 298
Adams, W. M. 6, 26
'aesthetic governmentality' 210, 233, 236, 243–244
aesthetics 232, 233, 235, 242, 245
Africa 73–75, 200, 216, 295, 314
Agamben, G. 141
agency 287, 288, 295, 303
Agrawal, A. 235
agroforestry 100
Ahlers, R. 35
Aidland 312–313
Alemán, Arnoldo 237
'alternative tourism' 199
Anaya, James 144
Anderson, B. 171
animals 14, 194
Anthropocene 286, 287, 289, 304, 316n1
anthropocentrism 195, 196, 199, 200, 201
antiessentialism 153, 157–158, 162, 163
'anti-politics' 183–184
apartheid 74, 87
Apthorpe, Raymond 312
Ashcroft, B. 74
Ashley, C. 73, 216
Asia 194, 200
assemblages 171, 176, 178–179, 182–184, 235, 287

Ateljevic, I. 306
Australia: ecotourism 191, 192–193; green politics 188–189; Ngarrindjeri Aboriginal community 29, 50, 53, 55–66; protected areas 195–196; threats to jobs 200; water 35; whale watching 193

Bailey, S. 11, 251
Baillie, G. 111
Bakhtin, Mikhail 108, 109, 112–114, 120, 124, 125n2, 125n3
Bali 9, 27, 28, 36–38, 40–44
Ban Ki-moon 218
Barke M. 275
Bart, F. 252
Baudrillard, J. 146
Baum, T. 33
Bauman, Z. 162
Bay Islands 9, 209, 271
Beakhurst, Graham 5
Beeton, S. 190, 191
benefit sharing 219, 220–225
Benjaminsen, T. A. 254, 255, 262
Benton, Ted 6
Berg, Peter 54
Berkes, F. 73
Berry, Thomas 55
Bhaktivedanta, A. C. 98
Bhatt, Narayan 94–95
Biermann, F. M. 76
biodiversity 5, 8, 10, 132; deep ecology 188; ecotourism 194–195; India 93, 104; neoliberalism 75
biopolitics 132, 134, 141, 144, 145
biopower 143

318 *Index*

Bird, N. 213
Bjørst, Lill Rastad 210–211, 284–301
Blaikie, P. 4–5, 12, 189, 215
bodily performances 109, 110, 112
Bohm, David 54
Boonabaana, B. 34
Borrini-Feyerabend, G. 73
boundaries, in protected areas 263–265, 266
Boykoff, M. T. 17
Brandist, C. 114
Brazil 245, 272, 275
Briassoulos, H. 52
Brin, E. 275
Britton, S. G. 255
Brockington, D. 8, 10, 75–76, 131, 254
Brookfield, H. 4–5, 12, 189, 215
Brown, J. 213
Brundtland Commission 16, 81–82, 156
Bryant, R. L. 11, 51, 251
Bryceson, I. 255
Buddhism 160, 165n5
Buechler, S. 35
Büscher, Bram 74, 76, 314
Buscher, M. 273, 274

Calma, Tom 61, 63
Camp Coorong 50, 58, 64–65
Campbell, L. M. 13
Campbell, Petra 144
CAMPFIRE 85
Cao, L. 156–157
capital accumulation 170, 181, 183, 184, 190, 192, 255
capitalism 1, 8, 13, 44, 62, 190, 191; capitalist nature 158–159, 160, 161, 162, 163; carbon 17; conservation economy 131, 132; contradictions of 309, 310, 314–315; Disneyist production 311; ecotourism 184; environmental imperatives 196–197; Rapa Nui 139; slum tourism 270; unsustainability of 305
Caribbean 194
Carlsson, L. 73
carnivalesque 108, 110, 112–114, 120, 124, 125n2
Carrier, J. G. 176
Carson, Rachel 188

caste 96, 98, 271
Ceballos-Lascuráin, H. 16
celebrities 111–112, 121–122
Central America 194
Chachage, C. S. L. 258, 259, 262
Chaderopa, Chengeto 29, 70–91
Chagga people 261
Chambers, Robert 312, 313
Chant, S. 39
Chavez, D. 82
Cheung, C. 33
Chile 134–150
China 132, 151–168, 184, 195
Chouliaraki, L. 110, 111
Christianity 120, 121
Christiansen, Lene Bull 108–127
chronotope 108, 109, 113, 119, 124
churequeros 239–240
citizenship 133, 188, 200; *see also* eco-citizenship
class 7, 8, 11, 33, 44, 304
Clean Out Loud 119
climate change 1, 9, 16–17, 202, 304, 305; Fiji 302; Greenland 210–211, 289; India 95–96; Maldives 309; minorities 44; Murray-Darling River Basin 60; water issues 35
Cockburn, Alexander 5
Cole, J. W. 137
Cole, Stroma 9, 27, 28–29, 31–49
College of African Wildlife Management (CAWM) 253, 263
colonialism 6, 7, 12, 27, 53–54, 74, 271, 314; Africa 216; Australia 29, 56; feminist perspective 33; internal 256; 'logic of elimination' 145; Rapa Nui 139; Tanzania 253
co-management 70, 73–88
commodification 163, 194, 216
Commoner, Barry 5
community 8, 12–13, 15, 25–29, 190–191, 200, 298; benefit sharing 219, 220–225; co-management 76, 80–81, 85–87; community development 190; conservation and control thesis 131; disempowerment of local communities 51, 87–88; ecotourism 195, 197, 198, 200; enclave tourism 255–256; Ngarrindjeri Aboriginal community

50, 53, 55–66; religious festivals 120; Roskilde Festival 109, 123–124; slum tourism 277–278; sovereignty 52; undermining of local 52–53
community associations 41
Community-Based Natural Resource Management (CBNRM) 213, 217, 219
community-based tourism (CBT) 12, 213, 216–217, 218, 219; ecotourism 189, 190, 191, 194–195, 198, 200; memories and imaginaries 315; South Africa 85
conflict 11, 16, 31, 51, 71, 209, 211, 234; Kilimanjaro National Park 260, 264; Ngarrindjeri Aboriginal community 56; Rapa Nui 137–138; water resources 42–43
Conran, M. 276
conservancies 219
conservation 6, 8, 11, 15–16, 71, 131–132, 170, 189, 313–315; China 155; cities 233; Community-Based Natural Resource Management 213; conservation economy 13–14; ecotourism 192, 193, 198, 200; 'fortress' 254, 263; governmentality 235–236; Namibia 210, 219; neoliberal 2, 75, 189–190, 191, 194–195, 197–198, 201, 202; Nicaragua 245; power relations 235, 265–266; protected areas 195; Rapa Nui 134; religiously motivated 93, 94, 95, 100–103, 104–105; sea turtles 196; Shangri-La 153, 156, 157; South Africa 72–73, 74, 80, 81–82, 83, 87; Tanzania 252, 253–254, 256–257, 260, 262–263, 265; technoscience 162; unintended consequences of 234
consumption 170, 176, 182–183, 192–195, 209, 311
Convention on International Trade in Endangered Species (CITES) 81
Cooper, M. 177
Coorong Wilderness Lodge (CWL) 50, 64–65
corporate social responsibility (CSR) 52, 218, 305, 314
Cosgrove, D. 256
Costa Rica 231, 237, 245; ecotourism 13, 191, 192, 195, 196, 198; gendered political ecology 28, 36, 38–39, 40–44; local community benefits of tourism 201–202
Crenshaw, Kimberle 33
Crippa, L. A. 138
Crutzen, Paul 310, 316n1
cultural flows 50, 63–64, 66n1
cultural globalization 215–216
cultural imperialism 54
'cultural turn' 31
culture 164, 177, 304, 309
'culture industry' 311–312
Cunningham, Eric J. 133, 169–187

DanChurchAid 119–123
D'Andrea, A. 273
Dasmann, Raymond 54
Davidov, Veronica 314
Dávila-Poblete, S. 34
Debord, Guy 172, 183, 312
decision-making 76, 77, 82–83, 85, 217
decolonization 54
deep ecology 188–189, 191, 194–195, 196, 201, 202, 203n2
deforestation: China 154, 155, 163; Guatemala 198; India 103; Japan 175
Deleuze, Gilles 54–55, 184
Delhi 244
Denmark 29, 108–110, 114–124
Dentith, S. 111, 113
depoliticization 97, 110, 211, 285, 286
Derrida, Jacques 141
deterritorialization 54–55
development 14, 16–17, 209–211, 215–217, 305, 312–313; co-management 87; Greenland 284–285, 292, 298; Namibia 218–225; Nicaragua 232, 241, 242, 244; poverty alleviation 217–218; Shangri-La 153; Tanzania 251, 257, 260; technoscience 162; unsustainable 40; *see also* economic development; sustainable development
Devine, J. 198
Diamond, J. 200
Dickens, Charles 272
Diekmann, Anya 211, 270–283
discourse 10, 28, 163–164, 235; 'aesthetic governmentality' 236; images and discourses of nature 170, 171–173, 179,

182–183; Live Clean, Live Healthy campaign 242; protected areas 195–196
'discursive ecological formations' 12
Disney 311, 312, 314
dispossession: Ngarrindjeri Aboriginal community 56, 61; Rapa Nui 135; Tanzania 255, 260, 263, 265
dominant cultures 256
Dominican Republic 194
Douglas, J. A. 8, 11, 12
Douglas, Mary 271
Dredge, Dianne 210–211, 284–301
Dressler, W. 76
drought 60, 61, 66, 210, 213, 223
Dubois, W. E. B. 310
Duffy, Rosaleen 8, 10, 14, 75, 77, 309
Durrant, M. B. 263

Earth Day (1970) 5
Easter Island (Rapa Nui) 132, 134–150
Echuca Declaration 64, 67n4
ecocentrism 195, 196, 199
eco-citizenship 188, 190, 191, 201
'ecological ethnicity' 58, 62
economic development 209, 303, 312; China 154, 156; co-management 76; ecotourism 192, 195, 198; Greenland 291, 292, 293; Namibia 218–219; neoliberal policies 32; Ngarrindjeri Aboriginal community 66; religious tourism in India 97
economic growth 44, 59, 136, 225, 256, 292
eco-politics 188–189
ecosophy 53
ecosystems 188, 189, 194–195, 202
ecotourism 8, 13–14, 16, 26, 162, 188–206, 210; Australia 64; consumption of nature 192–195; definition of 188; Ecotourism-Extraction Nexus 314; images and discourses of nature 171–173; indigenous tourism 27; Japan 133, 169, 170–173, 175–184; neoliberalism 75–76, 77, 197–198; protected areas 195–196, 199; Shangri-La 132, 152–155, 157, 161, 164; South Africa 72–73, 78–80
eco-utilitarianism 189, 190, 194, 196, 199, 200
Edmunds, David 7–8

education 243, 244
Egypt 146
elites 51, 81–82, 97, 104, 210, 237, 260
Elliot, J. 73
empathy 109, 110, 125
employment: Bali 37; Costa Rica 39; ecotourism 193; Greenland 292; Japan 182; Makuleke community 85; Namibia 210, 214; Nicaragua 245; Rapa Nui 136; Tanzania 251; threats to jobs 200
empowerment 64, 87
enclave tourism 255–256, 259, 265, 266
environmental degradation 5, 11, 26, 170, 211, 234, 305; cities 233; crisis of sustainability 304; driving forces of 246; marginalization 132; Murray-Darling River Basin 61; Namibia 224; Nicaragua 231; Rapa Nui 144; religious tourism in India 95–96, 97, 98–99, 105; Roskilde Festival 110; Shangri-La 155
'environmental discourse' 163–164
environmental non-governmental organizations (ENGOs) 26, 100, 101–103
environmentalism 198, 199; celebratory 109–110; religious tourism in India 93, 94, 95, 100, 103, 104; Roskilde Festival 109, 117–119, 124
Equitable Tourism Options (EQUATIONS) 51
equity 51, 192, 305
Escobar, A. 132, 153, 158–163, 191
ethics 133, 190, 194; of care 202; ethic of nature/ethic of use 201; slum tourism 273, 279
ethnicity 7, 8, 44; 'ecological' 58, 62; intersectionality 33, 42; Japan 177; power struggles 11; *see also* race
ethnography 273
Europe: green politics 188–189; migration 310–311; whale watching 194
exclusion 11, 71, 210, 215, 254, 260, 265
'Expatria' 313

fair trade 121–123
Faleomavaega, Eni 144
Faria, C. 33
favelas 272, 275

Index 321

feminist perspective 7–8, 32–33, 35
Ferguson, L. 32, 42
festival eco-voluntourism 29, 108–112, 116–117, 119–125
Fiji 1, 145, 302
Finland 191, 197
Fisher, Josh 210, 231–250
'flaneurism' 272
Fletcher, R. 13, 132, 164
flooding 155
foreign-owned businesses 258–260, 266
forest bathing 177–179
forests: Japan 169–170, 174–175, 177–179; Nicaragua 231; religious tourism in India 94–95, 97, 100, 102–103; Shangri-La 154–155; Tanzania 254–255, 256, 260–262, 263
'fortress conservation' 254, 263
Foster, John Bellamy 5, 6
Foucault, Michel 12, 27–28, 141, 143, 145, 146, 159, 236, 312
Frangiolli, Francesco 217
Frankfurt School 304, 311
Freire-Meideros, Bianca 270
Frello, B. 112
Friedman, B. 59
Friends of the Earth 292
Funck, C. 177
futuring devices 295–296

garbage 231–233, 235, 236, 238–240, 241–243, 245
Garrett, Peter 193
gaze 158–159, 170, 235
Gebhardt, K. 221
Geertz, Clifford 309
gender 7–8, 11, 28, 31–36, 39–44, 304
Getz, D. 83–84
Gezon, L. L. 9
Ghertner, D. A. 233, 236, 243–244
Gill, Stephen 60
Giroux, H. & S. 59
globalization 55, 184, 303; cultural 215–216; economic 189, 190, 198, 201; neoliberal 59, 60, 135
'globally circulating environmental theories' 104
Godfrey-Smith, W. 200–201
Goodman, M. 17

Goodwin, H. 214
Gössling, Stefan 9, 31–32, 35, 209–210, 271, 302
Gough, S. 199
governance: aesthetic 246; co-management 83–84, 85; Live Clean, Live Healthy campaign 245; Ngarrindjeri Aboriginal community 55, 56, 62; self-organization 215; Shangri-La 157, 163; sustainable tourism 52
government 312
governmentality 27–28, 145, 146, 235–236, 302; 'aesthetic' 210, 233, 236, 243–244; 'green' 12, 16, 158, 159; Rapa Nui 132, 134, 143
Graburn, N. 273
Gray, B. 73
Greece 9
green economy 155–156, 157, 159
'green governmentality' 12, 16, 158, 159
green parties 188–189, 202
green tourism 177
Greenland 210–211, 284–301
Greenpeace 189, 292
Grzimek, Bernhard 314
Guatemala 198
Guattari, Felix 53, 54–55, 184

Haberman, D. 96–97
Hall, C. M. 302
Hall, V. 33
Hannam, Kevin 211, 270–283
Hanson, A-M. S. 35
Hardin, Garrett 26
Hardt, M. 143
Harrison, D. 216
Hartman, F. 305
Harvey, D. 191
Hathaway, Michael 184
Haun, Beverly 135
Hawai'i 51, 141, 145
health 211, 232, 241–242, 244, 279, 280
Hemming, S. 57
Heron, B. 111
Higgins-Desbiolles, Freya 50–69
Higham, J. E. 194
Hillman, B. 160
Hindman, Heather 313
Hindmarsh Island Bridge conflict 56, 57

Hinduism 92, 98, 103–104, 271
history 6–7
Hitorangi, Santi 145
Hodge, William 135
Holifield, R. 286, 287, 298
Hollinshead, K. 28, 164, 302
Holroyd, Megan 210, 251–269
Honduras 9, 26–27, 35, 271
Honey, M. 254
hotels: Rapa Nui 137; religious tourism in India 95; Tamarindo 38; Tanzania 258, 259
Hotu Matu'a 140–141, 145
Howe, L. 37
human-environment relations 6, 11, 13, 161, 163
human rights 144, 211; *see also* rights
humanitarianism 109, 111, 112, 115, 116
Hunt, C. A. 192
hunting 74, 78, 79, 81, 86, 253
Hutnyk, J. 273
Hutton, J. 26
Huxham, C. 80
hybrid natures 153, 162–163

identity 11, 12, 235; Bali 37; ecological 306; 'environmental discourse' 164; Roskilde Festival 108
Igoe, Jim 75–76, 172, 179, 183, 309–316
images 170, 171–173, 179, 182–183, 310, 312, 314
imaginaries 176–177, 179, 182, 183, 235, 315
India 51, 244; religious tourism 29, 92–107; slum tourism 211, 270–283
indigenous people 7, 27, 197; Finland 191; indigenous knowledge 51, 160; Ngarrindjeri Aboriginal community 29, 50, 53, 55–66; Ovahimba 213–214, 216, 219–225; protected areas 195–196; Rapa Nui 132, 134–135, 137–146
Indonesia 36–38, 40–44
inequalities 11, 13, 188, 215, 298, 305; Actor-Network Theory 286; gender 32, 34, 35; Global South 216; India 96; Namibia 214; slum tourism 211, 271, 272, 273, 274; Tanzania 260
infrastructure 40, 311; ecotourism 192, 193, 197; India 95, 96; Managua 242

interconnectivity 54
intersectionality 32–33, 35, 41–42
Inuit 195–196
investors 43–44
islands 9, 209–210, 309

Jacobsohn, M. 220
Jamal, T. B. 83–84
Jantzen, Connie Yilmaz 121, 122
Japan 133, 169–187, 192
Jenkin, G. 55
Jensen, C. 287, 295–296
Jolie, Angelina 111–112
Jung, H. Y. 110

Kali Yuga 98–99, 100
Kaokoland 210, 214, 219–225
Kaumaitotoya, Banuve 302
Kelly, William W. 174
Kenya 1–2
Khadka, S. R. 9
Khandelwal, B. 100
Kilimanjaro 210, 251–252, 253, 256–266
Kinnaird, V. 33
kinship 143–144
Kinzer, Stephen 232
Kiribati 1
Kiso 169–170, 172–183, 184
Knight, Catherine 176
Knight, John 175
Knight, R. 132
knowledge 10, 215; 'authorizing' 179; colonialism 54; community 25; indigenous 51, 160; power relationship 82
Krishna 93, 94, 97, 98, 99–101, 103, 105
Kruger National Park (KNP) 29, 70, 71–72
Kütting, G. 9

land: conflict over 211; India 97; land rights 29, 72, 188; Makuleke community 77–78; Ngarrindjeri Aboriginal community 57–58; Rapa Nui 137, 138, 140, 141; Tanzania 253
'last chance' tourism 1
Latour, B. 287
Law, John 288
Lefebvre, Henri 309
Leite, N. 273

Leopold, A. 189
Li, Tania M. 159, 171, 179, 182, 183, 209
Lindhardt, Thure 121–122
Lisle, D. 280
Live Clean, Live Healthy campaign 210, 232–233, 236, 240–245
Liverman, D. M. 75
Locke, John 201
logging 154–155, 174–175, 263
London 272
'losers' 3, 13, 16, 82, 173, 192
Lovelock, James 26, 189
Luke, T. W. 159
Lusseau, D. 194
Luthy, Tamara 29, 92–107

Maasai 1–2, 254, 313
Madagascar 9
Makuleke community 29, 70, 71–73, 77–88
Maldives 1, 309
Mallya, U. 258, 259, 262
Managua 210, 231, 232, 233, 236–245
Manandhar, P. 9
Mansfield, B. 75
Marcus, George 145
marginalization 11, 14, 16, 71, 132, 170, 214; cities 233; colonialism 54; development processes 215, 216, 217–218; driving forces of 246; Makuleke community 83, 88; Marxist political economy 234; Ngarrindjeri Aboriginal community 56; Nicaragua 245; Ovahimba 220; Rapa Nui 132; Tanzania 210
Mariki, S. B. 262, 263, 264
Marxism 4, 5–6, 159, 234, 286, 287
Mathews, A. 104
Maxwell, S. 216
Mbonile, D. J. 252
McFarlane, C. 171
media 17
memory 142, 315
migrants 37, 41–42
migration 6, 38, 39, 138, 310–311
Millennium Development Goals 35, 217, 218
mining 200, 218–219, 284–286, 288–289, 291–298

Misana, S. B. 261–262, 264
Mizayaki Hayao 176
moai 135, 140, 141, 146
mobile ethnography 273
modernization 157, 174
Mol, Anne-Marie 288, 298
Møller, Jesper Switzer 116–117, 124
Mollett, S. 33
'moral economy' 135, 146
Morgan, Nigel 9, 306
Mosse, D. 104
Mostafanezhad, Mary 1–21, 111–112, 209–211
Mount Kilimanjaro 210, 251–252, 253, 256–266
mountain sports 180
Mukupa, Karen 121–122
multinational corporations 51, 263
multiplier effect 256, 258, 260
multi-scalar approach 8, 35, 71, 134, 170, 173, 280
Mumbai 270, 271, 273–281
Muñoz, Juan Sabolvarro 238–239
Murillo, Rosario 240–241
Murray-Darling River Basin (MDRB) 50, 57, 59, 60–64, 66, 67n4
Myanmar 1

Naess, Arne 188, 189, 200, 202
Namibia 210, 213–214, 216, 218–225
Narsaq 289–293, 294, 296, 297–298
national parks 197, 200; American 314; co-management 70, 75, 76, 77–88; Guatemala 198; Rapa Nui 137; Shangri-La 157, 160–161; Tanzania 251–252, 253–258, 260–265; *see also* protected areas
nationhood 62, 65
nature 5, 304; Actor-Network Theory 287; antiessentialist political ecology 157–158; 'anti-politics' 184; capitalist 158–159, 160, 161, 162, 163; changing views of 199; conservation and control thesis 131; ecotourism's consumption of 192–195; 'end of' 302–303; 'environmental discourse' 164; ethic of 201; hybrid natures 153, 162–163; images and discourses of 170, 171–173, 179, 182–183; interactions with

163–164, 202; Japan 176, 178, 179; organic 159–161, 162, 163; religious tourism in India 94, 95; 'rights of' 133; Shangri-La 153, 154; status of 94; technonature 161–162, 163; values 200; Western conceptions of 74–75, 313–314
nature reserves 155, 157, 234–235; *see also* protected areas
Negri, Antonio 145–146
Nelson, F. 260
neoliberalism 4, 17, 32, 51, 52, 210; alternative discourse to 44; Australia 59; co-management 88; conservation economy 131–132; depoliticization 286; dominance of neoliberal paradigm 197, 202; ecotourism 191, 192, 197–198; feminist perspective 33; globalization 55; 'green governmentality' 158; impact on low-income communities 191; nature management 70; neoliberal conservation 2, 189–190, 191, 194–195, 197–198, 201, 202; Nicaragua 232, 238, 241; as political rationality 59–60; 'Pro-Poor Tourism' 225; Rapa Nui 134, 136, 139; self-organization 215; slum tourism 270, 280; South Africa 75–77; Tanzania 254
neo-Marxism 13
Nepal 9
Neumann, R. P. 234–235, 264–265
New York 272
New Zealand 13–14, 27, 188–189
Ngarrindjeri Aboriginal community 29, 50, 53, 55–66
Ngorongoro Conservation Area 253
Ni, R. 156–157
Nicaragua 210, 231–250
Nicholls, Ron 50–69
Nieves Rico, M. E. 34
Nijman, J. 215–216
Nixon, R. 7
NOAH 292
non-governmental organizations (NGOs) 26, 51, 132; mining in Greenland 292; religious tourism in India 93, 94, 100, 101–103, 104; Roskilde Festival 109, 116; South Africa 79; Tanzania 256, 314; voluntourism 111
Nora, Pierre 141
Norum, Roger 1–21

Novelli, M. 221
Noy, C. 275
Núñez, Eliseo 233
Nunkoo, R. 34
Nyerere, Julius 241

objectivism 288
O'Brien, M. 188
Oceania 194
Ochieng, O. M. 80
Oelschlaeger, M. 133
O'Hare, G. 275
Okech, R. N. 258
Olwig, Mette Fog 108–127
organic nature 159–161, 162, 163
O'Riordan, T. 199
Ortega, Daniel 232, 233
Orwell, George 315
Othering 278–280
Outterson, K. 273
Ovahimba people 213–214, 216, 219–225

Pacific Islands 136, 146, 194
package tourism 259
Paoa, Petero Edmunds 138
Parajuli, P. 58, 62
patriarchy 33, 37, 41
Pearl Harbor (Pu'uloa) 145, 146
Peet, R. 2, 4–5, 7, 10, 12–13, 16, 25, 95, 266
Peluso, N. L. 265
Penna, S. 188
Pereiro, X. 27
Peterson, V. 32
Pinochet, Augusto 136
place 6, 7–8, 71, 105, 145, 281
plural approach 4–5
Poddar, J. 98, 99
Polanyi, Karl 146n3
political ecology 2, 3–6, 51–53, 152, 302–306; Actor-Network Theory 287; antiessentialist 157–158, 163; conceptual framework 10–14; contextualization 170; core elements 6–8; 'counter-work' 146n2; development 209–211; dominant narratives in 71; ecotourism 188, 189, 191–192, 198, 202; festival ecovoluntourism 112; Japan 173, 183, 184;

literature review 8–10; Live Clean, Live Healthy campaign 233, 244–245; power relations 215, 251, 265–266; protected areas 252; Rapa Nui 134–135; religious tourism in India 94–96; Shangri-La 153; slum tourism 271, 276, 277, 280–281; status of nature 94; strength of 246; urban 210, 233–236, 244, 286
political economy 4, 31, 32, 160, 215, 233–234
pollution 211, 271, 272, 296
popular festive eco-voluntourism 110–112, 116–117, 119–125
Porteous, J. D. 137
postcolonialism 12, 73–75
post-structuralism 4, 6, 10, 13, 158
poverty 6, 9, 36; Bali 42; feminization of 34, 39; Guatemala 198; Makuleke community 88; Namibia 210, 214; Nicaragua 239, 241; poverty alleviation 217–218; Rapa Nui 138; rural 216; slum tourism 270, 273, 275, 276–277, 280
power 10–11, 15, 31, 51, 71, 215, 265–266, 298; Actor-Network Theory 287; co-management 82, 83–84; communities 13; conservation and preservation 235; developing countries 251; ecotourism 184; ethical 226; feminist perspective 7–8; Foucault on 28; knowledge relationship 82; landscape 295; Marxist political economy 234; Namibia 214; Nicaragua 245; Rapa Nui 132; scales of power relations 306; slum tourism 281; tourism industry 33; water and power dynamics 34–35
Power Sports Inc. 180–182
Pretty, J. N. 81
Prime, Ranchor 101–102
Pritchard, A. 306
private sector 76, 198
privatization 59, 210, 211; India 97; Rapa Nui 136, 138; Tanzania 258; water 38, 39, 41–42
production 5, 209, 296, 311
property rights 7–8, 12, 75, 188, 195, 202
'Pro-Poor Tourism' (PPT) 14, 52, 210, 214–215, 216–217, 218, 225
protected areas (PAs) 132, 199, 252; co-management 70, 74, 75, 76, 77–88;

discourse of 195–196; ecotourism 198, 200; Rapa Nui 137; Shangri-La 155, 157; Tanzania 251, 253–256, 260–265, 266; *see also* national parks
Pudacuo National Park 157, 160–161
puritanism 110, 117–118, 119
Pu'uloa 145, 146

Qi, Z. 156

Rabelais, François 110, 113, 125n3
race 7, 8, 11, 44, 310–311; apartheid in South Africa 74, 87; intersectionality 32–33; Japan 177; *see also* ethnicity
racism 7, 54
Ramage, Patrick 193
Ramkissoon, H. 34
Rancière, J. 238
Rapa Nui 132, 134–150
rape 264
Rathje, Bill 238
Reality Tours and Travel 275–278
recycling 109, 162, 238, 239, 240, 242, 243, 245
reforestation 101–103, 134, 163, 175
Reid, D. 198
reinhabitation 54, 55
religious tourism 29, 92–105
Ren, Carina 210–211, 284–301
Renaissance carnivals 110–111, 113, 124
'rendering technical' 179
resistance 7, 28, 155, 265, 266
resources 6, 11, 13, 16, 52; co-management 70, 73–88; commodification of 216; community access and control 26–27; Community-Based Natural Resource Management 213, 217, 219; conflict over 209, 211, 234; ecotourism 192; indigenous people 27; Ngarrindjeri Aboriginal community 58; protected areas 195–196, 266; Tanzania 254, 257, 260–262, 263, 266; *see also* forests; land; water
restoration 5, 93, 94, 95, 97, 103
reterritorialization 54–55, 134
revenues 86, 193, 257, 260, 262–263, 266; *see also* benefit sharing
Ricoeur, Paul 131

326 *Index*

rights: citizenship 188; Ngarrindjeri Aboriginal community 56–57, 61, 63, 66; Rapa Nui 137, 138–139, 145; 'rights of nature' 133; *see also* property rights
Rigney, D. M. 57
Rigney, Grant 29, 50–69
Rigney, Matt 61–62
Ringer, G. 225
Rio de Janeiro 272
Robbins, Paul 3, 10, 13, 51, 94, 170, 188, 298; 'all producers of nature' 209; conservation and control thesis 81–82, 131, 252, 266; dominant narratives 71; environmental subjects and identities 108; governmentality 27–28; landscape 295; market economy 307n2; paradox of conservation 132; plural approach 5
Roberts, Gregory David 270
Rocheleau, Dianne 7–8, 33
Rose, D. B. 54
Roskilde Festival 29, 108–110, 112, 114–124
Rothfuss, E. 221
rural development tourism 312, 313
rural landscapes 176–177, 182–183, 235

Saarinen, Jarkko 210, 213–230
sacred groves 93, 94–95, 96–97, 99–100, 103, 104–105
Said, Edward 135, 145
Salazar, N. B. 176
Sandfær, Mogens 116–117
Sandinistas 232, 233, 237, 241, 243
Sandino, Augusto C. 238
'sanitary citizenship' 244
scale 6, 8, 35
Scheyvens, R. 87, 218
Scott, James 146n3, 174
Scott, W. 199
sea turtles 13, 191, 194, 196
Sebastien, L. 257
self-determination 26; Greenland 292; Ngarrindjeri Aboriginal community 56, 65, 66; Rapa Nui 132, 134–135, 137, 139, 144–145
Self-Discovery Adventure 180–181
Selin, S. 82
Selinger, E. 273

Serengeti National Park 253, 254, 314
Shangri-La 132, 151–168
Sharma, K. 275
Sharma, S. K. 9
Shelton, Eric 1–21, 131–133
Shinde, K. 103
'showpiece projects' 313
Simpson, A. 146
Sinclair, T. 216
Singer, Peter 188, 190
Singleton, S. 73
slum tourism 211, 270–283
Smirl, Lisa 313, 316n2
Smith, Neil 6, 9, 309
social control 264–265
social justice 6, 52, 77, 198; celebrity engagement with 112; co-management 73; Ngarrindjeri Aboriginal community 63
social movements 104, 188–189, 202
social relations 304, 305
solidarity 109, 110
Sørensen, Gerald 302
South Africa 7, 29, 70–91, 214
South America 194
sovereignty 50, 52, 56, 60, 62–63, 66
sports 180–182
Stables, A. 199
stakeholders 12, 15, 70, 305, 306; co-management 73, 80, 83–85; ecotourism 191
Steiner-Aeschliman, S. 117–118
Stengers, I. 288
Stewart-Harawira, M. 54
Stonich, Susan 9, 26–27, 31, 35, 39–40, 209, 271, 302
Stronza, A. 15
sustainability 5, 26, 52, 305–306; conservation and control thesis 131; crisis of 304; economic 211; ecotourism 190, 191, 192, 194–195; Ngarrindjeri Aboriginal community 62, 64; Roskilde Festival 118
sustainable development 70, 191, 225; co-management 74, 81; concept of 199; ecotourism 195, 197; neoliberal business model 197; Ngarrindjeri Aboriginal community 63; poverty alleviation 218; Shangri-La 153, 155–156, 157

sustainable tourism 12, 52, 133, 302; Actor-Network Theory 298; ecotourism 193; poverty alleviation 218; Rapa Nui 132, 134, 135, 144, 145
Swyngedouw, E. 32

Takeda, L. 26
Takikawa, Jiro 180, 182
Tamarindo 38–39, 40–44
Tanzania 210, 234–235, 241, 251–269, 314
technonature 161–162, 163
Thailand 51
Thomas, J. A. 184
Thompson-Carr, Anna 1–21, 25–30
Thorne, Frank 307n1
Tibet 201
Tibetan Buddhism 160, 165n5
Tiessen, R. 111
Tonga 193
tour companies 259–260, 275–276
Tourism Concern 36, 51
Tourism Investigations and Monitoring Team (TIM-Team) 51
tourism studies 2, 5, 8–10, 11, 13, 31–32, 51–53, 302–306, 309
'tragedy of the commons' 26, 52
trees 7–8, 100–103
trekking 201
Trevorrow, George 64–65
Trevorrow, Tom 58, 60
Tribe, J. 32
TripAdvisor 1, 274, 297, 298
Tsing, A. 14, 104
Tsunetsugu, Y. 178
Tucker, H. 34
Tuki Tepano, Rafael 138–139
Turner, R. L. 87

United Nations Educational, Scientific and Cultural Organization (UNESCO) 134, 137, 232, 240, 294
United States 272, 311
Universal Declaration on the Rights of Indigenous Peoples 27
unsustainable development 40
urban environments 234, 235, 236–245, 312–313; *see also* slum tourism

Urban Political Ecology (UPE) 210, 233–236, 244, 286
urbanization 96, 239
Urry, J. 273

value, instrumental and intrinsic 200–201
values 74, 200
Vangen, S. 80
Vilas, S. 75
violence 137–138, 139, 141, 144, 263–265, 266
Vizenor, Gerald 135
voluntourism 26, 29, 276; festival eco-voluntourism 29, 108–112, 115, 116–117, 119–125; growth of 111
voyeurism 270, 272–273, 280
Vrindivan 29, 92–103, 105

Walters, K. 99
Warren, K. 305
water 31, 33, 34–36, 211; Bali 9, 27, 38, 44; Bay Islands 271; conflict and activism 42–43; Costa Rica 42–43, 44; Fiji 145; Namibia 223; Ngarrindjeri Aboriginal community 50, 53, 57, 58, 60–64; religious tourism in India 95, 99; slum tourism 211, 271, 272, 281; Tamarindo 38–39; unsustainable development 40; water security 44
Watts, M. 2, 4–5, 7, 10, 12–13, 16, 25, 95, 266
Wearing, Michael 133, 188–206
Wearing, Stephen 133, 188–206
Weber, Max 117–118, 146n3
Wendt, Albert 140
West, P. 134, 144, 176
whale watching 27, 191, 193–194
White, C. 132
White, D. F. 298
White, Geoffrey 136
White, N. 73–74
wilderness 74, 131, 133, 176, 199, 221
wildlife-based tourism 74, 78–79, 82, 253
Williams, Raymond 302, 313–314
'winners' 3, 13, 16, 82, 173, 192
Wolf, Eric 5, 137
Wolfe, Patrick 145

women 7, 28, 32–36, 39–44; Bali 37, 41, 43; Costa Rica 39, 41, 42–43; female celebrities 112; Makuleke community 88; Ovahimba people 224; rape of 264
Wood, D. J. 73
World Bank 73, 218
World Heritage Sites: Greenland 289, 294; Nicaragua 232, 240; Rapa Nui 134, 135, 137, 139; Shangri-La 154
World Travel and Tourism Council 32
World Wildlife Fund (WWF) 101–102, 189, 292
worldviews 58, 59, 64, 65–66, 195, 199, 273
Worster, D. 133

Xu, H. 195

Yano, C. R. 176–177
Young, Forrest Wade 132, 134–150

Zackey, J. 155
Zanzibar 35
Zhang, Jundan 132, 151–168
Zimbabwe 85
Zweeten, M. 35